International Conference on Sustainable Agriculture (ICSA 2018)

IOP Conference Series: Earth and Environmental Science
Volume 327

Padang, Indonesia
13 - 14 November 2018

ISBN: 978-1-5108-9664-2
ISSN: 1755-1307

Printed from e-media with permission by:

Curran Associates, Inc.
57 Morehouse Lane
Red Hook, NY 12571

Some format issues inherent in the e-media version may also appear in this print version.

Institute of Physics
Dirac House, Temple Back
Bristol BS1 6BE UK

Phone: 44 1 17 929 7481
Fax: 44 1 17 920 0979

techtracking@iop.org

Additional copies of this publication are available from:

Curran Associates, Inc.
57 Morehouse Lane
Red Hook, NY 12571 USA
Phone: 845-758-0400
Fax: 845-758-2633
Email: curran@proceedings.com
Web: www.proceedings.com

TABLE OF CONTENTS

SUSTAINABLE AGRICULTURAL SYSTEM ORGANIC FARMING DIVERSIFIED

IOP Conf. Series: Earth and Environmental Science **327** (2019) 011001 doi:10.1088/1755-1315/327/1/011001

PREFACE

This proceeding contains the manuscripts which selected in International Conference on Sustainable Agriculture, that held on 13th – 14th of November 2018 at Andalas University, West Sumatera, Indonesia. This conference was organized by Faculty of Agricultural Technology, Andalas University in collaboration with Universiti Teknologi MARA, Malaysia and Chiang Mai University, Thailand.

The main topic of conference is sustainable agriculture and divided into three specified themes, they are Socio-Economics Value Chains in Agricultural System, Precision and Smart Technologies in Agricultural System, and Sustainable Agricultural System Organic Farming Diversified.

There were 71 papers that had been presented orally and poster during conference. 29 of 71 papers have been selected after reviewing by authorized reviewers. The selected papers are published in IOP conference series: Earth and Environmental Sciences.

Lastly, Our team would like to express our sincere thanks and appreciation to all the participants, colleagues as well as Andalas University administration for their indispensable support in organizing this scientific meeting and publisher (IOP Publishing) which accepted our manuscripts for publishing.

Dr. Ir. Feri Arlius, M.Sc
Chairman of the ICSA 2018

IOP Conf. Series: Earth and Environmental Science **327** (2019) 011002 doi:10.1088/1755-1315/327/1/011002

Peer review statement

All papers published in this volume of *IOP Conference Series: Earth and Environmental Science* have been peer reviewed through processes administered by the proceedings Editors. Reviews were conducted by expert referees to the professional and scientific standards expected of a proceedings journal published by IOP Publishing.

Systematic Literature Review on Evaluation of Quality Management Practices in Palm Oil Supply Chain: The Case of Upstream

MFSA Ghani[2], Zahid-Muhamad M[1], MFA Aziz[2] and S Syahlan[2]

[1, 2] Faculty of Plantation and Agrotechnology, Universiti Teknologi MARA Jasin, 77300, Melaka, Malaysia

Corresponding Author's E-mail Address: zahidmuhad@gmail.com

Abstract. Malaysia and Indonesia are classified among the world's leading producers and exporters of palm oil. In previous findings, the palm oil industry is linked together from upstream to downstream and consists of four sub-sectors which are growers, millers, refiners, and oleo chemicals. There are various quality management practices that are currently being implemented in palm oil industry. This research attempts to identify the current quality management practices in palm oil upstream supply chain and to develop an instrument to produce quality palm oil in upstream sector. This paper presents a systematic review on quality management practices in upstream sector and the final selection comprehends 32 articles selected and comprehensively analyzed to assess quality management practices in upstream processes of palm oil supply chain. This study adds value to the current knowledge by extending organizational readiness theories. This study is relevant, practical, and useful to both practitioners and academics by providing a holistic implementation roadmap for the upstream chain to guide the managers to implement quality management practices at the organisational level.

1. Introduction

Palm oil industry are linked to each other from upstream to downstream and consists of four sub-sectors which are growers, millers, refiners, and oleo chemicals [1]. Upstream industry consists of growers sector who is involved in the cultivation of oil palm such as smallholders and large plantations. The activity involved at this stage includes nursery establishment, planting of oil palm trees and production of fresh fruit bunches (FFB). The trees will bear fruit after 3 years and able to give productivity up to 25 years and it produce continuously, where the fruit bunches will be harvested every few weeks. The fresh fruit bunches (FFB) will then be taken to mill, where oil will be extracted with pressure and steam that produces Crude Palm Oil (CPO) and also Palm Kernel Oil (PKO) [2]. Midstream involve the millers alone.

Whereas for downstream industry, it consists of oleochemicals and refiners sectors such as palm kernel crushers, palm oil refiners, palm-based edible products manufacturers, specialty fats producers and oleochemical producers [1]. The extracted oil is taken from the mill to refineries to be refined into primary palm oil fractions such as stearin and olein [2]. The refined oil are then transported to downstream ingredients manufacturers of consumers goods who might further process it into more complex fractions which then be sold to retailers or consumers.

Hence, in order for these firms to operate efficiently along their supply chain, they need a better approach in their supply chain management (SCM). Figure 1 illustrates the value chain in this industry.

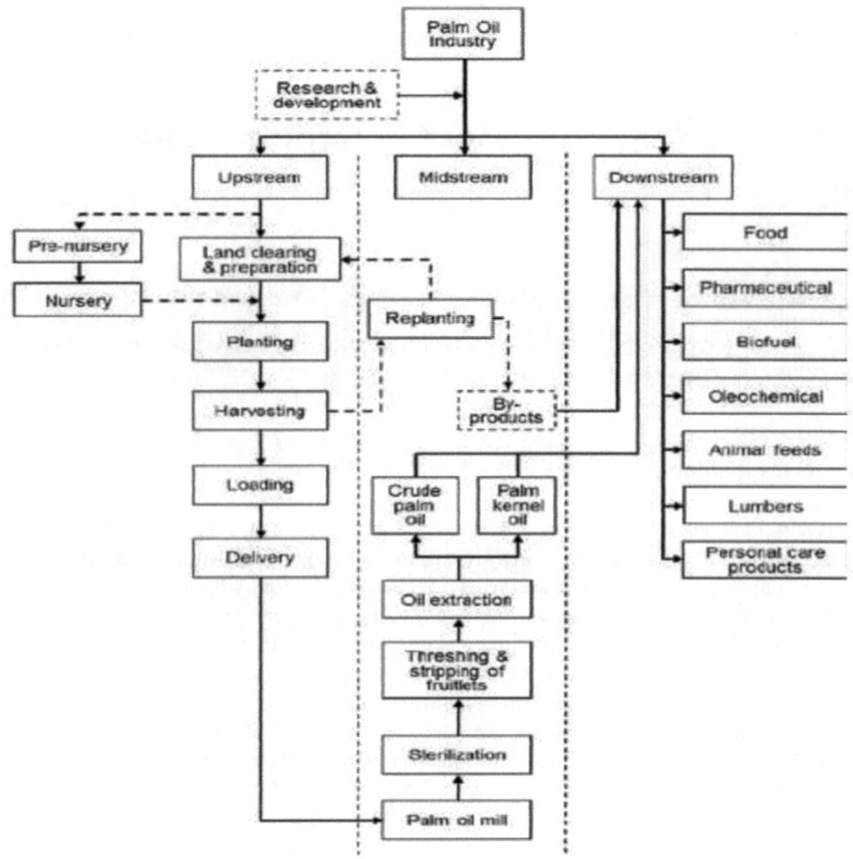

Figure 1. Process Flow of Oil Palm Upstream Sector

Based on above supply chain provided by [3], activities exist at upstream sectors that mainly affects the palm oil is planting, harvesting, loading and delivery to the mill and after received at mill it undergoes sterilization, threshing and stripping of fruitlets and finally extracted to produce crude palm oil (CPO) and palm kernel oil (PKO). According to [4], the main practice that is being referred by organizations involved in Palm Oil Supply Chain is the Codes of Practice for the Oil Palm Supply Chain (COPS). This is due to ensure that Sustainable Palm Oil is produced to meet requirement of quality palm oil, environmental protection and food safety [5].

The main objective of this research is to propose a research framework to investigate quality management practices for upstream supply chain in the palm oil industry. The scope of the study will only review the literature. The paper investigates the constructs of quality management practices for upstream chain in palm oil industry. The rest of the paper is organized as follows. The subsequent section summarized related literature, which is followed by the proposed framework to investigate key quality management practices in palm oil industry. Next, the research methodology been discussed. The final section presents the conclusion and the further study.

2. Methodology

This research is an exploratory in nature. Thus, it can be inferred that this study is a revelatory to the understanding of a phenomenon or problem. The method being used in this research is Systematic Literature Review (SLR). A systematic literature review is a method involving identify, evaluate and interpret all available research which is relevant to a particular research question, topic area, or situation of interest. Systematic review is a method of literature review which adopt a series of steps to ensure the appropriate transparency is brought in the process of analyzing lots of paper. It suggests that there are four phases needed to prove this research is relevant which is: planning, sampling, analysis and reporting [6].

The quality of a journal is reflected by its indexation value. Compared to non-indexed journals, indexed journals are considered to have higher scientific quality. Many popular indexation services have been developed over the past few years. These include Google Scholar, Science direct, SCOPUS, Scimago and much more database. There are many article databases containing millions of journals with various ratings so to ensure that only relevant journals were used in this paper, the journal's index used as references is checked through these indexation service provider.

Indexing helps the journal to achieve its main purpose of being accessible to a wide audience. A journal's reputation will be improved and referred as a reliable source of high quality information in one's field if it widely accessible. Searching through database is the first activity researchers will do as part of their research process, and mainly they will look forward to retrieve it from established and well-known databases. Thus, being indexed in a popular database in a particular field will help increase a journal's readership.

Jorge Hirsch proposed a research performance indicator known as The Hirsch index, or h index. Hirsch's index is an original and simple new measure that incorporates both quantity and visibility of publications [7]. According to [8], an h index of 20 after 20 years of scientific activity indicates a successful scientist, an h index of 40 after 20 years of scientific activity indicates outstanding scientists, likely to be found only at the top universities or major research laboratories, and an h index of 60 after 20 years, or 90 after 30 years characterizes truly unique individuals.

Building a search string by determining identifying the keywords is the first search strategy for SLR. It is the most important step as the database will determines which papers to be retrieved just by using the keywords. Following relevant works in the field of oil palm upstream [3], six key processes were selected for this review. Hence, the author considered the oil palm upstream process consists of nursery, land clearing and preparation, planting, harvesting, loading and delivery to mill.

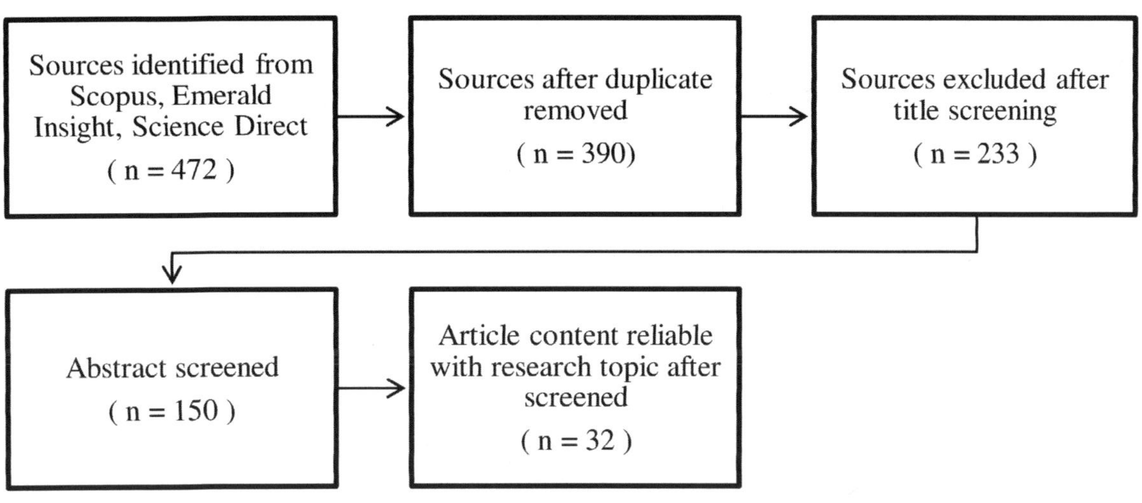

Figure 2. SLR Articles Screening Process

IOP Conf. Series: Earth and Environmental Science **327** (2019) 012001 doi:10.1088/1755-1315/327/1/012001

The sources were identified from various databases and a number of 472 articles were retrieved. The criteria is further screened by removing duplicate articles which reduces the number of articles into 390. After title screening, the articles were further narrowed to 233 and further narrowed into 150 after the abstract were screened. This number includes articles discussing about QMP, SCM, GAP, and other main practices and management system exists in the oil palm upstream supply chain area. Figure 2 shows the flow of article screening process that were being conducted.

Non-indexed journals are also used as one of the information sources relates to literature published in journals that are not being indexed on major bibliographic databases such as SCOPUS, Google Scholar and Science direct. Non-indexed journals were not being used as primary source in this research as it is only being used as side-reference to support the primary data from indexed articles.

3. Results

Oil palm involves a lot of process in its supply chain and to assess its quality management practices, each process that directly leads to the formation of the palm oil itself needs to be identified. This process were identified by reading journals, SOP's of big companies and Codes of Agricultural Practices (MPOB).

From extensive readings of past papers, it can be observed that there are only a limited number of journals about management of oil palm upstream activities that were found relevant with this study. In addition, there are very limited number of journals exists regarding quality management in palm oil upstream process.

Other findings is that even if there are good quality journals, lots of them are outdated and thus it needs to be excluded from this research's journal database management system. After screening of journals, the data were then critically analysed to extract only good management practices to be taken as a criteria, and from those criteria a checklist of upstream process in palm oil supply chain.

From the literature review, the process involved in oil palm upstream supply chain were simplified as in Figure 3. Those process were the main activity that were operating daily in an estate. The process begins with nursery establishment as the seedlings were grown in this process. Land clearing, land preparation and planting also linked to nursery where it involves only during replanting process after oil palm reach economics life of 25 years. Other operations were conducted daily from these main operations, there are sub-activity exists which contributes direct and indirect effects towards the product of oil palm which is the FFB itself.

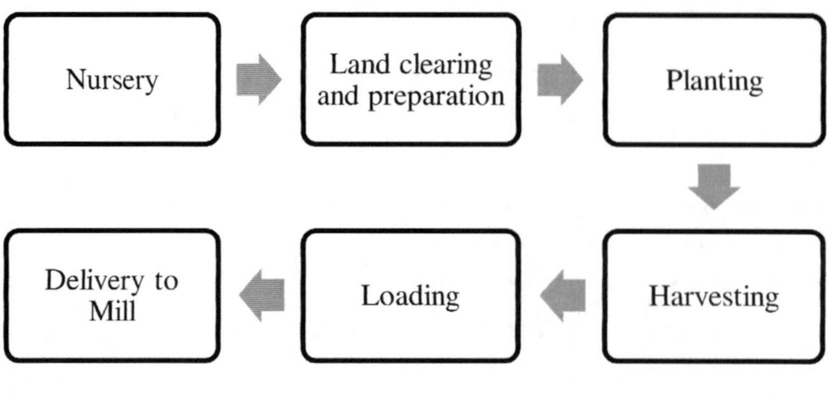

Figure 3. Process Flow of Oil Palm Upstream Sector

Table 1 lists all the articles reviewed which were relevant with each sub-process involved it the oil palm upstream process. Some of the articles explain about more than one process in the oil palm upstream activity.

Table 1. Articles Review

Articles Theme	Source
FFB post-harvest handling	[9, 10]
Transport of FFB	[11, 12]
FFB characteristics	[13, 14, 15, 16]
FFB quality assessment	[17, 18, 19, 20, 21]
FFB ripeness assessment	[22, 23, 24]
Harvesting round management	[25, 26]
Weed management	[27, 28]
Fertilizer management	[29, 30, 31, 32, 33]
Planting material	[34, 35]
Land selection	[36, 37, 38]

3.1. FFB post-harvest handling
According to [40], first process to be supervised is the post harvesting handling where harvester should handle the oil palm FFB with care after the FFB being harvested because the FFB could easily become damaged and bruised if it is not handled properly. All FFB and LF were collected from the palm base should be delivered free from any contamination of ground debris and stones to the collection points at the road side with minimum damage, delay, contamination and being sent to the mill together with harvested bunches. In other study by [9], in order to resist quality reduction it is recommended that FFB should be handled carefully. High impact when dropping and throwing the fruit by the harvester and fruit loader should be avoided. Meanwhile, the fruit collector and transporter should try to reduce the friction and bump among the fruit.

3.2. Transport of FFB
A study by [11] suggest that to reduce bruising during transportation of FFB to mill is by taking sterilizer cages to the field for immediate sterilisation process. For process of transport of FFB to the mill, harvested bunches should be transported by lorry, tractor-drawn trailers or rail cages. According to [12], they also states that the FFB and LF should be sent to the mill preferably within 24 hours after harvesting. Transport should be arranged in such a way that the FFB do not become contaminated with undesirable substances and quality is preserved. The general requirement is that the compartments should be dedicated to FFB, free from previous load residues, free from odour of previous loads, mineral oil and is dry. This should be checked before each consignment. The result of the cleaning actions should be checked visually and recorded. Another aspect to be prevented during transport is the penetration of rainwater and splash water. Even if the load compartments are empty, it needs to be covered in order to prevent rain penetration and contamination from bird droppings. In case of road transport, the outside of the means of transport including the chassis should be free from visible traces of previous cargoes [12].

3.3. FFB characteristics
A study by [13] states that a ripe FFB should have an OER higher than 21% and the FFA content less than 5%. The OC and FFA level are determined through analysis at the chemical laboratory [13]. The FFA level determines the quality of palm oil, because high presence of FFA leads to production of poor quality of oil [17]. The temperature and moisture content influences FFB physical and mechanical properties [39]. Previous studies also shows that the FFA content and oil quality will slowly be reduced if there was no immediate heating step taken [15]. A study was also found that a damaged FFB causes the lipase enzymes to be rapidly hydrolysed, contamination of FFB by microorganisms and FFA development on the FFB [16].

3.4. FFB quality assessment

According to [17], the FFB quality can be determined by it ripeness, FFA and OC level. To ensure quality FFB is graded physically right before it passes through process of oil extraction and current FFB quality assessment is very time consuming and prone to errors and inconsistencies [18].

The storage time of damaged oil palm fruits is also a major problem that could activate the lipase enzymes and eventually reduce the oil quality where a study proves that the length of the storage periods showed significant correlation with the FFA release rate within a 2 hour storage period [40].

3.5. FFB ripeness assessment

In practice, manual ripeness assessment is carried out by visual inspection [17]. The FFB colour varies from very dark purple for unripe to orange for ripe bunch. To harvest, the bunch ripeness needs to be analysed first. To determine the ripeness of bunch on a tree, observing the number of loose fruits on the ground is one of the method [24]. Another way to determine the ripeness is based on observing different colour intensities of FFB by referring to the colour ripening index. Using this index, a ripe bunch could be differentiated from other FFB categories such as the black bunches, under ripe bunches, overripe bunches, empty bunches and rotten bunches. According to indicator used, purplish black indicates an unripe bunches, reddish black indicates an under ripe bunches, red indicates a ripe bunches, and reddish orange indicates an overripe bunches. Another method available to determine the ripeness according to [41] is by measuring the fruit firmness using penetrometer based on colour, shape and feature. Other than that, measuring level of chemical type and parameters are also related to ripeness such as spectral image analysis and pH value.

3.6. Harvesting round management

According to study by [26], Plantations Company usually have harvesting round of 7-day, 10-day or 14-day interval while smallholders usually having harvesting intervals of 14 or 15-day sometimes up to 30 days. Having a shorter period of harvesting interval is important as they proved that the yield increase at 5–20% when reducing length of harvesting round from 14 to 10 days. Another study by [25] that were more focused on smallholders further proven the above research findings where it shows that harvesting once a month predicted the lowest yields in smallholder oil palm plantations. They find that short harvesting intervals around 7–10 days helps to improve fresh fruit bunch productivity of oil palm crops.

3.7. Weed management

Based on study by [28], weeds affect crop production and the importance of managing weeds is usually ignored by farmers and government officials. The effects of weeds on oil palm are difficult to be expressed due to their long economic life up to 20-30 years. The researchers reported that because of strong competition from weeds, yield losses in oil palm plantations are up 6% to 20%. Oil palm growers normally assume that weeds can easily be controlled by herbicides or mechanically. This misinterpretation prevents effective weed management. Overall, chemical control has been widely used as a weed management tool for oil palm sectors. There have been issues where the application of herbicide mainly affects the main crop being grown which is the oil palm itself. A findings from [27] proved the issue to be wrong based on their research data regarding the FFB amount per plant and the number of fronds for each plant. The results shows that as long as the herbicides used in the research such as glufosinate ammonium, paraquat, and glyphosate were not sprayed directly at the plant, they had no adverse effect on the vegetative growth of the oil palm.

3.8. Fertilizer management

A study by [29] states that the fertilizer input on oil palm is very much needed to sustain high yields and usually contributes to 40–65% of total upkeep costs of the field. For a duration of 25 years, fertilizer management within a plantation requires an annual plan for 25–30 hectare of each block. A specific fertilizer sequence is received by each block and each were different from the others. On a multiannual period, blocks may receive only mineral fertilizers, only organic fertilizers or they receive both mineral and organic fertilizer but alternately or known as mixed fertilizer sequence. Based on study by [33], N, P and K nutrient elements is one of the main factor that affect in oil palm seedlings and also mature palm growth and development that resulted from better nutrient management. Fertilizer application practice were conducted to ensure the growing plant get all the required nutrients for optimum growth. The most important nutrient element in terms of plant growth, carbohydrate content and physiology is Nitrogen [30]. Phosphorus is the second most important nutrient for plants that are needed for energy transfer and storage and low availability of phosphorus is a limiting factor to plant growth [31]. Potassium is the third important nutrient that needed for maintaining osmotic balance, phloem transport, and photosynthesis [32].

3.9. *Planting material*

Current issue arising is the yields of oil palm reached a stagnant level due to unimproved planting materials being used in the field [34]. They suggest that to improve productivity, generation of superior populations is much needed. This is vital to develop an elite planting material to be used for breeding purposes and as potential genetic resource. However, we need to estimate heritability among the fresh fruit bunch yield traits of different breeding populations because high heritability estimates resulting high genetic advance for FFB yield. At the end of the process of choosing a right planting material, the number of bunches (BN) and average bunch weight (ABW), is a major outcome that will be gained [35].

3.10. *Land selection*

To start an oil palm plantations it is good to begin with the selection of appropriate land or soil, to determine the land potential that can be developed to obtain the proper management of oil palm fruit productivity [36]. Oil palm can grow, develop and produce as expected when the available soil water is even and sufficient with areas of rainfall 2000-2500 m / year and dry months less than one month per year [38]. The temperature should be around 22◦C to 33 ◦C with relative humidity more than 85% [37].

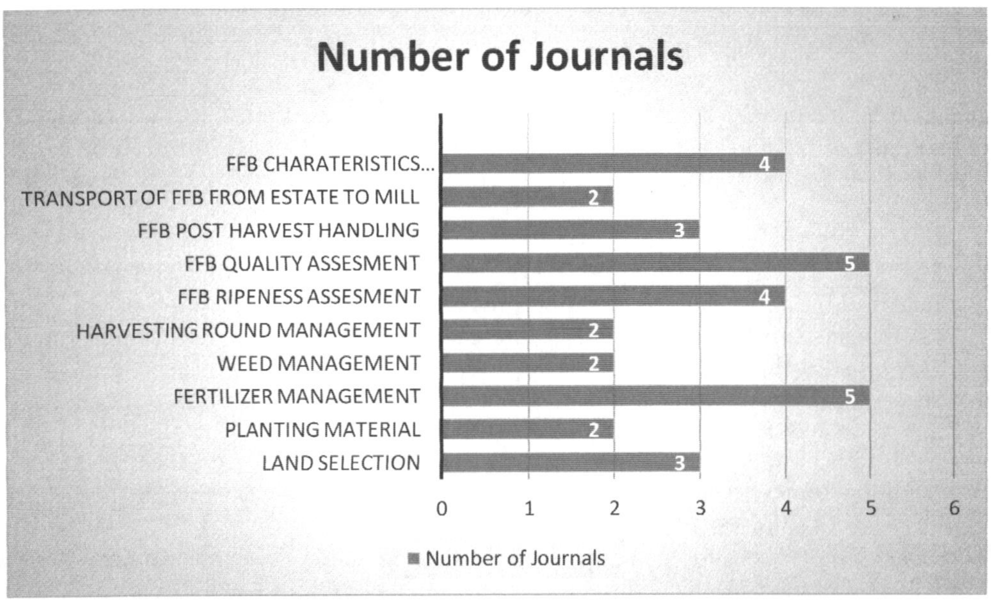

Figure 4. Graph Representation of Number of Articles Discussed about Management Practices in Oil Palm Upstream Sector

Based on the results obtained (Figure 4), the sector with highest number of article is fertilizer management and FFB quality assessment. This might due to the importance and direct effect of such factors towards the oil palm quality itself.

4. Results

To date, no instruments have been successfully produced and published to the public to assist oil palm growers to track down their quality management practices. So the best way to solve this problem is to produce such instrument and this paper had systematically reviewed past papers, extract key information and use them as a criterion in producing an instrument for assessing quality palm oil in the form of checklists. A checklist a simple standardized list of the required steps developed for a repetitive task to reduce mistakes that might cause its quality to deteriorate. Using checklists helps supervisors to do it right every time, and ensures they did not forget anything throughout the supply chain. Academically, this study proposes the set of constructs for studying quality management practices in palm oil industry. The framework is expected to be investigated in palm oil organisation in Malaysia.

References

[1] Abdullah, I., Wan Mahmood, W. H., Md Fauadi, H. F., Ab Rahman, M. N., & Mohamed, S. B. (2017). Sustainable manufacturing practices in Malaysian palm oil mills: Priority and current performance. Journal of Manufacturing Technology Management, 28(3), 278-298.

[2] Chong Tan, Y., & Oly Ndubisi, N. (2014). Evaluating supply chain relationship quality, organisational resources, technological innovation and enterprise performance in the palm oil processing sector in Asia. Journal of Business & Industrial Marketing, 29(6), 487-498.

[3] MPOB (2003). Oil Palm Fruit Grading Manual. Second edition. MPOB, Bangi. p. 4 (4.3.1). [Accessed 25 Feb. 2018].

[4] Cock J., Donough C.R., Oberthür T., Indrasuara K., Rahmadsyah, Gatot A.R., Dolong T., Increasing Palm Oil Yields by Measuring Oil Recovery Efficiency from the Fields to the Mills.

[5] Egghe, L. (2006). Dynamic h-index: the Hirsch index in function of time. Retrieved June 12, 2006

[6] Hirsch, J.E. (2005a). An index to quantify an individual's scientific research output. Proceedings of the National Academy of Sciences, USA, 102(46), 16569–16572.

[7] Krisdiarto, A. W., & Sutiarso, L. (2016). Study on Oil Palm Fresh Fruit Bunch Bruise in Harvesting and Transportation to Quality. Makara Journal of Technology, 20(2), 67-72.

[8] Mat Sharif, Z. B., Mohd Taib, N. B., Yusof, B., Bin Rahim, M. Z., Bin Mohd Tobi, A. L., & Othman, B. (2017). Study on Handing Process and Quality Degradation of Oil Palm Fresh Fruit Bunches (FFB). In IOP Conference Series: Materials Science and Engineering, 203 (1),[012027].

[9] Hadi, S., Ahmad, D., & Akande, F. B. (2009). Determination of the bruise indexes of oil palm fruits. Journal of food engineering, 95(2), 322-326.

[10] Ismail, N. Z. N., Jamian, R., Mohamed, M. S., Ab Rahman, M. N., & Deros, B. M. (2014). A conceptual approach of 5S to improving quality and environmental performance of Malaysian oil palm dealers.

[11] De Andrade, E. T., Teixeira, L. P., da Silva, I. M., Pereira, R. G., Piamba Tulcan, O. E., & de Andrade, D. O. (2013). Facilities for Obtaining Soybean Oil in Small Plants. INTECH.

[12] Sivasothy, K., Mohd Halim, R., Basiron, Y., 2005. A New System for Continuous Sterilization of Oil Palm Fresh Fruit Bunches. Journal of Oil Palm Research 17, 145- 151. [Accessed 5 March. 2018].

[13] Tan, C.H., Hasanah, M.G., Ainie, K., Chin-Ping, T., Abdul, A.A., 2009. Extraction and Physicochemical Properties of Low Free Fatty Acid Crude Palm Oil. Food Chemistry 113, 645–650. [Accessed 8 March. 2018].
[14] Foster Jr, S. T., Wallin, C., & Ogden, J. (2011). Towards a better understanding of supply chain quality management practices. International Journal of Production Research, 49(8), 2285-2300.
[15] Fatin, S.A., Rosnah, S., Yunus, R., 2014. Effect of Chopping Oil Palm Fruit Spikelet's on the Free Fatty Acid Content Release Rate and Its Mechanical Properties. International Journal of Research in Engineering and Technology 3(1), 511-516 [Accessed 10 March. 2018].
[16] Greenpalm.org. (2018). Green Palm: Palm oils versatility and efficiency. [Online] Available at: http://greenpalm.org/about-palm-oil/why-is-palm-oil-important [Accessed 25 Feb. 2018].
[17] Hai, T. C. (2002). The palm oil industry in Malaysia. WWF, Malaysia.
[18] Hassan, A., Mohammad, J. A., (2005) Oil Palm Bulletin (May 2005), p. 31-38, [Accessed 25 Feb. 2018].
[19] Fadilah, N., Mohamad-Saleh, J., Abdul Halim, Z., Ibrahim, H., & Syed Ali, S. S. (2012). Intelligent color vision system for ripeness classification of oil palm fresh fruit bunch. Sensors, 12(10), 14179- 14195.
[20] Lee, J. S. H., Ghazoul, J., Obidzinski, K., & Koh, L. P. (2014). Oil palm smallholder yields and incomes constrained by harvesting practices and type of smallholder management in Indonesia. Agronomy for Sustainable Development, 34(2), 501-513.
[21] Wibawa, W., Mohamad, R., Omar, D., & Juraimi, A. S. (2007). Less hazardous alternative herbicides to control weeds in immature oil palm. Weed biology and management, 7(4), 242-247.
[22] Lim, C. I., Biswas, W., & Samyudia, Y. (2015). Review of existing sustainability assessment methods for Malaysian palm oil production. Procedia CIRP, 26, 13-18.
[23] Dilipkumar, M., Chuah, T. S., Goh, S. S., & Sahid, I. (2017). Weed management issues, challenges, and opportunities in Malaysia. Crop Protection.
[24] Manghani, K. (2011). Quality assurance: Importance of systems and standard operating procedures. Perspectives in clinical research, 2(1), 34.
[25] Comte, I., Colin, F., Grünberger, O., Follain, S., Whalen, J. K., & Caliman, J. P. (2013). Landscape-scale assessment of soil response to long-term organic and mineral fertilizer application in an industrial oil palm plantation, Indonesia. Agriculture, Ecosystems & Environment, 169, 58-68.
[26] Maryanie, D. I., & Sutopo, W. (2013). Yuniaristanto,"A Holistic Feasibility Study Framework to Determine Valuable Chain in Palm Oil Industry". In Proceeding of the World Congress on Engineering 2013 (Vol. 1, pp. 3-5).
[27] Almodares, A., Taheri, R., Chung, M., & Fathi, M. (2008). The effect of nitrogen and potassium fertilizers on growth parameters and carbohydrate contents of sweet sorghum cultivars. J. Environ. Biol, 29(6), 849-852.
[28] Matola, A., Masamba, K., Mwangwela, A., & Mlotha , V. (2015). Quality evaluation of sunflower and groundnut oil produced by two cooperatives under the one village one product programme in cenrtal Malawi. African Journal of Agricultural Research, 1338- 1343.
[29] Fita, A., Bowen, H. C., Hayden, R. M., Nuez, F., Pico, B., & Hammond, J. P. (2012). Diversity in expression of phosphorus (P) responsive genes in Cucumis melo L. PloS one, 7(4), e35387.
[30] Tripler, C. E., Kaushal, S. S., Likens, G. E., & Todd Walter, M. (2006). Patterns in potassium dynamics in forest ecosystems. Ecology Letters, 9(4), 451-466.
[31] Mohidin, H., Hanafi, M. M., Rafii, Y. M., Abdullah, S. N. A., Idris, A. S., Man, S., ... & Sahebi, M. (2015). Determination of optimum levels of nitrogen, phosphorus and potassium of oil palm seedlings in solution culture. Bragantia, 74(3), 247-254.

[32] Okoye, M. N., Okwuagwu, C. O., & Uguru, M. I. (2009). Population improvement for fresh fruit bunch yield and yield components in oil palm (Elaeis guineensis Jacq.). American-Eurasian Journal of Scientific Research, 4(2), 59-63.

[33] Okwuagwu, C. O., Okoye, M. N., Okolo, E. C., Ataga, C. D., & Uguru, M. I. (2008). Genetic variability of fresh fruit bunch yield in Deli/dura x tenera breeding populations of oil palm (Elaeis guineensis Jacq.) in Nigeria. Journal of Tropical Agriculture, 46, 52-57.

[34] Mohamed Amine Ben Chikh El Fegoun. Upstream Supply Chain Analysis for Oil Palm, the Wall Street Journal, 2015. [Accessed 25 Feb. 2018].

[35] Idayu, I., & Supriyanto, E. (2014). Oil palm plantations management effects on productivity Fresh Fruit Bunch (FFB). APCBEE procedia, 8, 282-286.

[36] Mohdaly, A. E.-R., Seliem, K. A.-H., Abu EL-Hassan, A.-M. M., & Mahmoud, A. A. (2017). Effect of Refining Process on the Quality Charactheristic of Soybean and Cotton seed Oils. International Journal of Current Micrbiology and Applied Science, 207-222.

[37] Carr, M. K. V. (2011). The water relations and irrigation requirements of oil palm (Elaeis guineensis): a review. Experimental Agriculture, 47(4), 629-652.Morakinyo, T. A., & Bamgboye, A. I. (2015). Effects of age on some physical properties of oil palm fruitlets. Agricultural Engineering International: CIGR Journal, 17(3).

[38] Sharif, Z. B. M., Taib, N. B. M., Yusof, M. S. B., Rahim, M. Z. B., Tobi, A. L. B. M., & Othman, M. S. B. (2017, May). Study on Handing Process and Quality Degradation of Oil Palm Fresh Fruit Bunches (FFB). In IOP Conference Series: Materials Science and Engineering (Vol. 203, No. 1, p. 012027). IOP Publishing.

[39] Henson, I. E., & Harun, M. H. (2007). Short-term responses of oil palm to an interrupted dry season in north Kedah, Malaysia. Journal of Oil Palm Research, 19, 364.

[40] Morakinyo, T. A., & Bamgboye, A. I. (2015). Effects of age on some physical properties of oil palm fruitlets. Agricultural Engineering International: CIGR Journal, 17(3).

[41] Shahidi, F. (2005). Quality Assurance of Fats and Oils. Retrieved 8 June, 2018, from Bailey's Industrial oil and fat products: http://dx.doi.org/10.1002/047167849X.bio072

[42] Simarani, K. (2009). Effect of Palm Oil Mill Sterilization Process on the Physiochemical Characteristics and Enzymatic Hydrolysis of Empty Fruit Bunch. Asian Journal of Biotechnology, 57-66.

[43] Van Duijn, G. (2013). Traceability of the palm oil supply chain. Lipid Technology, 25(1), 15-18.

[44] Von Geibler, J. (2013). Market-based governance for sustainability in value chains: conditions for successful standard setting in the palm oil sector. Journal of Cleaner Production, 56, 39- 53.

[45] Turner , R. (2010). Fats and Oils quality charateristic, extraction and refining overview. Farm energy conference. Manchester NH.

[46] YCW, (2014), Home Quality and Specification Quality and Identity Characteristics of Palm Oil – Part 1 – Quality Characteristics, [online] Available at https://oilpalmblog.wordpress.com/2014/03/29/quality-and-identity-characteristics-of- palm-oil-part-1-quality-characteristics/ [Accessed 1 March. 2018].

Export Analysis of Major Commodities in Malaysia

N A Ismail[1], B A Talib[2], A Mokhtar[3]

[1]Faculty of Plantation and Agrotechnology, Universiti Teknologi MARA, Cawangan Melaka, Kampus Jasin, 77300 Merlimau, Melaka, Malaysia
[2,3]Faculty of Economics and Management, Universiti Kebangsaan Malaysia, 43600 UKM Bangi, Selangor, Malaysia

Corresponding author's e-mail address: amalina1520@melaka.uitm.edu.my

Abstract. This study aims to investigate the factors associated with the commodity export demand and to explore and quantify the contribution of the commodity export to economic growth. The technique of (VECM) Vector Error Correction is applied to estimate the export and economic growth model. It tests whether there has been a long-run relationship between commodity export volume and exogenous variables such as own price, substitute price, exchange rate, world population and industrial production index. The results indicated that there is a long run relationship among variables. The result suggests that exchange rate have a negative impact to the palm oil, rubber and tin where depreciation may reduce export of these commodities. The findings of the research also show that the commodities export has mixed effect on economic growth in Malaysia. Export of petroleum, palm oil and sawlog has a positive and significant relationship with economic growth in the long run while only petroleum and palm oil contributed to the economic growth in the short run. The evidence of impulse response function shows that exchange rate and FDI shock have a relatively higher significant impact on all exports of major commodities and economic growth respectively. This study implies that export of major commodities in Malaysia should be treated as an engine of growth. The effective management of exchange rate and export promotion policies can play a significant role in increasing commodity exports thus will improve Malaysia economic growth.
Keywords: Major Commodities, Export, Economic Growth, Vector Error Correction Model

1. Introduction

The export of major commodity in Malaysia plays one of the important roles to lead the domestic economy. However, an export commodity in Malaysia is caught in price instability and international interdependence. Moreover, the contribution of major commodity export earnings to total export revenue and GDP continues to decline. These issues are widely discussed by the industrial players and those directly and indirectly involved in the industries.

In the case of palm oil, it depends on the consumption pattern of the major markets; thus, any changes in consumption from these countries will affect the price of palm oil. During the first nine months of 2013, the average price of CPO had declined to RM 2332 per tonne. This was due to the reduction in external demand, unfavourable weather condition and increased production of oil and fats by major producing countries. These factors have led to the replacement of palm oil with other types of oil such as soybean, rapeseed and sunflower seed in the world market [1].

In addition, the rubber's price was also subjected to the world's market, and the surplus in supply has discouraged this industry for a long time. This sector's return to investment is influenced by the unremunerated price. Since the second half of 2012, Malaysia's smallholders had reduced their tapping activities following to the drop in the natural rubber price. This had forced them to find other alternative sources of income [1]. In 2013, the export volume of rubber had contracted 4.1% due to the weakening in demand, particularly from China and the US. The lower export price which averaged at RM 6.87 per kg in 2013 resulted in the decline of rubber earnings to RM 3 billion [2].

Major commodity was also heavily influenced by the uncertainty of the world market. Several notorious issues had surfaced as a result of the diminishing value of ringgit against the currencies of the nation's major partners. For example, when a price is quoted in US dollar, it would depend on the exchange rate. Therefore, if ringgit depreciates, the market price of commodity quoted in foreign currency would fall and encourage greater foreign demand for the commodity. However, according to [3] due to depreciation in Malaysia's currency, the high demand for natural rubber will lead to an increase in its price in the long term.

In addition, the economic slowdown of certain developed countries, especially in Europe, North America and Japan significantly affect Malaysia's economy. Malaysia's commodities remained to be at the mercy of the world's economy, and certain commodity exports are subject to this development.

Moreover, the country's major commodities also face competition from other commodities. As for rubber, Malaysia has to compete with Thailand and Indonesia which are relatively lower cost producers. Additionally, in view that synthetic rubber is a substitute for natural rubber the manufacturers are more inclined to utilise the cheaper material to maintain their competitiveness [3]. The palm oil is also facing stiff competition not only from other countries but also from other types of edible oil such as soybean, rapeseed, sunflower and corn [4]. This situation is even more unfavourable during bumper harvest.

Lastly, the contribution of major commodities to the country's export earnings and the gross domestic product was also declining. In 2014, the sector's contribution went down to only 10.2 per cent, signifying the growing importance of manufactured products as opposed to the commodity-based [5]. Moreover, weak global demand and excess supply of palm oil, rubber and petroleum also contributed to the decline. Although in 2010 the major commodities' export earnings share to GDP increased to 11.7 per cent due to firmed prices and favourable external demand, the sector's share declined from 16.7 per cent in 2010 to 13.4 per cent in 2014 due to economic factors.

The general objective of this study is to analyse the exports and their contribution towards GDP growth. The study focuses on these major commodities due to their importance to GDP, total employment and foreign exchange earnings.

2. Methodology

2.1 Sources of Data

The data used for this study covers from the year 1980 to 2014 were gathered from the Economic Report of Ministry of Finance Malaysia (MIF), Department of Statistic Malaysia (DOS), Malaysia Palm Oil Board (MPOB), Ministry of International Financial Statistics (IMF). Mineral Yearbook of US Geological Survey, Rubber Statistical Bulletin of International Rubber Study Group, United States Census Bureau and Federal Reserve Economic Data.

2.2 Export Demand Model

Theoretically, the number of exports demanded the given commodity be expected to be influenced by the price of exports, weighted average of the export prices of the country's trading partner and weighted average of the real income of the country's trading partner [6]. A new model has been developed based on previous research made by [7, 8, 9, 10, 11]. The export demand equation was shown below:

$$LOGQXD_t = \beta_0 + \beta_1 LOGPX_t + \beta_2 LOGPS_t + \beta_3 LOGEX_t + \beta_4 LOGWP_t + \beta_1 LOGIPI_t + e \qquad (1)$$

where,

Dependent variable: **LOGQXDT**: Quantity export demanded ('000 tons)

Independent variable: **LOGPX**: Price of export commodity (RM/ tons), **LOGPS**: Real price of substitute (US$/ tons), **LOGEX**: Exchange rate (RM/ US$), **LOGWP**: World population (million), **LOGIPI**: Industrial production index (2012=100)

Initially, if the variables confirm the existence of cointegration, then the VECM is estimated. In this study, the VECM model is as follows:

$$\Delta LOGQXD_t = \alpha_{0i} + \sum_{i=1}^{k} \alpha_{1i}\, \Delta LOGQXD_{t-i} + \sum_{i=1}^{k} \alpha_{2i}\, \Delta LOGPX_{t-i} + \sum_{i=1}^{k} \alpha_{3i}\, \Delta LOGPS_{t-i} + \sum_{i=1}^{k} \alpha_{4i}\, \Delta LOGEX_{t-i} + \sum_{i=1}^{k} \alpha_{5i}\, \Delta LOGWP_{t-i} + \sum_{i=1}^{k} \alpha_{6i}\, \Delta LOGIPI_{t-i} + \lambda_1\, (LOGQXD - \alpha - LOGPX - LOGP - LOGEX - LOGWP - LOGIPI)_{t-1} + u_{it} \quad (2)$$

where k is the number of lag, ΔLOG represents the change in natural logarithm.α1, α2, α3, α4, α5 are the parameters of the independent variables, λ_1 (LOGQXD_ - α - LOGPX-LOGPS-LOGEX -LOGWP-〚LOGIPI)〛 _(t-1) known as error correction term and uit is the stochastic error term. The optimum lags length k is chosen based on the minimum values of Akaike information criteria (AIC) and Schwarz information criteria (BIC).

2.3 Economic Growth Model

The economic growth function is assumed to depend on the export earnings of major commodities. A new model has been developed based on previous research made by [12, 13, 14]. The economic growth equation was shown below:

$$LOGRGDP_t = \beta_0 + \beta_1 LOGXEP_t + \beta_2 LOGXEPO_t + \beta_3 LOGXER_t + \beta_4 XES_t + \beta_5 XET_t + \beta_6 LOGFDI_t + \beta_7 LOGTL_t + \beta_8 WRGDP_t + e \quad (3)$$

where,

Dependent variable: **LOGRGDP**: GDP growth Malaysia (%). **Independent variable: LOGXEP**: Export earning of petroleum (RM million) **LOGXEPO**: Export earning of palm oil (RM million), **LOGXER**: Export earning of natural rubber (RM million), **LOGXES**: Export earning of sawlog (RM million), **LOGXET**: Export earning of tin (RM million). **Control variable**: **LOGFDI**: Foreign Direct Investment: **LOGTL**: Total Labour Force, **LOGWRGDP**: World GDP growth (%)

Initially, if the variables confirm the existence of cointegration, then the VECM is estimated. In this study, the VECM model is as follows:

$$\Delta LORGDP_t = \Theta_0 + \sum_{i=1}^{k} \Theta_{1i}\, \Delta LOGRGDP_{t-i} + \sum_{i=1}^{k} \Theta_{2i}\, \Delta LOGXEP_{t-i} + \sum_{i=1}^{k} \Theta_{3i}\, \Delta LOGXEPO_{t-i} + \sum_{i=1}^{k} \Theta_{4i}\, \Delta LOGXER_{t-i} + \sum_{i=1}^{k} \Theta_{5i}\, \Delta LOGXES_{t-i} + \sum_{i=1}^{k} \Theta_{6i}\, \Delta LOGXET_{t-i} + \sum_{i=1}^{k} \Theta_{7i}\, \Delta LOGFDI_{t-i} + \sum_{i=1}^{k} \Theta_{8i}\, \Delta LOGTL_{t-i} + \sum_{i=1}^{k} \Theta_{9i}\, \Delta LOGWRGDP_{t-i} + \lambda_1 (LOGRGDP - \Theta - LOGXEP - LOGXEPO - LOGXER - LOGXES - LOGXET - LOGFDI - LOGTL - LOGWRGDP)_{t-1} + u_{it} \quad (4)$$

where k is the number of lag, ΔLOG represents the change in natural logarithm. Θ_1, Θ_2, Θ_3, Θ_4, Θ_5, Θ_6, Θ_7, Θ_8, Θ_9 are parameters of the independent variables, λ_1(LOGGDP-Θ-LOGXEP-LOGXEPO-LOGXER-LOGXES-LOGXT-LOGFDI-LOGTL-LOGRWGDP)$_{t-1}$ known as error correction term, and u_{it} is the stochastic error term. The optimum lengths lag (k) is chosen based on the minimum values of Akaike information criteria (AIC) and Schwarz information criteria (BIC).

2.4 Empirical Analysis

From Table 1, the unit root testing of first difference data shows that all research variables have an ADF probability value either 5% or 1% significant level. This means that all research variables have stationary on the first difference so that the variables can be said to be integrated into degrees 1 or I (1). Thus, there exists the possibility that they share a long-run equilibrium relationship.

The choice of optimal lags for the VECM system is selected based on Akaike information criterion (AIC) and Schwarz information criterion (BIC) which is 1 for the model; petroleum, rubber, sawlog, tin, economic growth and 2 for palm oil model. Diagnostic statistics reveal that the residual is generally well behaved and in particular free from autocorrelation and heteroskedasticity problem.

Table 1. Unit *Root Test*

		Augmented Dickey-Fuller (t=1980-2014)			
Variables	Level	First Difference	Variables	Level	First Difference
LOGQXDPO	-3.8573**	9.4473***	LOGRGDP	-4.3013	-4.344291
LOGXPOP	-2.1362	5.4115***	LOGXEP	1.9114	5.7603***
LOGPOS	-2.0368	4.1325***	LOGXEPO	3.8667**	6.8108***
LOGQXDR	-2.618	6.0465***	LOGXER	2.659	5.7512***
LOGRP	-1.7703	3.0038**	LOGXES	3.6616**	5.83508***
LOGSRP	-3.7634**	5.9782***	LOGXET	0.7336	5.5758***
LOGQXDS	3.6616**	5.8350***	LOGFEDI	-2.9536	-7.3875***
LOGSP	2.1512	2.9984**	LOGTL	-1.6258	-4.981**
LOGSWP	1.4942	6.5148**	LOGWRGDP	-5.8538	-4.6177***
LOGQXDT	2.7049	5.7450***	LOGXP	-1.017546	3.3316**
LOGTP	1.9557	3.0049**	LOGPP	3.4433	3.1560**
LOGAP	5.1181***	3.9173**	LOGLNGP	1.2525	7.9640***
LOGEX	1.7855	6.3008***	LOGEX	1.7721	5.9498***
LOGWP	1.9341	5.7664***	LOGWP	1.6337	5.4558***
LOGIPI	1.7388	4.5072***	LOGIPI	1.5101	4.4413***

Note: *** 1%, **5%, *10% level of significance

2.5 Multivariate Cointegration Tests

Given the common integrational properties of the export demand, the next stage in the analysis is to test for the presence of the cointegration the five separate models of the major export commodities and also one economic growth model. In this analysis, we employed the Johansen and Juselius (JJ) procedure of testing multiple cointegrating vectors. The JJ procedure is established on a unified framework of estimating and testing cointegrating relations within the VECM formulations. Next appropriate statistic is provided to test a hypothesis for the number of cointegrating vectors and also test any restrictions upon the coefficients of the vector. Table 2 provides the results from the cointegration test. Both the maximum eigenvalue and trace statistics indicate there exist at most a single cointegrating vectors for the entire sample period since the null of r = 0 can be rejected at the 5% significant level. Given in the result is the presence of at least one cointegrating vector within each of the models.

Table 2. Johansen Cointegration Test Statistic

	Petroleum		Palm Oil		Rubber	
H_0	Trace	Max	Trace	Max	Trace	Max
r = 0	123.87*,***	58.52*,**	139.91*,***	67.64*,***	164.45*,**	84.69*,**
r = 1	65.35	22.64	72.28*	28.09	79.76*	39.20*,**
r = 2	42.71	16.67	44.19	18.27	40.6	17.45*,***
r = 3	26.04	14.49	26.92	13.49	23.11	14.65
r = 4	11.55	9.33	12.44	7.58	8.46	7.19
	Sawlog		Tin		Economic Growth	
H_0	Trace	Max	Trace	Max	Trace	Max
r = 0	101.28*	48.07***	112.88*,***	54.19*,**	227.15**	66.06*,**
r = 1	53.21	19.12*,***	58.69	33.52*,**	161.08**	46.35

r = 2	34.09	15.09	25.17	13.68	114.73	29.24
r = 3	18.99	10.99	11.49	7.41	85.49	23.91
r = 4	7.99	7.95	4.08	4.06	61.58	19.89

Note: *** 1%, **5%, *10% level of significance

Coefficient estimates and significance levels associated with the tests of zero loading restrictions appear in Table 3 for each export demand model. Based on petroleum export (LOGXDPO) result, in the long run, variables such exchange (LOGEX), world population (LOGWP) and industrial production index (LOGIPI). However the price of petroleum (LOGPOP) and liquefied natural gas price (LOGLNGP) are not statistically significant. The significant exchange rate as one of the determinants of the foreign demand of petroleum is consistent with theory; if the Malaysia ringgit depreciates then it will raise the competitiveness of the domestic commodities thus encourages export [15]. However, the coefficient of the world population does not meet the expected sign. This is also in line with the research of to [16] as the populace of a country moves to urban area their energy consumption in the transportation sector will be reduced, thus will give impact to the energy export demand.

Meanwhile, the price of palm oil (LOGPOP) is not a significant factor which determines the world demand market for Malaysia's palm oil market (LOGXDPO) but the exchange rate (LOGEX), price of soybean (LOGPOS) and industrial production index (LOGIPI) could explain the variation in palm oil export demand where they are significant at 1 per cent level. The result is parallel with the studies done by [8] that confirms the belief that soybean oil was indeed a substitute for palm oil. According to [17], the significant negative impact of exchange rate to palm oil export can be seen during the financial crisis 1997, whereas the consumer change in food preference and agriculture commodities which serve as an intermediate good only account a small percentage of the household expenditure which generally remained unchanged. Contrary with the theory, the world population has a negative relationship with quantity demand of palm oil export. This could be related to the decreased of palm oil consumption as a biofuel in the UK due to the policy changes such as Renewable Energy Directive (RED) and Greenhouse Gas Emission Target [28].

Variables such as synthetic rubber price (LOGSRP), exchange rate (LOGEX), world population (LOGWP) and industrial production index (LOGIPI) are significantly influenced rubber export in the long run (LOGXDR). Even though, the signs of the coefficient estimates on world population are positive as expected but negative for other variable. Support the findings that natural rubber and synthetic rubber have a complementary relationship [18]. As the currency of country depreciated, the rubber export volume will decrease due to the theory in demand of the agriculture product such rubber is inelastic [19]. Where it was found out that industrial production index and quantity demand for rubber export has a negative relationship because, as the import of rubber from Sri Lanka increased, India had reduced importing rubber from Malaysia [20].

Besides that, in the case of sawlog, all variables such sawlog price (LOGSP), exchange rate (LOGEX), world population (LOGWP) and industrial production index (LOGIPI) are theoretically significant in determining quantity demand of sawlog export (LOGDXT) except for sawnwood price (LOGSWP). Based on argued that the sawlog price and export demand of sawlog are expected to inversely related *ceteris paribus* [7, 11]. Related found a positive relationship between exchange rate and sawlog export in Malaysia [21]. According to [22], population growth is one of the main factors that influence the demand for forestry product.

In addition, for tin, the price of tin, aluminium price and world population has the strong positive response to the export demand of tin and negative response of exchange rate to the world demand for tin. The studies which has done support this finding that exchange rate and tin has a negative relationship with quantity demand for tin export [7, 23].

Table 3. Summarised from Result of VEC for Export Demand Model

Variables	Petroleum	Palm Oil	Rubber	Sawlog	Tin
LOGPP	-0.07	/	/	/	/

LOGLNP	0.05	/	/	/	/
LOGEX	1.17***	-6.2***	1.26***	4.54*	-0.9***
LOGWP	-0.68***	-1.61***	0.66***	4.9***	-0.79***
LOGIPI	-1.2***	4.67***	-0.04***	29.82***	-0.02*
LOGPOP	/	-0.42	/	/	/
LOGPOS	/	/	/	/	/
LOGRP	/	/	0.07	/	/
LOGSRP	/	/	0.09*	/	/
LOGSP	/	/	/	/	/
LOGSWP	/	/	/	-24.67***	/
LOGTP	/	/	/	-3.89***	-0.9***
LOGAP	/	/	/	/	0.12***

Note: *** 1%, **5%, *10% level of significance

Given the presence of cointegration among the variables, it would be interesting how these five major commodities export will give impact to the economic growth in the long run.

$$LOGRGDP_t = 0.72LOGXEP_t^{***} + 0.99LOGXEPO_t^{***} - 2.24LOGXER_t^{***} + 1.69LOGXES_t^{***} - 0.61LOGXET_t^{***} + 0.29LOGWRGDP_t^{**} - 0.87LOGFDI_t^{***} - 1.29LOGTL_t^{**} \quad (5)$$

In the long run, almost all the coefficients are individually significant at 1% significance. There is a positive relationship between the economic growth (LOGRGDP) and petroleum export (LOGXEP), palm oil export (LOGXEPO), sawlog export (LOGXES), rubber export (LOGXER), tin export (LOGDXT) and Malaysia's GDP growth (LOGRGDP). This is parallel to the study done by [26] where Malaysia is one of the oil exporting countries that heavily depends on oil and gas as the source of GDP growth. According to [25], since 1985, Malaysia had reduced its dependence on rubber, and tin export to diversifying its source of income as fall in the export earning caused by declined in rubber and tin export price give negative impact to the overall Malaysian economy. Besides that, sawlog was one of the value-added product and generated RM 47.9 billion total value added to the Malaysia economy [26]. The control variable world economic growth (LOGWRGDP) has a positive significant whereas FDI (LOGFDI) and total labour force (LOGTL) has negative significant to Malaysia's economic growth. Due to the unstable price of rubber and tin and also the closure of several mines in 1992, the employment opportunities were affected, thus give a negative impact to the Malaysia economy [25]. The finding that FDI has a negative relationship with GDP growth [27].

VECM conveniently combines the long run behaviour of the variables and their short run relations and thus can better reflect their relationships among variables. Table 4 shows the matrix coefficients of five different model. As can be seen, the resulting support for the export demand of commodity and the contribution of commodity export to economic growth in Malaysia as a short run process whereby all the independent variables shock has contributed to short-run shocks do not last then one year. Based on petroleum result, only exchange rate lag 1 is positively significant whereas variables at lag 1 such exchange rate, world population and industrial production index are significant in the short run. However, world population has a negative relationship while others two variables have a positive relationship with the quantity demanded of petroleum export. For the case of palm oil, variables such exchange rate at lag 1 and world population at lag 1 and lag 2 respectively have significant negative relationship with quantity demand of palm oil export. For the case of rubber, the results show that the world population and industrial production index in the previous year influencing quantity demand of rubber export in the short run. Besides that, only lagged coefficient of sawlog price is negatively significant with the quantity demand of sawlog export in the short run. In addition, for tin, the lagged coefficient of the exchange rate, world population and industrial production index are significant in the short run. Based on Table 5, the positive coefficient of one-year lag of petroleum and rubber is significantly different from zero. This means that in the more earnings from this, both type of commodity will lead to higher economic growth in the short term.

Table 4. Matrix of Short-Run Coefficient for Export Demand

	ΔLOGXDPO	ΔLOGXDOP	ΔLOGXDR	ΔLOGXSD	ΔLOGQXDT
ECT	-0.25	-0.43	-0.74	-0.06	-0.69
ΔLOGXDPO(1)	-0.32	/	/	/	/
ΔLOGXDPO(-2)	-0.11	/	/	/	/
ΔLOGPOP(1)	0.05	/	/	/	/
ΔLOGPOP(-2)	0.2	/	/	/	/
ΔLOGPOS(-1)	0.32	/	/	/	/
ΔLOGPOS(-2)	0.35	/	/	/	/
ΔLOGEX(-1)	-1.14*	0.69**	-0.23	0.08	-1.31**
ΔLOGEX(-2)	-0.34	/	/	/	/
ΔLOGWPC(-2)	-1.04*	0.29*	0.73**	-0.09	0.42**
ΔLOGWPC(-2)	-0.94*	/	/	/	/
ΔLOGIPIC(-1)	0.35	0.20	-0.01***	-1.26	-3.72**
ΔLOGIPIC(-2)	0.04	/	/	/	/
ΔLOGDOPC(-1)	/	0.14	/	/	/
ΔLOGPP(-1)	/	0.11	/	/	/
ΔLOGLNGP(-1)	/	0.14	/	/	/
ΔLOGXDR(-1)	/	/	-0.14***	/	/
ΔLOGRP(-1)	/	/	-0.02	/	/
ΔLOGSRP(-1)	/	/	0.01	/	/
ΔLOGXS(-1)	/	/	/	0.16*	/
ΔLOGSPC(-1)	/	/	/	1.06	/
ΔLOGSWP(-1)	/	/	/	0.08	/
ΔLOGDXT(-1)	/	/	/	/	0.42*
ΔLOGTPC(-1)	/	/	/	/	-0.32
ΔLOGAP(-1)	/	/	/	/	-0.08

Note: *** 1%, **5%, *10% level of significance

Table 5. Matrix of Short Run Coefficient for Economic Growth

	ECT	ΔLOGRGDP(-1)	ΔLOGXEP(-1)	ΔLOGXEPO(-1)	ΔLOGXER(-1)
Coefficient	-0.812	-0.1473	0.514**	0.941	0.604**

	ΔLOGXES(-1)	ΔLOGXET(-1)	ΔLOGFDI(-1)	ΔLOGTL(-1)	ΔLOGWRGDP(-1)
Coefficient	0.271	-0.584	-0.429	-0.31	-0.018

Note: *** 1%, **5%, *10% level of significance

2.6 Impulse Response Analysis

To fully appreciate the fully impacts that innovation in a variable of interest on remaining variables, we stimulate the generalised impulse- response functions using a level VAR specification. We opt for the generalised impulse response function developed by Pesaran and Shin to circumvent the results sensitivity to the variables alternative ordering in the traditional Cholesky decomposition. Figure 1 presents the plots of the impulse response function for five different model of major commodities export demand.

IOP Conf. Series: Earth and Environmental Science **327** (2019) 012002 doi:10.1088/1755-1315/327/1/012002

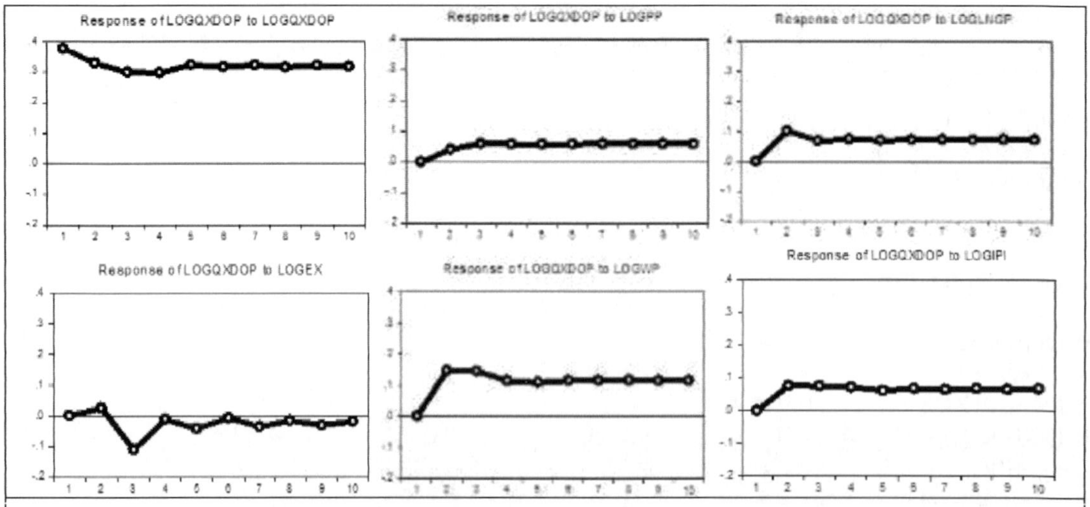

Figure 1 (a) Impulse Response Function Petroleum

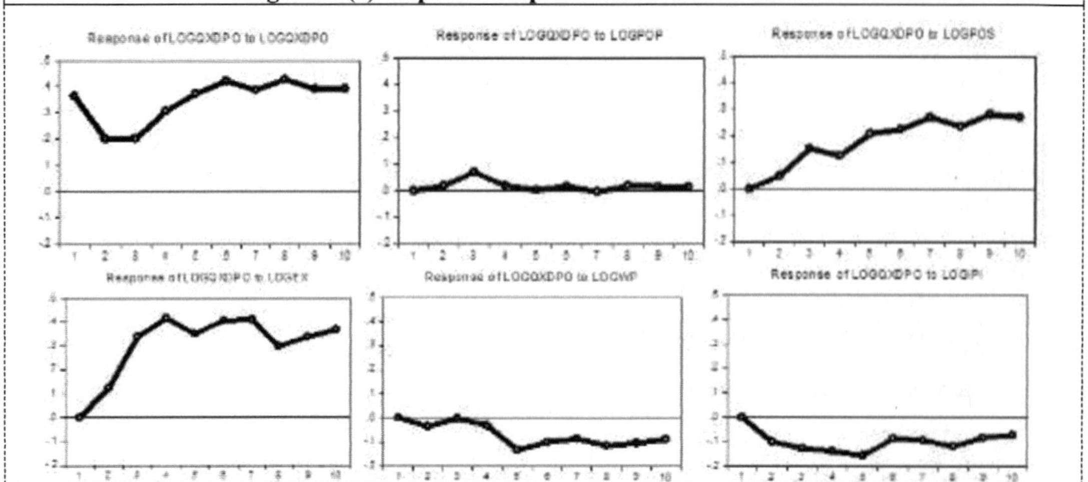

Figure 2 (b) Impulse Response Function Palm Oil

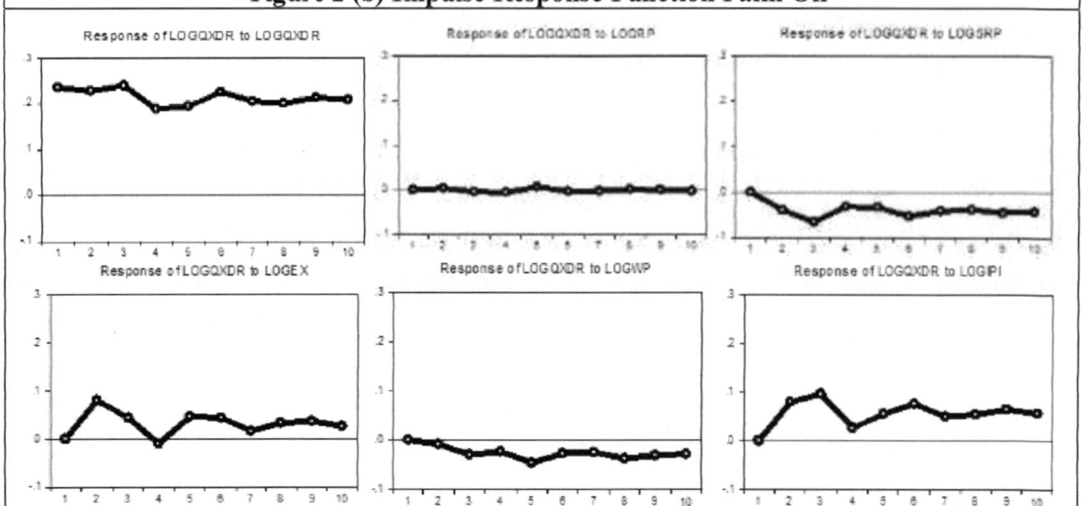

Figure 3 (c) Impulse Response Function Rubber

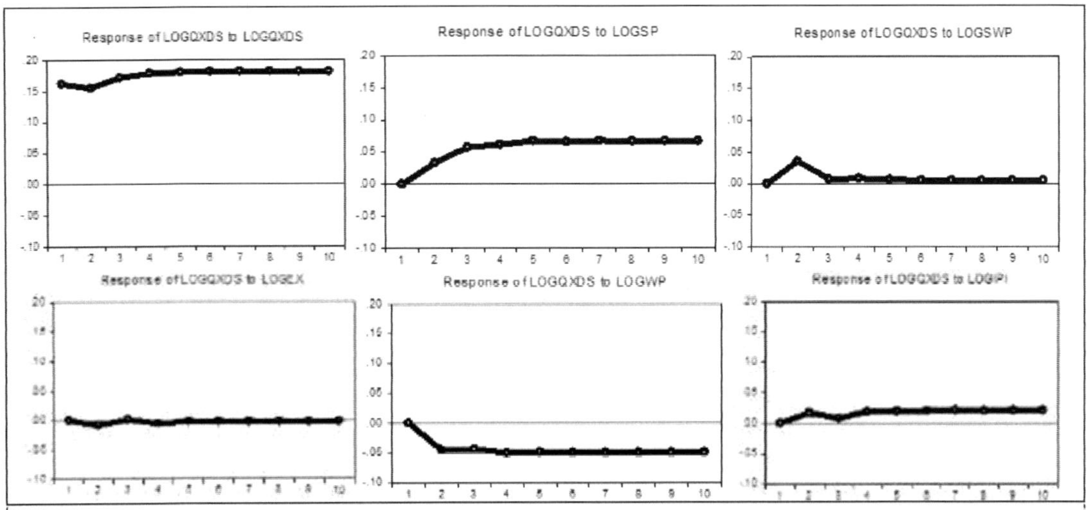

Figure 4 (d) Impulse Response Function Sawlog

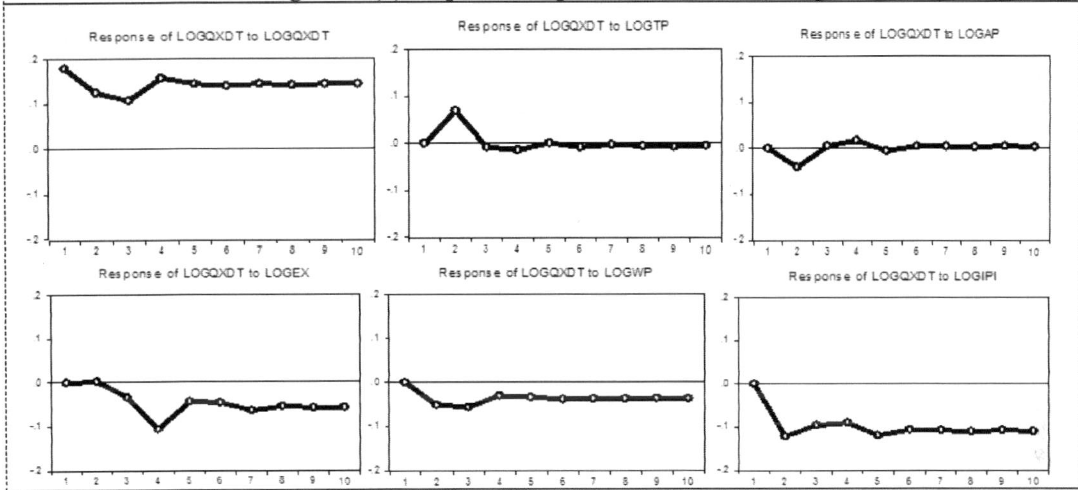

Figure 5 (e) Impulse Response Function Tin

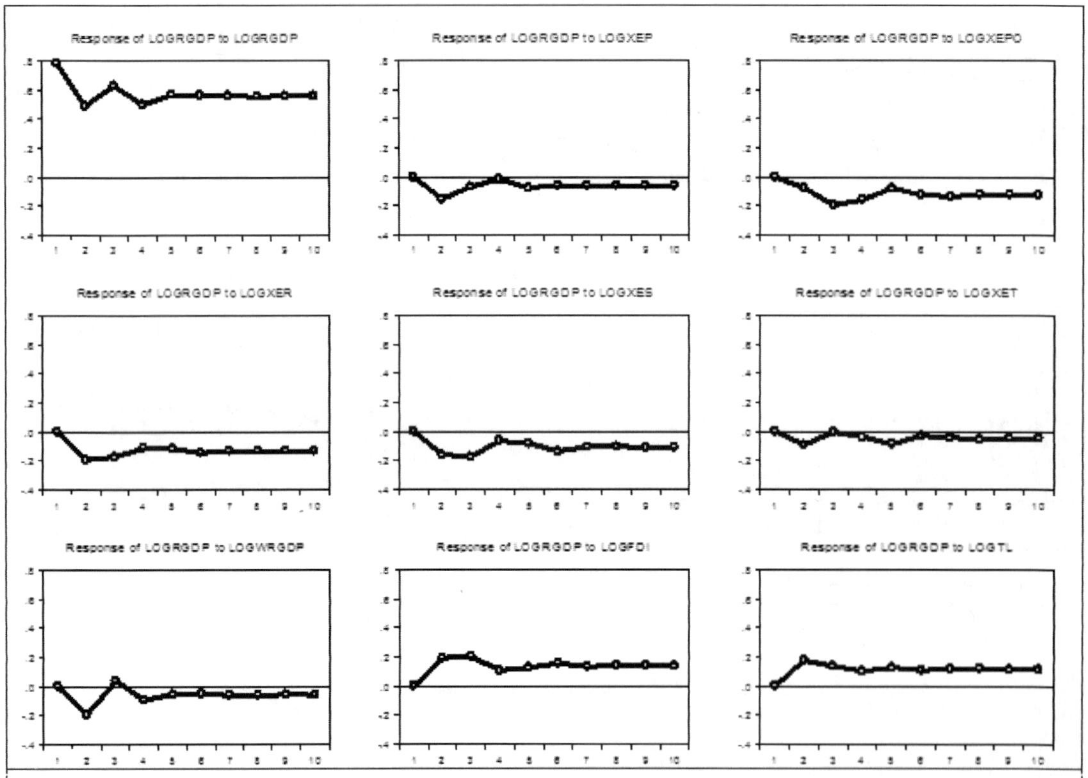

Figure 6 (f) Impulse Response Function Economic Growth

Figure 1 (a) shows in general petroleum's export demand significantly responded to all research variable shocks. The shocks increase in the following period until year 5, and after that it becomes stable. In contrast, there are positive shocks of the exchange rate to petroleum export in the first two years, but it continues to be negative fluctuated in the next eight years.

Figure 1 (b) shows that the quantity demanded of palm oil export responses positively to palm oil price, soybean price, exchange rate and it last for ten years. One standard deviation shock in palm oil export results in a negative response by industrial production index and world population after year 3.

Figure 1 (c) shows in the 2nd and 3rd-year exchange rate and industrial production index gives a positive response to rubber export and follows by fluctuating movement in the rest period while rubber price tends to be stable over the entire horizons. However, rubber export directly influenced negatively by synthetic rubber price and world population by one standard deviation.

Figure 1 (d) shows the positive impact of sawlog export itself, sawnwood price, sawlog price exchange rate and industrial production index by a standard deviation to sawlog export, but all variables tend to be stable after five years. However, the world population gives a negative response to sawlog export.

In figure (e) there seems to be the positive response of the tin export itself, aluminium price, tin price even though at first three years aluminium price responds negatively to tin export demand. For the case of exchange rate and industrial price index, their responses to the tin export are first positive and then turn negative after 2nd and 1st year respectively.

Figure (f) shows the response for petroleum export earning, palm oil export earning, rubber export earning, sawlog export earning, tin export demand towards Malaysia's economic growth is shown up and down movement and tend to balance after 6th year. However, FDI, economic growth and total labour respond positively fluctuated and become stable after the 8th year.

3. Conclusion

In this study, we employed VECM method to investigate the factors that influenced export demand of Malaysia's major commodities and to find out if these commodities export play any role in stimulating Malaysia's national income. Our result suggests that a long-run equilibrium relationship has existed for all five major export commodities and the economic growth model. Exchange rate, world population, and industrial production index influenced all exports of major commodities. All of them are also significantly influenced by their substitutes except for petroleum while only sawlog is have affected by its own price. Besides, it is evidence that major commodities export earning act as a stimulus to GDP growth in the long term. The respond of its own commodities export and price are positive whereas for other response variable, there are positive and negative effects existed when then impulse is major commodities export. FDI and total labour force give a positive impact when then impulse is economic growth. Thus Malaysian government should intensify efforts on the export of major commodities to generate revenue and foreign exchange needed to economy further. The improvement of simultaneous trade liberation and human capital policy will ensure increased investment by both local and foreign investor.

References

[1] Ministry of Finance Malaysia [MOF] 2013 *Economic Report* 2013/2014: 28-80
[2] Ministry of Finance Malaysia [MOF] 2014 *Economic Report* 2014/2015: 3/1-3/45
[3] Lembaga Getah Malaysia [LGM] 2014 *Natural Rubber Market Review* Malaysia Rubber Digest
[4] Mahfoor H, Mad Nasir S, and Ismail I L 2001 Challenges of Agribusiness: a Case for Malaysia *Agribusiness Management towards Strengthening Agriculture Development and Trade* (pp. 388-400) Thailand: Chiang Mai University
[5] Ministry of Finance Malaysia [MOF] 2015 *Economic Report* 2015/2015: 2/1-2/13
[6] Goldstein M and Khan M S 1978 The Supply and Demand for Exports: A Simultaneous Approach *Review of Economics and Statistics* **60** 275-289
[7] Mohamed Yusoff and Ahmad Zubaidi Baharumshah 1993 The Effects of Real Exchange Rate on the Demand for Exports *ASEAN Economic Buletin* **9**(3) 338-347
[8] Basri Abdul Talib and Zaimah Darawi 2002 An Economic Analysis of the Malaysian Palm Oil Market *Oil Palm Industry Economic Journal* **2**(1) 19-27
[9] Nourah A R A 2005 Modelling and Forecasting the Demand for Crude Oil in Asian Countries. *25th USAEE/ IAEE North American Conference: Fueling the Future: Prices, Productivity, Policies and Phropheis* (pp, 1-36)
[10] Ambiyah A 2011 Determinants of Palm Oil Export: Price and Income Elasticity Estimation. *Trends in Agricultural Economics* **4**(2) 52-57
[11] Abdul Rahim A S, Mohd Shahwahid H O, Mad Nasir S, and Awang Noor A G 2013 Market and Welfare Economic Impacts of Sustainable Forest Management Policy on Timber Market in Sarawak, Malaysia *Middle-East Journal of Scientific Research* **13**(8) 972-982
[12] Narayan S 2000 The Relationship between Commodity Prices and Output/Exports *Working Paper 2000/07*
[13] Diana Suziliana Maruzeki 2006 *Sumbangan Eksport Komoditi Terpilih Kepada Pertumbuhan Ekonomi Malaysia* TSm07 PE D5 Fakulti Ekonomi dan Pengurusan Universiti Kebangsaan Malaysia
[14] Ojo E J, Awe L T, and Olufemi O J 2014 Agricultural Export and Economic Growth in Nigeria: A Multivariate Johansen Cointegration Analysis *International Journal of Arts and Commerce*, **3**(3) 89-98
[15] Keong C C, Zulkornain Y, and Sen V L K 2005 Export-Led Growth Hypothesis in Malaysia: An Investigation Using Bound Test *Sunway Academic Journal* **2** 13-22
[16] Polimeni J M, Mayumi K, Giampetro M, and Alcott B 2008 *The Jevons Paradox and the Myth of Resource Efficiency Improvements* Earthscan
[17] Noor Zahirah Mohd Sidek, Mohammed Yusoff, Gairuzazmi Ghani, and Jarita Duasa 2011

Malaysia's Palm Oil Exports: Does Exchange Rate Overvaluation and Undervaluation matter? African Journal of Business Management **5**(27) 11219-11230
[18] Sakarindr P 1979 *An Econometric Study of Thai Rubber in the World Rubber Market* Iowa State University
[19] Tulasombat S, Bunchapattanasakda C, and Ratanakomu S 2015 The Effect ofExchange Rates on Agricultural Goods for Export: A Case of Thailand *Information Management and Business Review* **7**(1) 1-11
[20] Berendson B S M 1978 *Regional Models of Trade and Development* Interprint Ltd
[21] Deivani K 2000 *The Impact of the Recent Currency Crisis on Malaysia Export* University of Malaya, Malaysia
[22] Kanabahita C 2001 *Forestry Outlook Studies in Africa (FOSA)* Uganda Forestry Department, Ministry of Water, Lands and Environment, Kampala–Uganda
[23] Bojanic A N and Caudill S B 1992 The Demand for Bolivian Tin *Atlantic Economic Journal* **20**(4) 8
[24] Mohd Azfanizam Abdul Rashid 2015 *Oil and the Economy-Measuring the Relationship*
[25] Taylor R J 2007 Technical Progress and Economic Growth an Empirical Case Study of Malaysia *Edward Elgar Publishing Limited*
[26] Noor Aini Z, Nor Hazmira M, and Ismariah A 2014 Assessment of rubberwood value added in Malaysia's wooden furniture industry *International Journal of Economics and Management* **8**(1) 1-9
[27] Jamal O, Yaghoob J, and Tamat S 2014 Economic Growth, Foreign Direct Investment, Macroeconomic Conditions and Sustainability in Malaysia *Applied Econometrics and International Development* **14**(1) 214-223
[28] DEFRA 2013 *UK Consumption of Sustainable Palm Oil*

Landslide Susceptibility of Agricultural Area in Lembah Gumanti, Solok - West Sumatra Province

F Arlius[1], I Berd[1], and R M Sadewo[1]

[1] Department of Agricultural Engineering, Andalas University, Padang, Indonesia

Corresponding author's e-mail address: sipado@yahoo.com

Abstract. One of the factors that threaten food resilience in Indonesia is the natural disasters such as landslides. Cultivation of agricultural crops in hilly terrain with fairly steep slopes and high rainfall without regard to land conservation and good farmland cultivation practices will result in landslide disaster that could damage the irrigation and agricultural land. Soil movement is a consequence of natural dynamic phenomenon to reach new condition due to disturbance of balance in environment, either by natural or by human activity. Movement of the ground will occur on a slope, if there is a state of imbalance that causes the occurrence of a mechanical process, resulting in a part of the slope moving according to the force of gravity. The process of ground motion involves complex interactions between geological, geomorphological, hydrological, rainfall and landuse aspects. A research about landslide susceptibility of agricultural area has been conducted in Lembah Gumanti, Solok, West Sumatra Province. The research show that from the total area of 24,294 ha, high vulnerability zone is about 903 ha (4%), medium vulnerability zone is 6,913 ha (28%), and low vulnerability zone 16,478 ha (68%). Landslide vulnerability in agriculture area potentially arise on dry land farming (525 ha), paddy field (3 ha) and mixed garden (375 ha).
Keywords: Landslide; Food Resilience; and Solok

1. Introduction

From long time ago, farmers have settled and used hilly areas to fulfill their daily needs and support the family's economy by cultivating horticulture, plantations and food crops. The use of hilly land by farmers can lead to natural disasters such as landslides that will affect sustainable agricultural land. In this area the farmers mostly cultivate horticulture in hilly land with fairly steep slopes and high rainfall without regard to the principle of land conservation and cultivation procedures for sustainable agricultural land in the hilly area [1, 2, 3, 4].

The cultivation of agricultural land in West Sumatra also utilizes hilly land. One of the locations of the hilly farming system is in Lembah Gumanti District, Solok Regency. In this sub-district, there are four villages: Alahan Panjang, Sungai Nanam, Salimpat and Aie Dingin, where most of the population work as farmers that cultivate several types of horticultural crops. Landslides have occurred several times in this area which have caused land damage, such as loss of forest areas, damage to agricultural land and crop failure.

Identification of landslide susceptibility was considered necessary to avoid disasters and to design conservation systems for sustainable agricultural cultivation. The study entitled "Landslide Susceptibility of Agricultural Area in Lembah Gumanti, Solok - West Sumatra Province" has been carried out with the aim of finding out ar eas that have potential landslides in Lembah Gumanti District in Solok Regency in order to maintain sustainable agricultural land.

2. Methods

This research was conducted in Lembah Gumanti Subdistrict, Solok Regency, while data processing was carried out at the Laboratory of Agricultural Management and Geographic Information Systems, Faculty of Agricultural Technology, Andalas University, Padang.

The material used in this study is an administrative map of Gumanti Valley, DEM maps that are utilized for making slope maps, land use maps, soil type maps, and the last 10 years rainfall data and lithology maps, ArcGIS software, GPS and others. The initial step taken for this research is the preparation of a basic map that includes information on the location of the study, administrative maps, maps of soil types, land use maps, lithology maps, rainfall maps and DEM map analysis. The last 10 years of rainfall data form the basis for making rainfall distribution maps using the Polygon Thiessen method. Thematic maps that have been collected and created were scored and weighted according to the indicator parameters as stated [5].

Landslide vulnerability was categorized into 4 classes, namely: (1) High ground movement vulnerability zone, (2) Middle ground movement vulnerability zone, (3) Low ground movement vulnerability zone, (4) Very low ground movement vulnerability zone. Field observations were also made to validate the results of the analysis with the actual conditions in the field.

3. Results and Discussion

Lembah Gumanti is one of the sub-districts located in Solok Regency, West Sumatra, Indonesia. Lembah Gumanti Sub-District has a total area of 24,293 ha, where the area is divided into four villages, including Aie Dingin, Salimpek, Sungai Nanam and Alahan Panjang. The sub-district is about 65 km from the city of Padang, and located on the eastern slope of the Kerinci Seblat National Park. Geographically, the research area is located between 00o59'0''- 01o13'0" South Latitude and 100042'0'' – 100056'0'' East Longitude, with altitude is about 1,400 to 1,600 meter above sea level.

The annual rainfall was 2,837 mm, that was calculated from 2 rainfall stations that represented this research area. In Lembah Gumanti there are two types of soil, namely Gley Humus soil and Cambisol soil, where the Gley Humus soil is categorized as class of soil sensitivity to low landslides and Cambisol soil is classified as medium category. There are 4 rock types in this area, namely intrusive, sedimentary, volcanic and acidic volcanic rocks.

Intrusion rocks and acidic volcanic rocks are categorized as low sensitivity to landslides while sedimentary rocks are in the category of sensitivity to moderate landslides, and volcanic rocks have a sensitivity to high landslides. Sedimentary rocks have the largest area, where these rocks have a tendency to moderate landslides, so if these sedimentary rocks are in steep slopes, they can endanger so volcanic rocks, if sedimentary rocks and volcanic rocks are found in areas that are inflated and not accompanied by plants that have strong roots, landslides can occur if rainfall in the area is high.

Land use in the study area is generally dominated by forests, where protected forests have an area of 9,420 hectares and 2,514 hectares of forest, this is because the study area is in the Bukit Barisan area. The agricultural land in the study area is also categorized as large. where the agricultural land is divided into 4 namely mixed gardens, horticulture land, dry land agriculture and wetland agriculture.

Lembah Gumanti Subdistrict is more dominated by 15-45% slope class which covered 13,172 ha or more than 50% the area of Lembah Gumanti Subdistrict. Slope > 75% had an area of 1,025 ha, where this area needs to be considered as a conservation zone to avoid landslides that will negatively affect the environment.

To get a understanding about the level of landslide vulnerability in the study area, thematic maps (rainfall, soil type, lithology, land use and slope) that have been given scores and weights were overlayed together. Spatial analysis of landslide vulnerability showed that in the study area there were 903 ha of areas classified as areas with high landslide vulnerability, about 7,496 ha area with medium landslide vulnerability and 15,894 ha areas with low vulnerability (Tabel 1).

Refers to landuse in the study area, dry land is the area with the highest vulnerability of landslide in the study area, which is 525 ha, while for the vulnerability of medium landslide, dry land also occupies the widest area of 3,484 ha. Medium landslide vulnerability was more dominated by agricultural land such as dry land agriculture and mixed gardens. Dry land agriculture and mixed gardens has the largest area at the level of medium landslide vulnerability compared to low landslide vulnerability.

Table 1. Landslide susceptibility level in various landuse

Landuse	Landslied Susceptibility Level (ha)			Total
	High	Medium	Low	
Forest	0	1,759	10,176	11,935
Mix Farming	375	2,174	2,026	4,575
Holticulture	0	8	450	458
Settlement	0	56	219	274
Wet land	3	15	11	29
Dry Land	525	3,484	3,012	7,021
Total	903	7,496	15,894	24,922

The use of agricultural land in relation to surface flow and sedimentation, as well as land damage such as landslides, where such damage occurs in land clearing and processing of the land, land clearing is often carried out less in accordance with rules such as the application of conservation rules and the fields are steep slopes. In dry land agriculture, farmers cultivated horticultural crops such as shallots, tomatoes, and vegetables, while for mixed garden was located at this level of high landslide vulnerability covering an area of 375 ha and wetland agriculture such as rice field was about 3 ha. The high landslide vulnerability is dominated by agricultural land, where in this area need to be wary of both in terms of conservation and land use.

Mostly landslide occur in dry land and mixed farming area [6, 7], while cropping systems with agroforestry models are able to absorb water optimally and use it efficiently [8]. The concept of water balance in agroforestry places agroforestry in a strategic position to reduce the chance of the role of water in the occurrence of landslides. The type of vegetation recommended by the World Bank in the development and management of land in landslide areas prone to landslides is trees that can produce fruits and candlenut trees. Whereas in protected areas planted with vegetation or trees that are in accordance with local conditions such as acacia, pine, mahogany, johar, teak, pecan, and resin.

4. Conclusion

High level landslide susceptibility area in Lembah Gumanti Subdistrict covered an area of 903 ha, while medium level landslide susceptibility area was 7,496 ha and low level landslide susceptibility area was 15,894 ha. High level landslide susceptibility occured mostly in the dry land (525 ha) and mixed garden (375 ha), while other landuses contributed to medium level of landslide susceptibility.

References

[1] Arlius F, M A Tjandra, Delvi Y 2017 Land Suitability Analysis for Development of Arabica Coffee in Solok Regency (in Indonesian) *Jurnal Teknologi Pertanian Andalas* **21/1** pp 70-78

[2] Hoang L, Roshetko J, Huu T, Pham, Pagella T, Nguyen P 2017 Agroforestry-The Most Resilient Farming System for the Hilly Northwest of Vietnam *International journal of Agricultural System (IJAS)* **5**. 10.20956/ijas.v5i1.1166

[3] Wymann von D S, Romeo R, Vita A, Wurzinger M, Kohler T 2013 *Mountain Farming Is Family Farming: A contribution from mountain areas to the International Year of Family Farming 2014* (Rome: Italy: FAO, CDE, BOKU, pp 100)

[4] Sukanlaya C and A Terry R 2018 Identification of agricultural systems in the mountains of Northeast Thailand *Journal of Mountain Science* **15**(2) 419p

[5] Wahyunto. 2007 *Vulnerability of Landslides to Agricultural Land in the Citarum River Basin, West Java (in Indonesian)* (Bogor: Balai Penelitian Tanah)
[6] Hasnawir, Kubota T, Sanchez-Castillo L, Soma A S 2017 The Influence of Land Use and Rainfall on Shallow Landslides in Tanralili Sub-watershed, Indonesia *Journal- Faculty of Agriculture Kyushu University* **62** pp 171-176.
[7] Hadmoko D S; Lavigne F; Sartohadi J; Gomez C, and Daryono 2017 Spatio-temporal distribution of landslides in Java and the Triggering Factors *Forum Geografi* **31**(1)

IOP Conf. Series: Earth and Environmental Science **327** (2019) 012004 doi:10.1088/1755-1315/327/1/012004

Analysis of Energy Consumption in Rubber Cultivation in Malaysia: A Case Study

H A Zulekipli[2] and D E Pebrian[1]

[1, 2] Faculty of Plantation and Agrotechnology, Universiti Teknologi MARA Cawangan Melaka, Kampus Jasin, 77300 Merlimau, Melaka, Malaysia

Corresponding author's e-mail: darius@uitm.edu.my

Abstract. Efficient energy use in rubber cultivation is one of the important strategies that can facilitate the sustainability and competitiveness of this crop as one of the main contributors in Malaysian economy from agricultural sector. Therefore, a case study was conducted to analyze the efficiency energy in Malaysian rubber cultivation. Data and other relevant information were gathered through a combination of oral interviews and field observations at rubber plantations located in the Rubber Research Institute of Malaysia Mini Station (RRIMINIS), Jasin, Melaka and Lakai, Negeri Sembilan, Malaysia. Mathematical calculations and computer spreadsheet were used to analyze the collected data. The findings showed that the rubber cultivation in the study areas consumed 16080.86 MJ/ha of energy. The largest amount of this energy, which accounts for 65.93% was consumed by fertilizing operation, followed by planting, weeding, collecting latex, land preparation, spraying hormones, tapping and pruning with share of 15.92%, 9.66%, 4.03%, 3.93%, 0.44%, 0.06%, and 0.04%, respectively. Energy output/inputs ratio in the rubber cultivation was 0.83. Decisively, the energy inputs used in the rubber plantations were less effective since the energy output/inputs ratio is lower than 1. Thus, energy saving-technologies and techniques are highly recommended to be introduced in rubber cultivation in Malaysia in order to achieve sustainability of economics and productivity of rubber planting industry in the country.

1. Introduction

Rubber (*Hevea brasiliensis),* is of one the major crops in Malaysian agricultural sector. The existence of rubber planting industry has been around in very long and even over a century in this country. With a total of planted areas of 1,007,000.87 ha either run by estates and smallholding as reported by MRB [1], Malaysia is recorded as one of the largest producers of natural rubber (NR) in the world. The total exports natural rubber (NR) amounting to 1,193,891 tons in 2017, which shows an increase of 176,285 tons or 14.77 percent when compared to the year 2016. In addition, Malaysian rubber industry additionally provided about 4.41% contribution rates to the country exports in 2017[1]. With those achievements, this rubber plantation industry continuously forms the backbone and becomes one of the key sectors of the Malaysian economy, which contribute to the prosperity of the country through agriculture sector.

Similar to other sectors, agricultural activities required energy as an important input for its production processes. In fact, agriculture has a close relationship with energy since the farming has functions both an energy user and provider of energy in the form of bio-energy as well. Nowadays, modernization of agricultural production operations has led to increasing energy consumption in agricultural sector.

According to Bekhet & Abdullah [2], among the reasons that may lead to increasing energy consumption in the agricultural sector are labor shortages in the sector in line with the government's policy to reduce dependence on foreign labor and limited available land for agricultural activities due to the rapid development of residential and industrial areas. Efficient use of energy is compulsory for sustaining agricultural production economically and environmentally, because it provides financial savings, preserve fossil resources and reduce air pollution as mentioned by Pervanchon et al. [3] and Pimentel [4]. Therefore, it is important to identify the pattern of energy consumption to help the economics and environmental planning for both agriculture and energy sectors.

In order to sustain its important role in Malaysian agricultural sector, rubber planting industry in the country should be able to maintain its competitiveness through applications of efficient and effective inputs in the cultivation process with the purpose of obtaining the maximum output as possible. Therefore, the efficient use of energy in rubber cultivation has to be looked into seriously to meet these needs. As suggested by Refsgaard [5], among the ways of achieving this goal is through determining the cultivation methods that need smaller energy input with higher energy productivity as suggested by Refsgaard [5]. In other words, rubber has to be cultivated in a manner where the energy inputs are used efficiently and effectively. This is also consistent with Pimentel & Patzek [6], who said that utilizing effective energy in agriculture is one of the conditions for sustainable agricultural production as it leads to financial saving, better fossil fuel preservation and lesser air pollution.

Looking at the past research literature, many studies on the rubber industry in Malaysia have been reported by researchers such as Mustafa et al. [7], Ratnasingam et al. [8], In [9], Othman [10], Ismail [11] and Tugiman [12]. These past studies were focused on processing, economics and future prospect of this crop. Therefore, the findings were not relevant to be acquainted with the understanding of the energy use in rubber cultivation. Special attempts have been made to study energy use in various crops production systems in Malaysian agriculture. Nazri and Pebrian [13], for example, reported energy consumption in pineapple cultivation. Azwan et al. [14] examined energy utilization in Malaysian oil palm mechanization operation. Pebrian et al. [15] reported on workers' workload in oil palm cultivation. Pebrian et al. [16] also audited human energy inputs for oil palm and rice cultivations in Malaysia. Prior to that research, Gevao et al. [17] investigated energy use in Malaysian rice cultivation. The only one study with regards to energy consumption for rubber industry in Malaysia was reported by Saidur and Mekhilef [18]. They audited and analyzed the energy in the Malaysian rubber producing industries. Again nonetheless, this study was emphasized on energy use in downstream activities which focused on the rubber industry operations including tire manufacturing. Reviewing above-mentioned matter indicated that there is no study aiming at energy consumption on upstream activities such as rubber cultivation in Malaysia that has been published in the research literatures.

Therefore, a study on energy consumption in rubber cultivation in Malaysia is necessary to enrich the understanding on when, where and how much energy inputs are being expended in this crop cultivation in the field, and conclusively to recognize the opportunities to save the energy inputs for economical of operations. This paper initiates to analyze energy inputs-output used in rubber cultivation in Malaysia. Energy inputs-output analysis is employed to assess the effectiveness of energy use for a crop production system. The distribution of energy inputs, energy efficiency and energy pattern in the rubber cultivation were also explored. However, this study was limited to analyze the energy consumption in fields operations only. Energy analysis in nursery stage operations was not included. Specifically, this study could contribute to the development of a database for energy consumption in rubber cultivation in Malaysia, and it can be a guideline in applying related strategies and policies in the rubber planting industry in the country. Generally, this study also would be beneficial to enrich the database of energy consumptions in various crops production systems in Malaysia.

2. Material and Methods

2.1. Data collection

The data was collected through field observations and oral interviews at rubber plantations in RRIMINIS Jasin, Melaka and Lakai, Negeri Sembilan, Malaysia. Both the locations were governed by the Malaysian Rubber Board (MRB). The cultivated areas surfaces were considered as the undulating terrains. Data were obtained from field observations of agronomic practices such as land preparation, planting seedlings, weeding, fertilizing, pruning, tapping, collecting latex at twelve experimental plots. Agricultural inputs used in the operations were recorded. Active ingredient content of agrochemical inputs such as fertilizer and herbicides were also noted. Oral interviews were employed to gather general data from the farm management.

2.2. Field operations description

Field operations observed in rubber cultivation are land preparation, planting seedlings, weeding, fertilizing, pruning, tapping and collecting latex. Land preparation was conducted to remove the debris, plant residues and weeds. Appropriate herbicide such as glyphosate was used to chemically eradicate the growth of the weeds on the land. Planting distance of 6 x 3 m and a planting density of 512 to 555 trees per hectare was applied in planting the seedlings. Prior to the planting seedlings commence, a 37.3 kW tractor mounted with post hole digger was operated to prepare the planting holes. The tractor also transported the seedlings from nursery to the planting areas. Weeding was undertaken by a combination of manual and chemical methods. In areas with immature trees, weeding was carried out using hand-held tools to eliminate the weed from the ground. Meanwhile, chemicals with knapsack sprayer were utilized in areas of mature rubber. Fertilizing was carried out manually by workers. Type fertilizer applied was compound fertilizer or NPK fertilizer in granular form. Throughout the operation, the worker brings along the fertilizer inside a bucket and uses his hand to broadcast the fertilizer to ground nearby the trees. A tractor 37.3 kW was used to deliver the fertilizer bags from the fertilizer store to the assigned areas. Pruning was done manually using hand-sharp steel to cut dead leaves or unwanted part of trees with the purpose of achieving better circumference of the tree for higher rubber production. Tapping the trees was executed manually using a special tapping-knife to cut the trunk of trees to harvest the latex. Tapping begins when the trunk of the tree reached a circumference of 50 cm and a height of 125 cm from the bud union. A diagonal cut downward at approximately 30°of the slope was made to wound the bark of the tree to flow the latex. A small receiving cup was put below the wounded bark to receive the flow of latex and store it temporarily prior to collecting it throughput the areas. Collecting latex was performed manually. This activity was carried out after a week of tapping activity. The worker walks in the field and collect the rubber latex from the tree. The tractor was used to transport the collected latex to the main collecting point in the field.

2.3. Data Analysis

Energy use from both direct and indirect energy sources was analyzed based on the field operations involved in rubber cultivation. The number of energy inputs/output used for the operations was recorded in specified established units, such as human labor in hour and fuel in liter, while machinery, agrochemicals, seedlings, and latex yield were in kilogram units. Energy equivalent in mega Joule (MJ) per hectare was computed based on the amount energy inputs/output used, multiplied with the coefficient energy equivalent in agricultural operations published from the previous research literatures as shown in Table 1. The fuel consumption was measured by filling the tractor's fuel tank before and after performing each operation as suggested by Alcock [19], Pebrian and Yahya [20], Nazri and Pebrian [13]. Amount of the fuel to be topped up once each operation completed was considered as fuel consumption of the operation. The energy utilized by machinery was computed by using f equation 1 by Moerschner and Gerowitt [21] and Nazri and Pebrian [13]

$$Em = \frac{(WmxCmxHxN)}{Wo} \quad (1)$$

Where *Em* indicates the energy of machinery in MJ/ha, *Wm* indicates weight of machine in kg, *Cm* indicates coefficient of energy for machine in MJ/kg, *H* indicates working hours in hha^{-1}, *N* indicates number of application in unitless, and *Wo* indicates wear-outlife of machinery in hours.

The total inputs equivalent can be computed by summation of energy equivalences of all inputs in MJ units. Energy ratio (energy use efficiency) and energy productivity were calculated based on formulas by Nazri and Pebrian [13], Mandal et al. [22] and Singh et al. [23] in equation 2 and 3 by using the computed energy equivalents of the inputs and output.

$$EE = \frac{Eo}{Ei} \quad (2)$$

$$EP = \frac{Lo}{Ei} \quad (3)$$

Where *EE* indicates energy use efficiency in unitless, *Eo* indicates energy output in MJ/ha, and *Ei* indicates Energy inputs in MJ/ha, *EP* indicates energy productivity in unitless, *Lo* indicates latex output in kg/ha.

Table 1. Energy equivalents for different inputs and outputs in agricultural production

	Unit	Energy equivalent (MJ/unit)	Reference
Input			
1)Labor	h	1.96	Safa and Tabatabaeerfar [24]
2)Machinery			
Tractor 50 hp	kg	109.00	Pimentel [25]
Petrol engine, 5hp	kg	109.00	Pimentel [25]
High-pressure sprayer	kg	109.00	Pimentel [25]
3)Fertilizer			
Nitrogen (N)	kg	61.53	Pimentel and Patzek [6]
Phosphorus (P_2O_5)	kg	12.56	Pimentel and Patzek [6]
Potassium (K_2O)	kg	6.70	Pimentel and Patzek [6]
4)Chemical			
Hydrated lime	kg	1.17	Pimentel and Patzek [6]
Micronutrient*	kg	20.90	Anon [26]
Ethrel	kg	255.00	Anon [26]
Paraquat	kg	459.00	Anon [26]
Glyphosate	kg	453.00	Anon [26]
5)Fuel			
Diesel	L	56.31	Safa and Tabatabaeerfar [24]
Petrol	L	46.30	Safa and Tabatabaeerfar [24]
6)Suckers	kg	1.90	Ricaud [27]
Output			
7)Fruits	kg	1.90	Singh and Mittal [28]

Lastly, energy was classified into direct and indirect inputs, and renewable and non-renewable sources. In categorizing the direct and indirect energy inputs, therefore, agrochemicals, fertilizer, seeds and machinery was categorized as direct energy inputs, while human labor and fuel as indirect energy inputs. In grouping renewable and non-renewable energy sources, hence, fuel, fertilizer, agrochemicals and machinery were grouped into non-renewable energy, whereas human labor and seeds as renewable energy.

3. Results and Discussion

3.1. Energy consumption for field operations

Total energy consumption for field operations in rubber cultivation in the study areas was16080.86 MJ/ha as shown in Table 2. Fertilizing consumed 10602.65 MJ/ha of energy, or billed for 65.93% of total energy was the largest portion of energy consumption in in rubber cultivation. It followed by planting of 2559.34 MJ/ha (15.92%), weeding of 1552.84 MJ/ha (9.66%), harvesting of 647.38 MJ/ha (4.03%), land preparation of 632.37 MJ/ha (3.93%), spraying hormone of 70.67 MJ/ha (0.44%), tapping of 9.24 MJ/ha (0.06%), and pruning of 6.37MJ/ha (0.04%). Energy expenditure in fertilizing operation came from the computation of energy of fertilizers, labor, tractor and fuel. Energy consumption in planting operation was originated from the sum of energy of planting material (seeds), labor, tractor and fuel. Energy spent for weeding operation was a summation of the energy of labor along with hand-held tools and equipment used. In the areas of immature rubber, weeding was performed manually using a hoe to pull the weeds. The operation was laborious and spent a lot of time to complete the operation because the distances in between rows are very narrow as the rubber cultivation in the study area implemented operation was carried out by a worker using knapsack sprayer to deliver herbicides for eradicating the weeds. Energy for collecting latex and land preparation came from the sum of energy of labor, tractor and fuel. Energy for spraying hormone was obtained from the total amount of energy of labor, chemical and equipment used, meanwhile energy for tapping and pruning operations were computed from the energy of labor only since the operation is done manually.

Table 2. Distribution of energy consumption on field operation basis

Field operation	Energy (MJ/ha)	Percentage (%)	Rank
Fertilizing	10602.65	65.93	1
Planting	2559.34	15.92	2
Weeding	1552.84	9.66	3
Collecting latex	647.38	4.03	4
Land preparation	632.37	3.93	5
Spraying hormone	70.67	0.44	6
Tapping	9.24	0.06	7
Pruning	6.37	0.04	8
Total	16080.86	100	

3.2 Distribution of Energy Inputs Based On Energy Resources

The distribution of energy inputs in rubber cultivation on energy resources basis is shown in Table 3. The highest share of energy inputs equivalent was fertilizer usage amounting to 9957.27 MJ/ha, or 61.92% of the total equivalent energy in rubber cultivation. Huge consumption of energy inputs for fertilizer was due to the cultivation requires various types of fertilizers. The fertilizer used in the cultivation comprised of 123 kg of nitrogen, 123 kg of phosphorus, 98 kg of potassium and 28 kg of magnesium. Among the fertilizer components themselves, nitrogen consumed 7568.19 MJ/ha of energy input as the largest energy consumption, while the lowest was magnesium with consumption of 187.60 MJ/ha.

Fuel with energy inputs equivalent of 2449.49 MJ/ha (15.23%) was in the second largest energy consumption in rubber cultivation. This sequence are significantly followed by energy inputs for planting material or seeds with energy input equivalent of 1900 MJ/ha (11.82%), and chemical application of 1558.85 MJ/ha (9.65%). Human labor and machinery with energy input equivalent of 160.05 MJ/ha (0.93%) and 65.4 MJ/ha (0.40%), respectively, was considered as the lowest energy inputs.

Fuel energy input was considered as the second highest energy consumption due to the high mobility of the tractor that was used to assist transportation of other inputs for land preparation, planting, weeding, fertilizing operation, spraying hormone, tapping and collecting latex. Planting material or seed also consumed huge energy input due to a large amount of seeds planted for supplying the planting material for the areas. A total of 4000 kg of seeds were planted in a hectare of rubber plantation in the study area. Such amount of seeds can be considered as a large amount of energy input among the largest inputs in rubber cultivation. For the chemical, glyphosate and ethrel contributed to equivalent energy of 1558.65 MJ/ha. Labor and machinery instigate were in the lowest energy inputs in the study as operations were done manually. Lack involvement of machinery along with the low energy equivalent for labor (1.09 MJ/ha) are believed become factors that contributed the lowest energy to these inputs.

3.3. Energy Use Efficiency

Concerning yield per hectare, the total energy output of latex from rubber cultivation was 7108.6 kg/ha. Such yield was produced by energy inputs equivalent to 16080.85 MJ (Table 3). Thus, the energy productivity shows that the amount of product from the return of energy inputs use was 0.44 MJ/kg, which means for production 0.44 kg of the yield of latex consumed energy 1 MJ. As shown in Table 3, the energy use efficiency in rubber cultivation was 0.83. The energy use efficiency was computed based on the ratio of energy output to energy inputs. The computed value was then referred to a definition energy use efficiency by the International Agency (2016), who mentions that something or activity that can be considered having more efficient energy use if it delivers more services for the same energy input, or gives same services for less energy input. With that reference, conclusively, the energy use efficiency of 0.83 in rubber cultivation in the study area was less efficient since the ratio is lesser than 1. The energy use efficiency in rubber cultivation was much lower than the energy efficiency use of other crops cultivations in Malaysia such as pineapple of 3.6 by Nazri and Pebrian [13] and rice of 8.86, excluding inputs for irrigation by Gevao et al. [17]. the energy use efficiency in rubber cultivation in Malaysia was also lower than that of the other similar tree crops production in Turkey such as apple (1.10), apricot (1.24 to 3.37), citrus orange (1.25), lemon(1.06) and mandarin fruits (1.17) reported by Nuray [29], Gezer et al. [30], Esengun et al. [31], and Ozkan [32]. Generally, these findings are inline with the statement by Schäfer [33], who said that the energy input for crop cultivation varies in a wide range depending on habitat, crop species and variety, the intensity of production, and employed tools and machinery.

3.4 Energy Use based on Direct, Indirect and Renewable, Non Renewable

Energy use was grouped based on direct and indirect and also renewable and non-renewable forms. Grouping energy forms by Nazri and Pebrian [13] was used to divide inputs. The direct energy was defined as the energy that produced physically in farms such as labor and fuel, meanwhile the indirect energy referred to energy that was developed outside of farm such like fertilizers, chemical and machinery. These inputs were obtained from manufacturers. Renewable energy includes labor and seeds, while fuel, chemical and machinery are non-renewable energy. The results of the computation of energy use based on direct, indirect and renewable, non-renewable energy are presented in Table 4. Indirect and non-renewable energy forms were dominantly used in rubber cultivation with portions of 84% and 87%, respectively. The results reflected that the rubber cultivation in Malaysian is still highly depending on off-farm energy and low-returns energy from the farm. The percentage of the direct energy of 16% in rubber cultivation was lower than the similar energy in pineapple cultivation, which accounted for 23.56% [13]. The share of 84% indirect energy of in rubber cultivation was bigger than that of 76.44% in pineapple cultivation by Nazri and Pebrian [13]. In relation to renewable energy, its portion in rubber cultivation was 13% or lower than that of the pineapple cultivation of 20.68%. In the context of non-renewable energy, the portion of 87% in rubber cultivation was also bigger than that of 79.32% in pineapple cultivation. The results reflected that the rubber cultivation in Malaysian is still highly depending on off-farm energy and low in returns energy from the farm.

IOP Conf. Series: Earth and Environmental Science **327** (2019) 012004 doi:10.1088/1755-1315/327/1/012004

Table 3. The distribution of energy inputs used in rubber cultivation

Input	Unit	Amount of input used per hectare	Energy equivalent (MJ/unit)	Energy equivalent (MJ/ha)	Percent-age, (%)
Labor	H	76.55		150.05	0.93
Land Preparation	h	1.86	1.96	3.65	
Planting	h	15.62		30.62	
Pruning	h	3.25		6.37	
Weeding	h	14.5/15.06		57.94	
Fertilizing	h	8.5		16.66	
Spraying	h	3.53		6.92	
Hormone	h	4.71		9.23	
Tapping Harvesting	h	9.52		18.66	
Machinery	kg	1800		65.40[a]	0.40
Tractor 37.3 kW	kg	1792.8	109.00	65.13[a]	
Knapsack sprayer 16 L	kg	7.2		0.27[a]	
Fertilizer	kg	372		9957.27	61.92
Nitrogen(N)	kg	123	61.53	7568.19	
Phosphorus	kg	123	12.56	1544.88	
(P2O5)	kg	98	6.70	656.60	
Potassium(K2O) Magnesium	kg	28	6.70	187.60	
Chemical	kg	3.55		1558.65	9.69
Ethrel	kg	0.25	255.00	63.75	
Glyphosate	kg	3.30	453.00	1494.90	
Fuel	L	43.50		2449.49	15.23
Diesel	L	43.50	53.61	2449.49	
Seeds	kg	1000		1900	11.82
		1000	1.90		
Total energy inputs	MJ/ha			16080.86	100
Output				13335.34	
Yield of latex	kg/ha	7018.6	1.90	13335.34	
Energy use efficiency (energy output/inputs ratio)				0.83	
Energy productivity	MJ/kg			0.44	

[a]Based on calculation using equation (1).

Table 4. Energy use based on direct, indirect and renewable, non-renewable.

Type of energy	Energy equivalent, (MJ/ha)	Percent (%)
Direct energy	2599.54	16
Indirect energy	13481.32	84
Renewable energy	2050.05	13
Non-renewable energy	14030.81	87

4. Conclusion

This paper has successfully presented and analyzed the energy consumption in rubber cultivation in Malaysia through a case study. The computed energy use in rubber cultivation in the study area was 16080.86 MJ/ha. On the basis of operation, the findings of the study indicated that fertilizing consumed the highest share of energy use in rubber cultivation, which account for 65.93% of total energy use. While the lowest consumption was in the pruning operation. This study found that the ratio of energy output/inputs the rubber cultivation was 0.83. Conclusively, the energy inputs used in the rubber plantations were less effective since the ratio is lower than 1. This ratio also showed that the rubber cultivation in Malaysian is still highly depending on off-farm energy and low in returns energy from the farm. Thus, technological improvements, which are able to save the energy use and produce energy on farm should be introduced in Malaysian rubber cultivations. Anyhow, it is certain that the study contributed to enrich the database of energy consumption for major crops in Malaysian agricultural sector.

Acknowledgment
The authors are very grateful to the Rubber Research Institute of Malaysia Mini Station in Jasin, Melaka and Lakai, Negeri Sembilan, Malaysia for providing the study area for this research project.

References

[1] MRB (2018). *Natural Rubber Statistics 2018*. Malaysian Rubber Board (MRB). 206 Jalan Ampang, Kuala Lumpur. Malaysia Another reference

[2] Bekhet, H.A, & Abdullah, A., (2010). Energy use in agriculture sector: input-output analysis. *International Business Research,* 3 (3): 111-121

[3] Pervanchon, F., Bockstaller, C., & Girardin, P. (2002): Assessment of energy use in arable farming systems by means of an agro-ecological indicator: the energy indicator. *Agricultural Systems*, 72: 149–172.

[4] Pimentel, D 2009. Energy Inputs in Food Crop Production in Developing and Developed Nations. *Energies,*2: 1-24.

[5] Refsgaard, K., Halberg, N., & Kristensen, E. S. (1998). Energy utilization in crop and dairy production in organic and conventional livestock production systems. *Agricultural Systems, 57*(4), 599–630.

[6] Pimentel, D., & Patzek, T. W. (2005). Ethanol production using corn, switchgrass and wood; biodiesel production using soybean and sunflower. *Natural Resources Research, 14*(1), 65-76.

[7] Mustafa, G., Latif, I. A., & Egwuma, H (2016). An empirical analysis of supply response of rubber in malaysia. *American Journal of Agricultural and Biological Sciences*, 12 (4): 148.156.

[8] Ratnasingam, J., Ramasamy, G., Ioras, F, Kaner J., & Wenming, L (2012). Production Potential of Rubberwood in Malaysia: Its Economic Challenges. *JNotulae Botanicae Horti Agrobotanici Cluj-Napoca*, 40(2):317-322.

[9] In, S. (2012). *Forecasting the Price of Natural Rubber in Malaysia*. Master of Arts (Economics) Thesis. California State University, Sacramento, CA, USA.

[10] Othman, A 2008. Future of Malaysian Rubber Products Industries. IPICEX 2008. 17-21 November 2008, Shah Alam, Selangor

[11] Ismail, M.H.M (1989). *Malaysian Natural Rubber Industry: An Econometric Analysis on the Elasticity of Supply and Demand Approaches.* MS thesis. University of North Texas. Denton, Texas, USA.

[12] Tugiman, S (1979). Transfer of technology within the natural rubber industry in Malaysia. *Media Asia,* 6 :145-148

[13] Nazri, M & Pebrian, D.E., (2017). Analysis of energy consumption in pineapple cultivation in Malaysia: a case study. *Pertanika Journal Science & Technology*, 25 (1): 17 – 28.

[14] Azwan, M B, Norasikin, A, Abd Rahim, S, Norman, K., & Salmah, J (2016). Analysis of energy utilisation in Malaysian oil palm mechanisation operation. *Journal of Oil Palm Research*, 28: 485-495.

[15] Pebrian, D E, Yahya A and Siang T C 2014 Workers' workload and productivity in oil palm cultivation in Malaysia. Journal of Agricultural Safety and Health 20 (4): 234-254.

[16] Pebrian, D E, Yahya A, Nawi N M, Siang T C and Bockari-Gevao S M 2013 Monitoring and

Auditing Human Energy Input for Oil Palm and Rice Production in Malaysia *Philippine Agricultural Scientist 96(4):282-282.*

[17] Gevao, S. M. B., Ishak, W. I. W., Azmi, Y., & Chan, C. W. (2005). Analysis of energy consumption in lowland rice based cropping system of Malaysia. Songklanakarin. *Journal of Science and Technology, 27*(4), 819-826.

[18] Saidura R.,& .Mekhilef, S (2010) Energy use, energy savings and emission analysis in the Malaysian rubber producing industries. *Applied Energy*, 87 (8); 2746-2758.

[19] Alcock, R. (1986). Tractor-Implement Systems. Westport, Connecticut: AVI Publishing Co.

[20] Pebrian, DE and Yahya, A (2013). Mechanized System for in-field oil palm fresh fruit bunches collection-transportation. Agricultural Mechanization in Asia, Africa and Latin America, 44 (2):7-14.

[21] Moerschner, J., & Gerowitt, B. (2000). Direct and indirect energy use in arable farming - An example on winter wheat in Northern Germany. In B.P. Weidema & M.J.G Meeusen (Eds.), *Agricultural Data for Life Cycle Assessments*. The Hague, Agricultural Economics Research Institute (LEI). Report 2.00.01, Volume 1. February 2000. 195 p.

[22] Mandal, K.G., Saha, K.P., Gosh, P.L., Hati, K.M., & Bandyopadhyay, K.K. (2002). Bioenergy and economic analyses of soybean-based crop production systems in central India. *Biomass Bioenergy, 23*, 337–345.

[23] Singh, M.K., Pal, S.K., Thakur, R., & Verma, U.N. (1997). Energy input–output relationship of cropping systems. Indian J. Agric. Sci., 67(6), 262–266.

[24] Safa, M. & Tabatabaeefar, A. (2002). Energy consumption in wheat production in irrigated and Dry Land Farming. In Proc. Intl. Agric. Engg. Conf., Wuxi, China, Nov. 28-30.

[25] Pimentel D. (1980). Handbook of energy utilization in agriculture. Boca Raton, FL: CRC Press.

[26] Anon. (2004). Units, Equivalents and energy Constants. Cooperative Extension Service. Institute of Food and Agricultural Sciences, University of Florida.

[27] Ricaud, R. (1980). Energy input and output for sugarcane in Louisiana. In D. Pimentel (Ed.), *Handbook of Energy Utilization in Agriculture* (pp. 135-136). Boca Raton, FL., CRC Press.

[28] Singh, S and Mittal, J.P. (1992) *Energy in Production Agriculture*. Mittal Publications. New Delhi, India.

[29] Nuray, K (2009). Energy use and input-output energy analysis for apple production in Turkey. *Journal of Food, Agriculture & Environment,.7*(2), 419-423.

[30] Gezer, I., Acaroglu, M., & Haciseferogullari, H. (2003). Use of energy and labour in apricot agriculture in Turkey. *Biomass Bioenergy, 24*(3), 215–219.

[31] Esengun, K., Gunduz, O., & Erdal, G. (2007). Input-output energy analysis in dry apricot production of Turkey. *Energy Conversion and Management, 48*(2), 592-598.

[32] Ozkan, B., Akcaoz, H., & Karadeniz, F. (2004a.) Energy requirement and economic analysis of citrus production in Turkey. *Energy Conversion and Management, 45*, 1821-1830.

[33] Schäfer, W (2006) Energy crops and renewable energy: overall and process efficiency. *Forskningsnytt* (4): 11-13.

IOP Conf. Series: Earth and Environmental Science **327** (2019) 012005 doi:10.1088/1755-1315/327/1/012005

Non-destructive Evaluation of Fat Content of Coffee Beans Solok Radjo Using Near Infrared Spectroscopy

F Yuwita[1], Ifmalinda[2] and M Makky[2]

[1]Magister Student of Biosystem and Bioproses Engineering, Andalas University, Padang, Indonesia,
[2]Department of Agricultural Engineering, Andalas University, Kampus UNAND Limau Manis Padang 25163 West Sumatera, Indonesia

Corresponding author's e-mail address: fitriyuwita.s@gmail.com

Abstract. Prediction fat of coffee beans in West Sumatra is still done manually by chemical methods, time-consuming, and costly. With the development of food technology, the estimation of the content of the beans to do with how quickly and efficiently using NIRS technology. NIRS can analyze the chemical content of foodstuffs nondestructive. This study uses Arabica beans Solok Radjo two maturity levels are 4 (ripe) and 5 (over ripe), 30 samples with weight of samples 6 grams. NIRS instrument used is the science and technology FT-IR-1516 with a wavelength of 1000-2500 nm. Fat tests conducted in the laboratory Instrumentation Technology Center for Agriculture. In this study, the data spectral coffee beans processed using PCA (Principal Component Analysis) after correlated with fat content processed by the method of PLS (Partial Least Square). The prediction results of calibration best fat obtained r = 0.999, R^2 = 0.999, SEC = 0.008%, and RMSEC = 0.008%, validating the value of r = 0.998, R2 = 0.999, SEP = 0.009%, and RMSEP = 0.008%, of the model fat coffee beans can be identified by non-destructive evaluation using spectral analysis.

1. Introduction

Indonesia is known as one of the coffee producing countries and the world's largest exporter of coffee beans. West Sumatra in 2015 produces coffee as 34.059 tons of coffee, with planting areas of 42.925 Ha [4]. One of the producing coffee in West Sumatra is districk of Solok, famous with coffee names Minang Solok Radjo. Solok Radjo coffee at the moment it feels have been identified in Indonesia and Abroad, Solok Radjo coffee in 2016 becomes the winner and get the highest score in West Sumatra coffee festival as a quality coffee speciality, this was proved in 2013 the coffee was exported to the United States have as much as 4 ton, Australia 22 tons, to Italy 2 tons and 1.5 tons to Thailand [4].

Postharvest handling is key to success to improve the quality of Indonesian coffee beans. The estimation of the quality of the coffee is usually done through laboratory testing (destructive), where coffee beans are crushed and the juice is taken and analyzed by standard methods of chemical common in laboratories. However, the chemical method to spend a long time and cost a fortune, so it is not suitable to be applied in industries that require a very fast method and not destructive (non-destructive) to analyze the quality of the coffee. The detection of food quality can be realized quickly and efficiently through technological development Nir Spectrometer (NIRS). NIRS can be analyzed more quickly does not use chemicals and can analyze the chemical content of materials without damage.

Some research has been done before, as it has done [7], which analyzed the caffeine content of coffee powder. The results showed the caffeine content of coffee 6% higher when compared with UV detection. LC-UV method selected as the reference method for the calibration of NIRS system. Analysis on 83 samples of liquid coffee extract resulted in SEE (standard error of estimate) 0.34% while for SEP (standard error of prediction) of 0.07% with a correlation coefficient of 0.86. [14] has also been conducting research NIRS to predict that whole coffee beans caffeine content Gayo Arabica coffee beans, according to the determination of the chemical composition of coffee more efficiently and effectively carried out in the form of coffee beans than coffee powder.

The use of NIRS has many advantages compared to traditional testing is a simple preparation of samples, can analyze samples in the form of powder and seeds, rapid detection process, do not use chemicals. NIRS research on coffee has much to do but have not been to a coffee Minang Solok Radjo because Indonesia has a wide range of coffee varieties that have different characteristics. Therefore, the development of NIRS research with the object of coffee Solok Radjo need to be supported and developed to facilitate detection of the chemical content of the coffee beans quickly and accurately.

2. Materials and Methods

In this study, the samples' fat is determined through electromagnetic radiation absorbance, measured using NIR spectroscopy. 1 kg grams of coffee beans Solok Radjo with two maturity they are maturity level 4 (ripe) and maturity lever 5 (overripe) were harvested by manual picking from the stem. The beans were cleaned and placed into 30 ml glass tubes. The samples were compacted by placing the tubes on vortex for 10 seconds. More beans were added and procedure repeated until the tubes were fully filled. The sample's tubes were then placed into NIR spectroscopy and its spectral (1000-2500 nm) were recorded at normal scan speed using average of 10 scans method. Recorded signals were pre-processed with spectroscopy software (Unscrambler® X.1, Camo, USA).

The measured actual samples' fat by primary method [1]. As many 3 grams of sample (W0) is inserted into a filter paper that had been closed both ends with fat free cotton and inserted into the fatty sheath that has measured its permanent weight (W1) and connected with the soxhlet tube. Fatty sheath tube inserted into the soxhlet extractor chamber and added fat solvent (n-hexane). Then the solution at reflux for 3 hours. Solvents fat in the fat flask was distilled until all the fat solvent evaporated. Furthermore, fat tube dried in an oven at a temperature of 105 °C and tube dried in the desiccator until its weight is constant (W2). The calculation of fat content can use the formula:

$$\text{Fat (\%)} = \frac{W_2 - W_1}{W_0} \times 100\% \tag{1}$$

where W0 is weight of samples (g), W1 is weight empty fat tube (g), W2 is weight tube with fat (g).

The absorbance data was then analyzed using software (Unscrambler® X.1, Camo, USA) to correlate with its fat. PCA method was used to identify significant correlation between certain electromagnetic radiation wavelengths with sample's fat. Model to determine fat of the samples according to its electromagnetic radiation absorbance was developed using PLS, were in maturity level 4 using pretreatment normalization. Pretreatment normalization method is part of the normalization obtained by dividing each row of the matrix of data averaging so as to reduce the interference of factors that are not visible [6].

The samples fat as dependent variables, and the PCs from PCA analysis as factors. Data was partitioned into two groups; the first 2/3 (20 data) was used to train the model, and the rest of 1/3 (10 data) was used to validate the model. Model performance was analyzed using Sum of Squares Error (SSE), and Relative Error (RE).

3. Results and Discussion

3.1. NIRS spectrum and Chemical Ingredients Raw Coffee Beans Solok Radjo

IOP Conf. Series: Earth and Environmental Science **327** (2019) 012005 doi:10.1088/1755-1315/327/1/012005

Spectrum *raw* NIRS is a raw spectrum from the data given NIRS without pretreatment, from raw spectra can find the amount of noise generated and to know the chemical constituents in the beans. Chemical content indicated by the peaks and valleys of the resulting wavelengths on the spectrum. NIRS spectrum of the raw coffee beans are Arabica Solok Radjo between 1000-2500 nm wavelength can be seen in Figure 1.

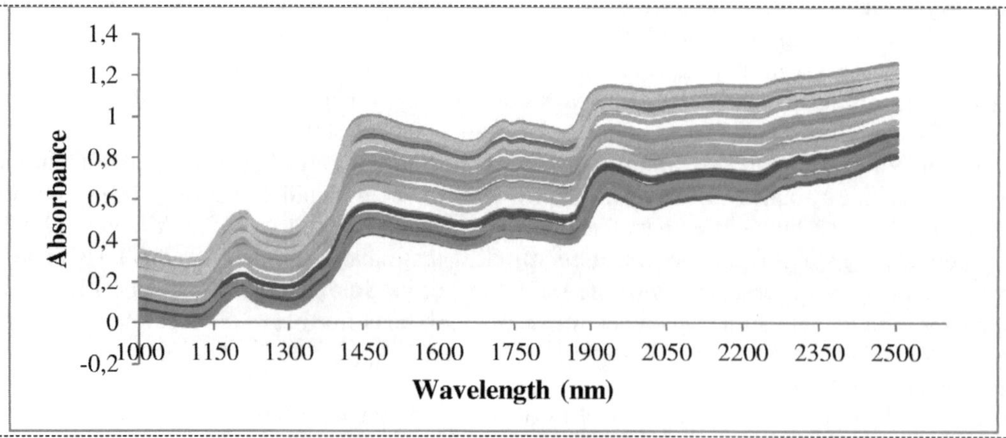

Figure 1. Raw NIRS Spectrum Arabica Coffee Solok Radjo.

Coffee beans spectrum has peaks and valleys at a particular wavelength, peak spectral absorbance NIRS in Solok Radjo Arabica coffee beans is between wavelengths 1128-1140 nm, 1202-1221 nm, 1392-1418 nm, and 1731-1766 nm, it shows information where the content of certain substances found in coffee. This is in accordance with the opinion [2], that the chemical components contained in the material to be tested affect the existence of the peaks and valleys of the spectrum, in addition to the peaks and valleys of the spectrum is also influenced by the physical properties of the material to be analysed.

3.2. NIRS spectrum classification Solok Radjo Coffee Beans on Two Levels Maturity

Classification of two coffee beans Solok maturity level Radjo seen by taking each of the spectrum of maturity level 4 (ripe) and 5 (overripe), this is to look at the differences of each spectrum. The results of the classification ofspectrum of coffee beans in the maturity level 4 (perfectly cooked) and 5 (overripe) can be seen in Figure 2.

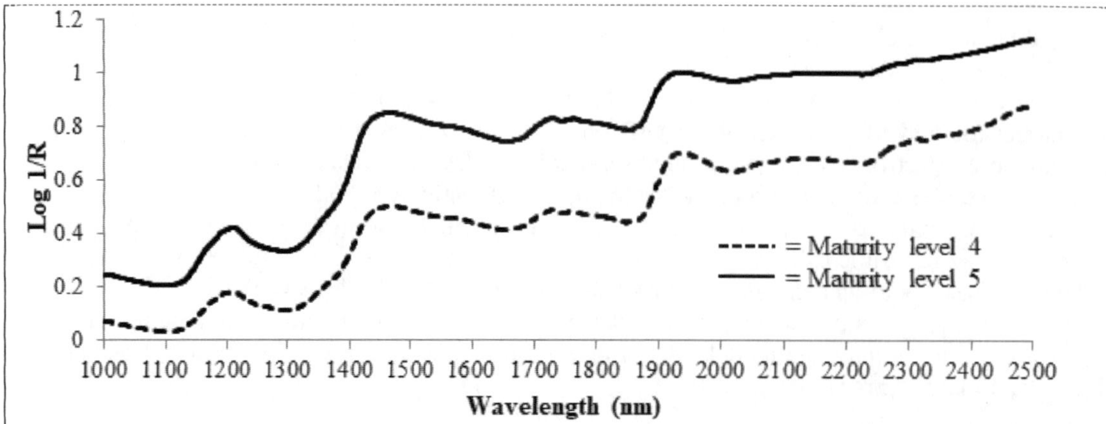

Figure 2. Classification of NIRS Spectrum of Solok Radjo Coffee Beans

Figure 2 show of spectrum results indicate that the spectrum of a typical coffee bean has the same pattern, but has a absorbance absorption values were somewhat different, this is due at the time of data collection is still no light is passed out materials as well as the difference of the chemical component contained by coffee. [12] said that the difference in the spectrum caused by several kinds disorders like rays of light from the sample, the difference in temperature, density and particle size and other disturbances coming from the detector, AD converter, or amplifer.

Analysis to see the difference in the maturity level of the coffee beans Solok Radjo using PCA (Principal Component Analysis). Conduct an analysis of factors to get a new variable in smaller amounts, PCA can be used as a basis [8]. PCA analysis performed to determine whether the maturity of the beans can be directly distinguished from each other. [10] states that the PCA is able classify an appropriately different materials such as classifying according to its kind. The results of PCA analysis Solok Radjo coffee beans can be seen in Figure 3.

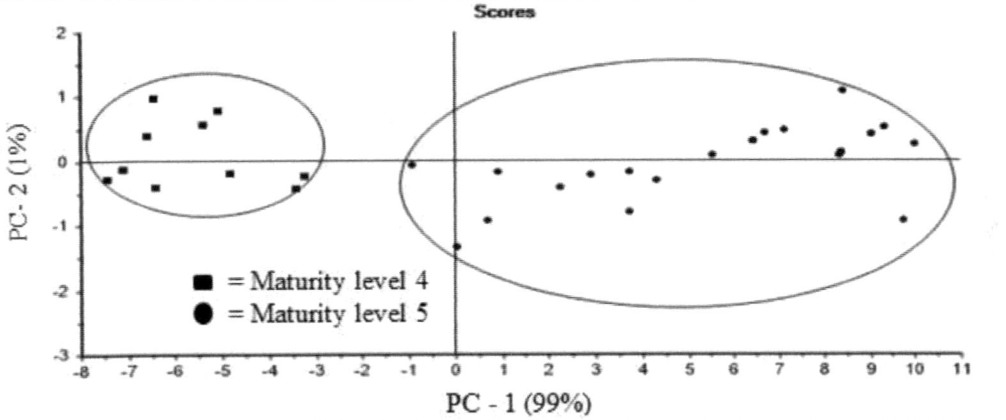

Figure 3. Results of PCA Analysis NIRS Solok Radjo Coffee Beans

The results of PCA analysis on raw (original data) coffee beans without pretreatment spectrum as shown in Figure 3, shows that they are separated or grouped well. Coffee beans in the maturity level 4 (ripe) are in the same pile as well as coffee beans at maturity level 5 (overripe). PCA can prove the difference spectrum at each maturity level Solok Radjo coffee beans. PCA results prove that NIRS can segment data based on the maturity level of the coffee beans. This shows the spectrum of coffee beans can be classified as good with a percentage of 100% success. Wavelength NIRS measurement results that are relevant to determining the substance content in coffee Solok Radjo show in Figure 4.

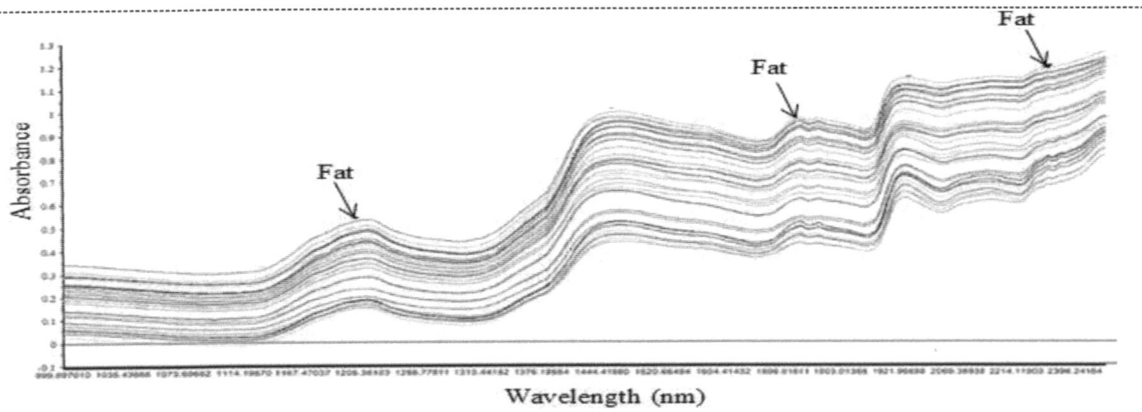

Figure 4. The Relevant Wavelength Solok Radjo of Coffee Beans

The result of the loading of the plot on raw coffee bean Solok Radjo spectrum shows the presence of peaks and valleys that indicates the presence of chemical substances content in the coffee. Based on the analysis of spectral peaks based on the description information table Cen and He [3] the fat is located at a wavelength of 1379-1388 nm, 1720-1733 nm, and 2300-2346 nm.

3.3. Prediction Model Calibration and Validation Results Fat Content Coffee Beans Solok Radjo Using PLS (Partial Least Squares Regression) Method.

Calibration and validation of coffee Solok Radjo models using software Unscramble X.1 with Partial Least Square Regression method. PLS is a relatively new approach to linear regression and algorithms. PLS calibration suitable for incorporation in a number of samples with chemical data destructively tested in the laboratory and data NIRS [16]. Model calibration is the relationship between the data absorbance NIRS and chemical data coffee Solok Radjo, the model calibration are well defined with parameters SEC, SEP, Bias, the difference in value of the SEC with the SEP, the correlation coefficient between the actual value and the predicted number of factors or variables latent etc. [9].

The fat content in Arabica coffee contained in a protective wax coating on oil seeds and coffee. Fat in coffee is one of the chemical composition of coffee shape coffee flavor. Total fat content in arabica coffee between 12-18% which is contained in a layer of protective wax beans [15]. The increase in free fatty acids during storage will cause rancidity in the coffee powder so that it will affect the taste and the quality of the coffee powder. [13] fat content is marked by a chain of chemical bonds CH.

Fat content prediction data processing coffee beans for two different maturity levels using methods Partial Least Squares Regression (PLS) can be seen in Figures 5a and 5b.

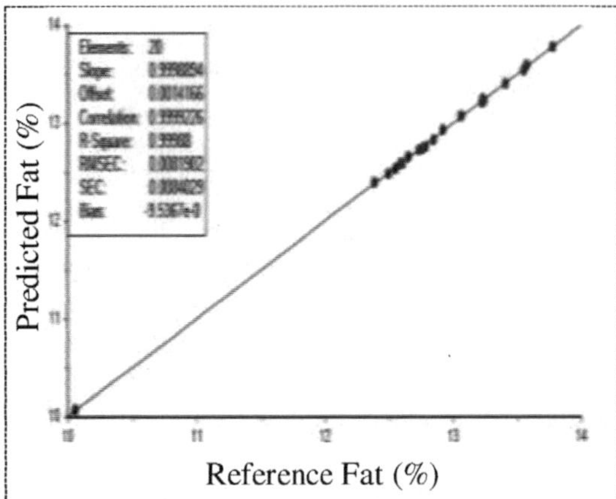

Figure 5a. PLS model calibration for samples' fat content coffee beans solok radjo at maturity level 4 (ripe)

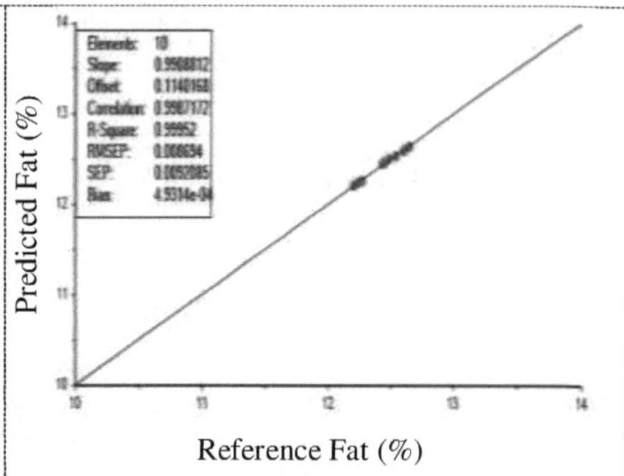

Figure 5b. PLS model validation for samples' fat content coffee beans solok radjo at maturity level 4 (ripe)

Results of the PLS prediction estimation can be seen from the results of the SEC and SEP error values obtained are still small, The resulting high r value, and the value of the error (RMSEC and RMSEP) is small, [16] is said to have good models when the error value generated RMSEC smaller than the standard deviation of the actual data. The results of the calibration predictive value r = 0.999, R^2 = 0.999, SEC = 0.008%, and RMSEC = 0. 0.008%, while the yield for the validation of the value of r = 0.998, R^2 = 0999, SEP = 0.009%, and RMSEP = 0.008%. [11] states that the determination value above 0.91 can be said to be good models. Use of PLS can predict the fat content of the coffee beans in the maturity level 4.

Spectral data processing coffee beans Solok Radjo at maturity level 5 (overripe) can be seen in Figure 6.

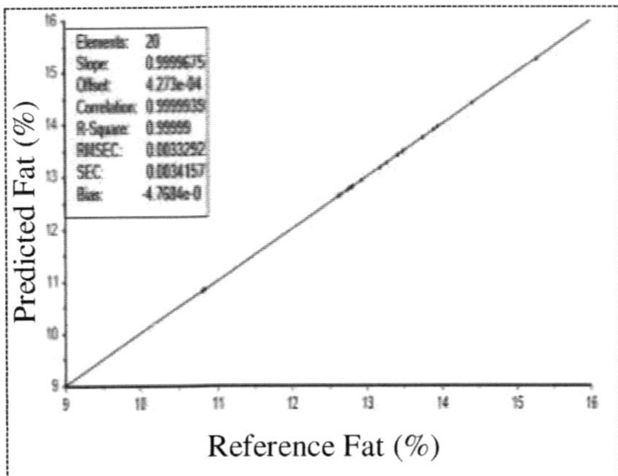

Figure 6a. PLS model calibration for samples' fat content coffee beans solok radjo at maturity level 5 (overripe)

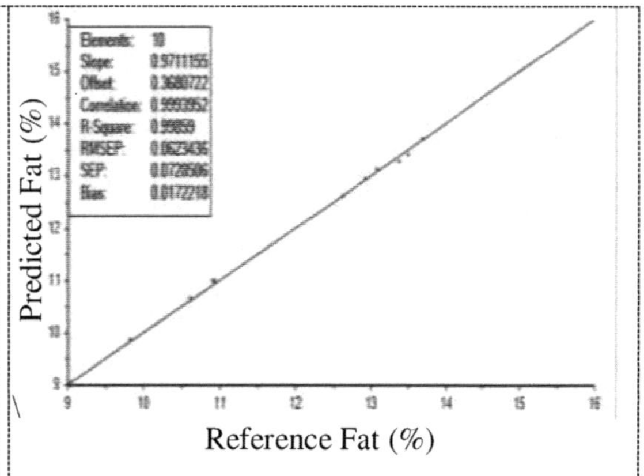

Figure 6b. PLS model validation for samples' fat content coffee beans solok radjo at maturity level 5 (overripe)

Figure 6 shows the results of a plot of data calibration and validation fat coffee beans Solok Radjo which were obtained by the predictive value of the calibration is r = 0.999, $R^2 = 0.999$, SEC = 0.003%, and RMSEC = 0.003%, while for validation is r = 0.999, R2 = 0.998 , SEP = 0.072%, and RMSEP = 0.062%. The results of PLS estimation without the fat content of coffee beans to show that the value of r obtained is big enough that that is above 0.91 and for the error value (RMSEC and RMSEP) obtained good report because less than zero.

4. Conclusion

In this study, nondestructive methods can be used to test the fat content of the coffee beans Solok Radjo. Model with absorbance values that are absorbed from the beans by using PLS. NIRS spectrum wavelength used is 1000-2500 nm and fat Solok Radjo arabica coffee is at a wavelength 1379-1388 nm, 1720-1733 nm, and 2300-2346 nm. Upon validation the model obtained value $R^2 = 0.999$ and RMSEP = 0.008%. Conclusion this research, NIRS can be used to estimate the fat content of coffee beans and as information for merchants to increase the quality of the coffee beans that will be marketed.

References

[1] AOAC International 1995 *Official Methods of Analysis of AOAC International* AOAC International. Secs. 942.15. Washington USA

[2] Blanco M and Villarroya I 2002 *NIR spectroscopy a rapid-response analytical tool* Trends in analytical chemistry **21** 240-250

[3] Cen H, and He Y 2007 *Theory and application of near infrared reflectance spectroscopy in determination of food quality* J. Trends in Food Sci & Technol **18** 72-83

[4] Directorate General of Plantation. 2016. *Statistics Indonesia Plantation* Jakarta

[5] Fajarudin 2016 *Will be Solok Coffee Export 18 Tons* The bow May 6 Padang

[6] Gaydow V, Kister J, Dupuy N 2011 *Evaluation of multiblock NIR / MIR PLS predictive models to detect adulteration of diesel / biodiesel blends by vegetal oil* Journal of Chemometrics and Intelligent Laboratory Systems **106** 190-197

[7] Huck C W, W Guggenbichler GK Bonn 2005 *Analysis of caffeine, theobromine and theophylline in coffee by near infrared spectroscopy (NIRS) compared to high-performance liquid chromatography (HPLC) coupled to mass spectrometry* J Analytica Chimica Acta **538** 195-203

[8] Iriawan N and Astuti S P 2006 *Statistics Data Processing with Easy to Use Minitab 14* Yogyakarta (ID) Andi

[9] Lammertyn J, Peirs A, De Baerdemaeker J and Nicolai B M 2000 *Light Penetration Properties of NIR Radiation in Fruit with Respect to Non-Destructive Quality Assessment* J Postharvest Biol Technology **18** 121-132

[10] Lee V S, Tue-Ngeun P, Traisathit P, Prasitwattanaseree S, Nimmanpipug P and Chaijaruwanich J 2009 *FTNIR and chemometric tools for the classification of thai wines*. journal of science and technology **3** 446-458

[11] Mouazen A M, Saeys W, Xing J, De Baerdemaeker J and Ramon H 2005 *Near infrared spectroscopy for agricultural materials: an instrument comparison a near infrared spectroscopy* **13** 87-97

[12] Ozaki Y, McClure W F and Christy A A 2007 *Near-Infrared spectroscopy in food science and technology* New Jersey (USA) John Wiley & Sons, Inc

[13] Rita H, Marliah A and Rosita F 2011 *Study of Three Varieties and Two Against Fermentation Method Quality Arabica coffee beans (Coffea arabica L.) Gayo, Bener Rousing* Proceedings of the Seminar The National Association of Expert Food Technology Indonesia (PATPI) North Sumatra

[14] Rosita R 2016 *Determination of chemical constituents Gayo Arabica coffee beans are non-destructively by near infrared spectroscopy* [Thesis] Bogor Graduate School of Bogor Agricultural University

[15] Wei F and Tanokura M 2015 *chemichal change in the components of coffee beans during roasting* J Healt and Disease Doi.org/10.1016/B978. **12** 409-517

[16] Williams P and Norris K 1990 *Near-infrared technology in the agricultural and food industries.* American As-sociation of cereal chemical, Inc. St. Paul USA 146

The Different of Processing Method Against Polyphenol and Antioxidant Activity of Moringa Herbal Tea

A Asben [1], B Rini[1] and L Aini [2]

[1]Faculty of Agricultural Technology, Andalas University. Limau Manis-Padang, West Sumatera, Indonesia 25163
[2]Student of Agricultural Industrial Technology Postgraduate Program, Andalas University, West Sumatera, Indonesia 25163

Corresponding author's e-mail address: alfi_asben@yahoo.com

Abstract. Moringa leaves are known contain many types of antioxidants such as ascorbic acid, flavanoids, phenolic compounds and carotenoids. Moringa leaves have the potential to be processed into herbal tea. This study investigated the different of tea processing methods against polyphenol, antioxidant activity and panelist level on moringa leaf herbal tea, and also to get the appropiate processing method of moringa herbal tea. The study was conducted based on 4 types of tea processing methods including black tea, green tea,oolong tea and fragrant tea with 3 replications. The data were reportedas average. This study showed that the tea processing method gives a different to the chemical components and teacharacteristics produced. The results of the processed moringa leaf tea with different tea processing methods showed that the appropriate of processing method was fragrant tea method with the value of moisture content 6.36%, total polyphenol 2.43%, antioxidant activity 71.58%,and the sensory test of color, aroma and taste were 4.13 (like), 4.13 (like) and 3.2 (regular), respectively.
Keywords: Moringa leaf, herbal tea, antioxidant, polyphenol

1. Introduction

Tea is a drink made from processed young leaves of tea plants (*Camellia sinensis*) which ranks second after water as the most consumed beverage in the world and is known to have many health benefits. Tea leaves contain caffeine compounds which are known to have a refreshing effect on the body. Tea is rich in vitamins such as vitamin C and B especially thiamin and riboflavin which can help the process of absorption of protein in the body and is believed to relieve headaches. In addition, polyphenol compounds in tea function as antioxidants that can slow down and prevent the oxidation process by free radicals so that cell damage can be avoided and can help prevent various chronic diseases [1].

Increasing public awareness of health encourages increasing consumption patterns of herbal products. Not only in Eastern countries, WHO data shows that demand for herbal products in European countries in the period 1999-2004 is estimated to reach 66% of world demand. This also encourages research on herbal products to grow rapidly [2].

Herbal tea comes from the processing of flowers, seeds, leaves, skin and roots of plants other than tea plants (*Camellia sinensis*). Basically, the processing of herbal tea is almost the same as processing tea in general as well as the way it is presented. Herbal tea can be consumed in single or herbal mixtures. Herbal tea is a product that is currently widely consumed and is preferred by the community because the raw material is easier to obtain and the benefits are more varied depending on the raw materials used. Some herbal teas that are now known by the public such as cocoa leaf tea, girlfriend water leaf tea, bay leaf tea and leaf tea [2].

Moringa oleifera is a type of plant that is often found in Indonesia, knows no season, can grow in a variety of climates and in some regions it is usually processed for consumption. *Moringa oleifera* is known as different in every country including benzolive, drumstick tree, kelor, marango, mlonge, mulangay, nebeday, saijhan, and [3]. Analysis of the nutritional content of *Moringa oleifera* leaves shows that Moringa leaves have extraordinary nutritional content. Leaves of moringa are believed to cure various diseases Even in the history of moringa plants have been used by Indian tribes as medicine since 5000 years ago [4].

The chemical content of moringa leaves is phenol, hydroquinine, steroid flavonoids, triterpenoids, alkaloid tannins and saponins. Various studies have proven that the bioactive content in Moringa leaves has the potential as a medicinal compound including anti-inflammatory, antifungal, anticancer, and antioxidant. Moringa leaves are supported as a good source of nutrients because they contain proteins including all essential amino acids, iron, calcium, vitamins, carotenoids, other complementary minerals and other phytochemical compounds [3].

Moringa leaves have the potential as tea because they have good antioxidant activity. Moringa leaves contain antioxidant compounds such as flavonoids, vitamin C, and vitamin [5]. Based on the research conducted by [6], in 1 gram of green tea Moringa leaf powder brewed in 100 ml of water, Moringa leaf green tea contains 42 mg EGCG (epigallocatechin-3-gallate). In the same ratio, green tea from the Camelia sinensis plant contains 70.2 mg EGGC (epigallocatechin-3-gallate). EGCG (epigallocatechin-3-gallate) is the most commonly found type of polyphenol constituent in tea.

To facilitate its use in the community and increase the shelf life, Moringa leaves can be processed into herbal Moringa leaf tea. Based on the method of processing, tea in Indonesia is divided into three groups, namely black tea (fermented tea), green tea (unfermented tea) and fragrant tea. Black tea is made with a full enzymatic oxidation process whereas green tea is made without enzymatic oxidation, while fragrant tea is green tea which is mixed with fragrances such as jasmine flowers to produce a distinctive taste. Jasmine flowers are chosen because they have a good aroma and are favored by consumers so they are commonly used in making fragrant tea. While in Taiwan there is one more type, namely oolong tea (semifermented tea) which is the result of a transitional processing between green tea and black tea [7].

The different methods of processing tea based on the enzymatic oxidation process will affect the physical and chemical properties of tea leaves due to the enzymatic oxidation process resulting in the overhauling of certain compounds. Catechins can be oxidized to thearubigin compounds that can give a brownish red color to the tea and theaflavin which can provide fresh taste [8].

Based on research conducted by [9], it is known that processing methods can affect the quality of the cocoa leaf tea produced. Green tea from cocoa leaves turned out to have the highest antioxidant activity and in terms of organoleptic cocoa leaf green tea is the most preferred tea by panelists compared to cocoa leaf tea which is processed by the method of making black tea, oolong tea and fragrant tea. Whereas in Moringa leaves, the effect of tea processing method on the condition of pre-processing tea and organoleptic reception of the produced Moringa leaf herbal tea is unknown.

Based on the description above, the author conducted a study on "The Different of Processing Method Against Polyphenol and Antioxidant Activity of Moringa Herbal Tea" to determine the relationship between different tea processing methods to the chemical components and panelists level of Moringa tea.

2. Materials and Methods

2.1. Materials

The material used in this study was Moringa leaves obtained from Pasir, Pariaman city. The ingredients for reagents used were distilled water, methanol solution, DPPH (0.2 M) solution, reagenfolin-ciocalteu, Na_2CO_3, gallic acid, 0.05 N ammonia chloroform, H_2SO_4, and Mayer, Bouchardat and Ragendorf reagents. The equipment used in this study included a blender (Miyako BL152GF), oven (Maspion MOT600), aluminum cup, aluminum foil, erlenmeyer, UV-Vis spectrophotometry (Shimadzu UV-1800), desiccator, funnel, teat, measuring cup, dropper cups, test tubes, and equipment for sensory testing.

2.2. Design and Data Analysis
The research is explorative to find out the type of herbal tea that is the most preferred Moringa leaf organoleptically based on processing methods. The study was conducted based on 4 types of tea processing methods including black tea, green tea, oolong tea and fragrant tea with 3 replications and the data werereportedas average.

2.3. Processing Methods of Moringa Tea
The stages of making Moringa green tea: Moringa leaves are picked and sorted. Then moringa leaves were carried out at a temperature of 70^0C for 20 minutes. Then grinding using a blender until tea powder is obtained. Then drying in an oven with a temperature of 80^0C for 30 minutes.The product is ready to be packed [10].

The stages of making Moringa black tea: Moringa leaves are picked and sorted. After that the moringa leaves are carried out for 16 hours at room temperature (27-30^0C). Then grinding using a blender until tea powder is obtained. After that enzymatic oxidation is carried out at room temperature (27-30^0C) for 18 hours (until the entire leaf turns black. Then drying in an oven with a temperature of 80^0C for 30 minutes.The product is ready to be packed [10].

The stages of making Moringa oolong tea: Moringa leaves are picked and sorted. Then the moringa leaves are carried out for 16 hours at room temperature (27-30^0C). Then grinding using a blender until tea powder is obtained. After that enzymatic oxidation is carried out at room temperature (27-30^0C) for 9 hours (until the entire leaf turns black. Then drying in an oven with a temperature of 80^0C for 30 minutes.The product is ready to be packed [9].

The stages of making Moringa Fragrant tea: Moringa leaves are picked and sorted. Then rub it in the oven at 90-100^0C for 1-2 hours until the moisture content is around 0%. Moisture is then carried out by giving water until the water content is 30-35% of the weight of the material. Then do fragrances by mixing tea powder with jasmine flowers with ratio of 1: 1. Then mix and stir the tea that has been moistened with jasmine for 18 hours. Then drying in an oven with a temperature of 80^0C for 30 minutes.The product is ready to be packed [10].

2.4. Observation
Observation on Moringa leaves includes observing water content [11], total polyphenols [12] and qualitative testing of alkaloids [13].

The observation of herbal tea leaves produced by Moringa leaves in the form of observing the chemical content including water content [11], antioxidant activity [1], and total polyphenols [12] then observing tea steeping water was tested against the level of preference of the panelists [15]. The best products are chosen based on the best content of polyphenols and antioxidants and the level of preference of the panelists.

3. Results and Discussion

3.1. Raw Material
Moringa leaves have an ovoid leaf shape, 1 - 2 cm long, 1-2 cm wide, blunt tip and base, flat edge, pinnate and compounded reinforcement arrangement in one stalk. Young moringa leaves are light green with a soft and weak texture while old Moringa leaves are dark green with a rather stiff and hard

texture [16]. The analysis carried out on the raw material of young leaves of Moringa plants includes water content, ash content, total polyphenol content and qualitative testing of alkaloids. The average value of the chemical components of the raw material for Moringa leaves can be seen in Table 1.

Table 1. shows that the moisture content of the fresh leaves of *Moringa oleifera* is 73.94%. The moisture content of the Moringa leaves is slightly lower than the moisture content of Moringa leaves according to the Ministry of Health (1995), which is equal to 75%.

Table 1. The Average Value of Chemical Components of Moringa Leaf Raw Materials

Analysis Component	Result ± SD
Moisture Content (%)	73,94 ± 0,62
Polyphenol Content (%)	1,24 ± 0,26
Alkaloid (+/-)	+

Note: SD (Standard Deviation)

The total polyphenol in fresh Moringa leaves is 1.24%. The total polyphenols in Moringa leaves are higher than the total polyphenols found in cocoa leaves which are equal to 0.8% [9].

After qualitative testing of alkaloid compounds, it was obtained that the positive fresh Moringa leaves contain alkaloid compounds. In accordance with the research conducted by [17], the phytochemical test results on Moringa leaves showed the presence of alkaloid compounds, flavonoids, phenolics, triterpenoids / steroids and tannins. According to research conducted by [18], Moringa leaves contain alkaloid compounds, tannins, saponins, steroids, triterpenoids and trigonelin compounds that can increase the quantity of breast milk.

3.2. Moringa Herbal Tea

After Moringa leaves processed into 4 types of tea including black tea, oolong tea, green tea and fragnant tea, then the analysis of the four types of herbal Moringa tea was carried out. The analysis included water content, total polyphenols, antioxidants, and panelists' preference test for Moringa herbal tea. The average value of the herbal tea content of Moringa leaves with various processing methods can be seen in Table 2.

Tabel 2. The Average Value of Chemical Components Moringa Herbal Tea

Processing Method	Analysis Component		
	Moisture Content (%)	Polyphenol Content (%)	Antioxidant Activity (%)
Black Tea	7.31±0,23	1.14±0.07	32.77±0.58
Oolong Tea	7.10±0.79	1.42±0.09	34.09±2.31
Green Tea	6.36±0.36	1.61±0.04	44.87±2.51
FragrantTea	6.36±0.18	2.43±0.26	71.58±1.80

Note: SD (Standard Deviation)

3.2.1. Moisture Content. The data in Table 2. show that the moisture content in herbal Moringa tea ranges from 6-7%. This is in accordance with the maximum limit of moisture content tea in Indonesian National Standards which is 8% so that the moisture content of herbal Moringa tea with various processing methods can meet the requirements. The difference in moisture content in tea products can be caused by differences in processing in each method. Green tea undergoes twice the process of decreasing moisture content, namely in the process of drying at 70°C for 20 minutes and the drying process at 80°C for 30 minutes. Black tea and oolong tea have greater moisture content than green tea and fragrant tea because the process of black tea and oolong tea is carried out at room temperature and the drying temperature of tea in each method is the same, namely 80°C for 30 minutes. The drying process of green tea at 70°C for 20 minutes can reduce the moisture content of tea leaves to around 60-70% [10]. Fragrant tea has the lowest moisture content, allegedly because fragrant tea undergoes several drying processes, namely drying during the process of making green tea, grinding process until the moisture content of tea powder approaches 0% and the drying process when processing fragrant tea. Moisture content in food is a very important component because it will affect the physical properties and shelf life of a food ingredient. If the moisture content of a food is too high, then the food will be easily overgrown by mold/fungus so that the shelf life is shorter. Whereas if the moisture content of a food is too low, the food will become fragile, affecting the physical properties of food. This can reduce consumer acceptance of these foods [19].

3.2.2. Polyphenol Content. The data in Table 2 shows that Moringa fragrant tea has the highest total polyphenol content which is equal to 2.43%. Green tea is 1.61%, oolong tea is 1.42% and black tea is 1.14%. The total content of polyphenols in herbal tea with Moringa leaves is higher than the total polyphenols found in cocoa leaf tea which ranges from 0.3-0.7% [9]. This value is still much lower than the total content of polyphenols in green tea made from tea plants according to Indonesian National Standards number 3945-2016 which is a minimum of 15%. According to research conducted by [6], young Moringa leaves dried to 13% moisture content containing EGCG (epigallocatechin-3-gallate) of 0.76%.

The oxidation process for making black tea and oolong tea is 18 hours and 9 hours respectively. The oxidation process of herbal tea with Moringa leaves is much longer than the oxidation process of tea made from tea plants which only ranges from 45-90 minutes. During the tea oxidation process, most of the catechins are converted to theaflavin and tearubigin. Theaflavin gives a touch of fresh taste in steeping tea and gives a reddish yellow color. Whereas tearubigin plays a role in giving a brownish red color to steeping water of black tea. The longer the oxidation process takes place, the more polyphenol compounds are broken down into other compounds so that the levels of polyphenols are lower [8].

Jasmine flowers contain phenolic compounds [20]. The process of mixing jasmine flowers in the processing of fragrant tea affects the total polyphenols in fragrant tea. This is because the process of imbibition of phenolic compounds found in jasmine flowers into herbal tea leaves Moringa so that the total value becomes higher than other Moringa herbal tea.

The phenol compound is a compound containing a hydroxyl (-OH) group that is bound directly to the aromatic hydrocarbon ring group. Classification of phenol compounds contained in plants are simple phenol, benzoquinone, phenolic acid, acetophenone, naphthoxinone, xanton, carinar bioflavonoids, stilben, derivatives tyrosine, cinnamic hydroxy acids, flavonoids, lignans, and tannins [21].

3.2.3. Antioxidant Activity. Antioxidants are compounds that can counteract or reduce the negative effects of oxidants in the body by complementing the lack of electrons possessed by free radicals. DPPH radicals are widely used as radical models for antioxidant testing. Phenol compounds in plants can capture radicals. The mechanism of phenol compounds in reducing DPPH radicals is through proton donation. In the form of purple DPPH radicals, this color will become light yellow after receiving protons [1].

Moringa leaves are a rich source of beta-carotene, protein, vitamins, calcium, potassium, iron and natural sources of antioxidants. Moringa leaves are known to have good antioxidant activity because they are rich in ascorbic acid, flavanoids, phenolic compounds and carotenoids [22].

The data in Table 2. Shows that in a concentration of 10,000 ppm, the highest antioxidant activity was in fragrant tea, which amounted to 71.58%. Moringa leaf black tea has antioxidant activity of 32.77%, oolong tea by 34.09% and green tea is 44.87%.

The longer the enzymatic oxidation process in the tea processing process, the antioxidant activity will be lower. Green tea and fragrant tea processed without enzymatic oxidation processes have higher antioxidant activity than black tea and oolong tea which have lower antioxidant activity because of enzymatic oxidation processes. This is because the oxidation process results in the loss of several antioxidant components due to the enzymatic oxidation reaction [1]. In line with total polyphenols, antioxidant activity of Moringa leaves fragant tea is higher than other teas due to imbibition of phenolic compounds from jasmine flowers to Moringa leaf herbal teas. According to research conducted by [20], the antioxidant activity of jasmine flower extract was 71.48%.

The high antioxidant activity in fragrant tea can occur due to the maillard reaction. Maillard reaction is a reaction that occurs between an amino group from a free amino acid, a residue of a peptide chain or protein with a carbonyl group from a carbohydrate if both are heated or stored for a relatively long time. The processing of fragrant tea uses high temperature in the process of rubbing which is 100^0C for 1 hour and continued with drying at a temperature of 80^0C for 30 minutes can cause a reaction between carbonyl groups and amino groups that produce Maillard reaction products namely melanoidin compounds. This melanoidin compound has benefits especially in the food industry, namely increasing the intensity of brownish color, increasing the content of antioxidants and increasing aroma [23].

3.2.4. Sensory Evaluation. To find out the consumer acceptance of Moringa herbal tea products produced by various processing methods, the sensory evaluation of Moringa herbal tea produced included color, aroma and flavor of steeping Moringa herbal tea. The sensory evaluation of Moringa herbal tea with various processing methods can be seen in Table 3.

Tabel 3. The Sensory Evaluation of Moringa Herbal Tea

Processing Method	The Value of Sensory Evaluation		
	Color	Aroma	Taste
Black Tea	4.20	2.96	2.77
Oolong Tea	3.10	3.23	2.83
Green Tea	3.16	3.03	2.97
Fragrant Tea	4.13	4.13	3.20

Note: SD (Standard Deviation)

The data above shows the results of sensory evaluations on a scale of 1 to 5. The scale in this study are (1) Very Dislike, (2) Dislike, (3) Ordinary, (4) Likes and (5) Very Like.

The data in Table 3 show that the level of acceptance of panelists on the color of herbal tea with Moringa leaves with various processing methods is on a scale of 3 (normal) to 4 (likes). The color of steeping Moringa herbal tea which is most preferred by panelists is the color of steeping black tea with an average value of 4.2 (likes). The longer the enzymatic oxidation process, the color of steeping herbal tea leaves Moringa becomes darker. Green tea has a steaming color that is slightly greenish yellow, steeping oolong tea is reddish yellow while black tea has a steaming color that is brownish red. The color difference is due to the enzymatic oxidation process, the catechin compounds are overhauled into theaflavin and tearubigin. The aflavin gives a touch of fresh taste in steeping tea and gives a reddish yellow color while the tearubigin plays a role in giving a brownish red color. Color changes in Moringa leaves can also occur due to the overhaul of the chlorophyll compound into black feophytin [8].

Steeping colored fragrant tea is much darker than green tea but brighter than black tea. Color changes in fragrant tea can be caused because in the process of processing, fragrant tea warms up with high temperatures during the process of grinding. Color changes that occur in dried food products are Mailard and caramelization reactions. Changes that occur due to mailard reactions are reactions that occur between carboxyl from carbohydrates with primary amino groups from proteins.

The data in Table 3 show that the level of acceptance of panelists on the aroma of herbal tea with Moringa leaves with various processing methods is on a scale of 3 (normal) to 4 (likes). The aroma of Moringa leaf herbal tea which is most preferred by panelists is the aroma of fragrant tea with an average value of 4.13 (likes). This is due to the aroma of jasmine produced by fragrant tea. In addition, according to [23], melanoidin compounds formed due to maillard reactions can have an effect on increasing the aroma of tea.

The aroma of tea products plays a very important role. The term aroma is interpreted as an odor sensation caused by the chemical stimulation of volatile compounds which are smelled by different oily nerves in the nasal cavity when food enters the mouth. The sensation or stimulation will always cause delicacy, which can then affect the level or acceptability of the panelist or consumer of a particular food product [19].

The taste is different from the smell and more often involves the five senses of the tongue. Taste is a parameter that is no less important in determining whether a product can be accepted by consumers or not. Components that can cause a sense of dependence on its constituent compounds. Generally food stuffs do not only consist of one taste but a combination of various kinds of flavors that are integrated so as to create a complete taste of food [19].

The data in Table 3 show that the level of acceptance of panelists on the taste of Moringa herbal tea with various processing methods is on a scale of approximately 3 (normal). This is because the steeping of Moringa leaves herbal tea has a tight taste. The flavor of the steamed herbal tea leaves Moringa leaves can be formed due to the content of polyphenol and alkaloid compounds in Moringa leaves. In addition, during the organoleptic testing of herbal tea, Moringa leaves are not added sugar so that most panelists did not like steeping the Moringa leaf herbal tea. However, Moringa leaf fragrant tea has the most preferred taste by panelists with an average value of 3.2 (usual). Overall, the most preferred Moringa leaf herbal tea products by panelists can be seen in the organoleptic radar graph found in Figure 1.

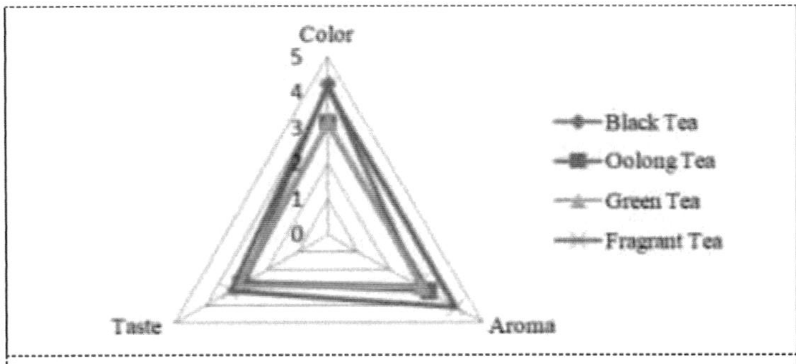

Figure 1. Organoleptic Radar Chart of Moringa Herbal Tea

From Figure 1 it can be seen that the whole Moringa fragrant tea is most preferred by consumers because it has the largest organoleptic radar area. The color organoleptic value of Moringa fragrant tea is on a scale of 4.13 (likes), the organoleptic value of aroma of fragrant tea from Moringa leaves is on a scale of 4.13 (likes), and the organoleptic value of steeping flavor of Moringa leaves is on a scale of 3, 2 (normal).

4. Conclusion

The tea processing method gives a difference to the chemical components and the characteristics of the Moringa leaf herbal tea produced. The longer the enzymatic oxidation process in tea takes place, the lower the antioxidant and polyphenol content in tea. Fragrant tea processing method is a herbal tea product with Moringa leaves with the best characteristics with a water content of 6.36%, total polyphenols 2.43%, antioxidant activity 71.58%, where the color organoleptic value of steeping Moringa fragrant tea is on a scale of 4.13 (likes), the organoleptic value of aroma of fragrant tea from Moringa leaves is on a scale of 4.13 (likes), and the organoleptic value of steeping flavor of Moringa fragrant tea is on a scale of 3.2 (normal).

References

[1] Rahmawati, Nur D. 2015. Aktivitas Antioksidan dan Total Fenol Teh Herbal Daun Pacar Air (Impatiens balsamina) dengan Variasi Lama Fermentasi dan Metode Pengeringan. [Artikel Publikasi]. Surakarta : Fakultas Keguruan dan Ilmu Pendidikan. Universitas Muhammadiyah

Surakarta.

[2] Hambali, E. M. Z. Nasution dan E. Herliana.2005.*Membuat Aneka Herbal Tea.*Jakarta. Penebar Swadaya.

[3] Fahey, J.W. 2005. *Moringa oleifera: A Review of the Medical Evidence for Its Nutritional, Therapeutic, and Prophylactic Properties.* Part 1. Trees for Life Journal. (http://www.malunggay-propagation.com/Jed_F ahey_text_GB.pdf): 3-27.

[4] Mishra, Satya P., Pankaj Singh dan Sanjay Singh. 2012. *Processing of Moringa oleifera Leaves for Human Consumption.* Bull. Env.Pharmacol. Life Sci. Volume 2 [1] December 2012: 28-31.

[5] Rahmawati. 2015. Pengaruh pemberian seduhan daun kelor (Moringa oleifera Lamk) terhadap kadar asam urat tikus putih (Rattus norvegicus). [Skripsi]. Semarang: Program Studi Ilmu Gizi Fakultas Kedokteran. Universitas Diponegoro.

[6] Sugianto, Ajeng Kinanti. 2016. Kandungan Gizi Daun Kelor (Moringa Oleifera) Berdasarkan Posisi Daun dan Suhu Penyeduhan.[Skripsi]. Bogor: Fakultas Ekologi Manusia. IPB Alegantina, S., Isnawati, A. dan Widowati, L. 2013.*Kualitas Ekstrak Etanol 70% Daun Kelor (Moringa oleifera Lamk) dalam Ramuan Penambah ASI.* Jurnal Kefarmasian Indonesia 3(1):1-8

[7] Sativa, Oryza. 2006. Kajian Proses Pembuatan Teh Herbal dari Campuran Teh Hijau (Camellia sinensis), Rimpang Bangle (Zingiber cassumunar Roxb.) dan Daun Ceremai (Phyllanthus acidus (L.) Skeels.). [Skripsi]. Bogor: Fakultas Teknologi Pertanian. IPB.Aminah, S., Ramadhan, T. dan Yanis, M. 2015. *Kandungan Nutrisi dan Sifat Fungsional Tanaman Kelor (Moringa oleifera).* Buletin Pertanian Perkotaan 5(2):35-42.

[8] Towaha, Juniaty. 2013. *Kandungan Senyawa Kimia pada Daun Teh (Camellia sinensis).* Warta Penelitian dan Pengembangan Tanaman Industri 19(3): 12-16

[9] Saputra, Abdi. 2016. *Pengaruh Pengolahan Daun Kakao (Theobroma Cacao. L) Terhadap Mutu Minumana Teh Kakao*. [Skripsi]. Padang: Fakultas Teknologi Pertanian. Universitas Andalas.

[10] Hartoyo, Arif. 2003. *Teh dan Khasiatnya Bagi Kesehatan*. Yogyakarta.Kanisiuns.

[11] Andarwulan, N., Kusnandar, F., Herawati, D., 2011. *Analisis Pangan.* Jakarta. Dian Rakyat.

[12] Dhianawaty, D. dan Ruslin. 2015. Kandungan Total Polifenol dan Aktivitas Antioksidan dari Ekstrak Metanol Akar Imperata cylindrica (L) Beauv. MKB. 47(1):60-64.

[13] Marsora, Astika. 2013. Analisis Fisikokimia dan Fitokimia Ekstrak Kental Daun Binahong Anredera cordifolia (Tenore) Steen.). [KTI]. Padang: Akademi Farmasi Ranah Minang.

[14] Rahmawati, Nur D. 2015. Aktivitas Antioksidan dan Total Fenol Teh Herbal Daun Pacar Air (Impatiens balsamina) dengan Variasi Lama Fermentasi dan Metode Pengeringan. [Artikel Publikasi]. Surakarta : Fakultas Keguruan dan Ilmu Pendidikan. Universitas Muhammadiyah Surakarta.

[15] Soekarto, S. T. 1985. Penilaian Organoleptik untuk Industri Pangan dan Hasil Pertanian. Jakarta. Bhratara Karya Aksara.

[16] Aminah, S., Ramadhan, T. dan Yanis, M. 2015. *Kandungan Nutrisi dan Sifat Fungsional Tanaman Kelor (Moringa oleifera).* Buletin Pertanian Perkotaan 5(2):35-42.

[17] Putra, I. W. D., Dharmayudha, A. A., Sudimartini, L. M., 2016. *Identifikasi Senyawa Kimia Ekstrak Etanol Daun Kelor (Moringa oleifera L) di Bali. Indonesia* Medicus Veterinus5(5) : 464-473.

[18] Alegantina, S., Isnawati, A. dan Widowati, L. 2013.*Kualitas Ekstrak Etanol 70% Daun Kelor (Moringa oleifera Lamk) dalam Ramuan Penambah ASI.* Jurnal Kefarmasian Indonesia 3(1):1-8.

[19] Winarno, F. G. 1991. *Kimia Pangan dan Gizi*. Jakarta. Gramedia Pustaka Utama

[20] Wang, H. F., Yih, K. H., Yang, C. H., dan Huang, K. F. 2017. Anti-oxidant activity and major chemical component analyses of twenty-six commercially available essential oils. Journal of Food and Drug Analysis 25: 881-889.

[21] Dhianawaty, D. dan Ruslin. 2015. Kandungan Total Polifenol dan Aktivitas Antioksidan dari Ekstrak Metanol Akar Imperata cylindrica (L) Beauv. MKB. 47(1):60-64.

[22] Siddhuraju, P., Becker, K., 2003. Antioxidant properties of various solvent extracts of total phenolic constituents from three dif erent agro-climatic origins of drumstick tree (Moringa oleifera Lam.). J Agric Food Chem 15: 2144–2155 (http://pubs.acs.org/doi/abs/10.1021/jf020444%2 B).

[23] Waskito, Y. A. P., Al-Baarri, A. N., Abduh, S. B. M. 2014. Intensitas Warna Kecoklatan, Sifat Antioksidan, dan Goaty Aroma pada Proses Glikasi Whey Susu Kambing dengan L-psikosa, L-tagatosa, dan L-fruktosa. Jurnal Aplikasi Teknologi Pangan 3(4) : 160-163.

Individual Effect of UV-C Radiation and Hot Water Treatment on Harvested Quality of Chili

K Fahmy[1], O C Chatib[1], D Cherie[1], P Maulida[1], A A Sirtin[1], N Susanto[2], Nofiarli[3], F Violalita[4], K Nakano[5], P Jitareerat[6]

[1] Department of Agricultural Engineering, Faculty of Agricultural Technology, Andalas University, Kampus Limau Manis, Padang-25163, Indonesia
[2] Department of Agricultural Products Technology, Faculty of Agricultural Technology, Andalas University, Kampus Limau Manis, Padang-25163, Indonesia
[3] Tropical Fruit Crops Research Institute, Ministry of Agriculture, Indonesia, Jl. Raya Solok Aripan Km 8 Solok West Sumatra, 27351-Indonesia
[4] Department of Food Technology, Agricultural Polytechnic State of Payakumbuh Jalan Raya Negara Tanjung Pati Km 7 – 26271, Indonesia
[5] The United Graduate School of Agricultural Science Gifu University, 1-1 Yanagido, Gifu, 5011193 JAPAN
[6] Division of Postharvest Technology, School of Bioresources and Technology, King Mongkut's University of Technology Thonburi, 126 Bangmod, Thungkru, Bangkok, 10140, Thailand

Corresponding author's e-mail address: khandrafahmy@ae.unand.ac.id

Abstract. The aim of study was to determine the individual effect of UV-C radiation and Hot Water Treatments (HWT) in preserving quality of chili packed in Modified Atmosphere Packaging (MAP) under ambiet temperature storage. Chili was radiated by UV-C for 15 min and for 20 min, with HWT at 45 °C for 15 min and at 53 °C for 4 min, and untreated as control. After treatments, all fruits were packaged and stored at room temperature (25 °C) for 1 month. The results that untreated chilli could be stored for 7 days, treated chilli with UV-C 15 min and HWT at 45 °C for 15 min could be stored for 14 days, while chilli treated with UV-C for 20 min and HWT at 53 °C for 4 min could be stored for 17 days. Among all treatments, chili treated with UV-C for 20 min suppressed the disease severity dan maintained the harvested quality of chilli. This study suggests that combination UV-C irradiated with HWT under low temperature might be extended the shelf life of chilli.
Keywords: Chilli, UV-C, Hot Water Treatment, Modified Atmosphere Packaging, Shelf-life

1. Introduction

Chilis are a horticultural product that use largely as ingredient in Indonesian food. The high demand of chili has been encouraged their trading not only in dosmetic market but also in oversea. However, in high harvesting season, chili supplies in the market become abundant and this condition affects on its marketability. Marketed to other regions is constrained by the lack of technology in quality preservation especially during storage and transportation process. However, chili is perishable commodity and easy infected by anthracnose disease which makes chili rotten qucikly. Fungi is the main cause of disease in chili, and it become major prevalent and widespread postharvest disease [1, 2].

Many treatments have been used to control of chili anthracnose pathagens in chili. Application of chemical fungicides has been used to control chili anthracnose disease, but these treatment cause the fungicide-resistant pathogens [3] and the risks to consumers and environment [4, 5, 2). Therefore, safe alternative treatments have become an essential requirement for the management of postharvest disease of perishable commodity.

UV-C has also emerged as an alternative method to avoid chemical fungicides. UV-C irridation increased the disease resistance against anthracnose and improved the harvested quality of perishable commodity. It has been reported that UV-C irradition reduced disease incidence and severity, and improve quality of mango compared with untreated fruit [6]. HWT also has potential treatment as a non chemical method for controlling postharvest disease. Application of HWT delayed of decayed in pepper fruit compared without treatment pepper fruit [7].

Moreover, Modifed Atmophere Packaging (MAP) has been subject of the many studies for maintaining the quality of perishable commodity during storage and transportation. The ability of MAP to reduce CI is thought as reduction in O2 and elevation of CO2 inside the package as well as a higher humidity inside packaging. It has been reported that MAP preserved the quality of 'Jiro' persimmon for export market. The optimal of bulk MAP was also designed for long-term storage of 'Fuyu' persimmon [8].

In order to control postharvest disease and prolong of storage of chili, the individual influence of UV-C irradition and HWT on chili rot disease and quality were investigated. In this study, chilis were stored at enviroment temperature (25°C) which is actual temperature for storage and transported of chili to the market.

2. Materials and Methods

2.1 Plant Materials and Storage Conditions

Chilli (Capsicum annum L.) at commercial maturity were harvested from Solok Regency, West Sumatra Province, Indonesia and transported immediately to laboratory. Fruits were sorted and selected on the basis of uniform size and absence of visual defects. Chilis were devided on five groups. First group was irradiated using UV-C Germicidal lamp 30 watt (peak emission 254 nm) for 15 min to provide doses 7 kJ/m2. The UV-C radiation doses was measured using UV–Light meter (YK-37UVSD, Taiwan). Second group was irradiated with UV-C for 20 min to provide doses 9 kJ/m2. Thrid group was treated with HWT at 45 °C for 15 min in water bath (WNB14-Memmert, Germany). Fourth groups were treated with HWT at 53 °C for 4 min. The last goup was untreated chilli as control. After treatment, chilis were packaged in in Low Density Poliethylene (LDPE) about 100 g per packaged, and stored at room temperature (25 °C) for 1 month.

2.2 Disease Severity

Disease severity of chili was evaluated by visual observation of fruit based on surface area of surface decayed in term of 0–4 index: 0 (no signs of decay; 1 (<25%); 2 (25–50%); 3 (51–75%) and 4 (>75%). Disease severity between 0 and 4 calculated as:

$$DS = \frac{[\Sigma\ (DI)\times(number\ of\ fruit\ at\ index)]}{(Total\ of\ fruit)} \quad (1)$$

where DS and DI were disease severity and disease index, respectively.

2.3 Quality Evaluation

The weight of chilli from each experimental condition was measured. Weight loss of each fruit in the chamber was calculated as a percentage of initial fresh weight. Skin colour was measured dusing a lobivon colour measurement (MRM200, China) yielding parameters L*, a* and b*. Skin colour measurement was made at the base, middle, and bottom of the chili. The results were expressed as the

IOP Conf. Series: Earth and Environmental Science **327** (2019) 012007 doi:10.1088/1755-1315/327/1/012007

hue angle ($h^o = \tan^{-1} b^*/a^*$). The firmness was measured by using fource gauge (FGS-Japan) expressing as F (N) represent the force exerted on a sample under compression. Titrable acidity was measured using titrimetric method using with 0.01 N iodine solution. pH was measured using pH meter to determin the pH change during storage.

2.4 Malondialdehyde Content
Malondialdehyde (MDA) was determined according to the method of Hodges et al (1999) with some modifications as mentioned in Fahmy and Nakano (2014b).

2.5 Total of Microbes
Chilis from each packet of treatments were evaluated to determine microbial load as per the standard plate count method. Total plate count agar was used to determine yeast and molds count. A 5 g sample was diluted with 9 mL of 0.86% NaCl solution, homogenized and was then further diluted to get serial dilutions. Serially diluted samples were plated onto total plate count agar. 1 mL of each dilution was transferred to sterile petri plate and 15–20 mL of sterile agar media at temperature 40–45 °C was poured, mixed well and allowed to set. Plates were incubated at 37 °C for 48 h. At the end of the incubation period the colony forming unit (cfu) were counted and multiplied by appropriate dilution factor to obtain total plate count.

2.6 Statistical Analysis
The design was completely randomised with 3 replications and each replicates consisted 100 g of chilis packaged in LDPE. Statistical significance was determined by subjecting the mean values to analysis of variance and means were compared by Tukey's test at the 5% level of significance using R 3.5.1 (R Foundation).

3. Results and Discussion

3.1 Disease severity
Disease severity (DS) increased rapidly on untreated chili and chili treated with HWT at 45 °C for 15 min on day 7, while chilis treated with UV-C for 15 min and 20 min, and HWT at 53°C for 4 min the DS were suppressed, and then increased gradually for all treatments tested untl the end of storage. From the result, chili had different shef life among the treatments. Untreated chili (control) had shelf life for 10 days, chili treated with UV-C for 15 min and HWT at 45 °C for 15 had shelf life for 14 days, while chili treated with UV-C for 20 min and HWT at 53 °C for 4 min had shelf life for 17 days. Sripong et al. (2015) reported that UV-C and HWT suppressed the dieasease severity on mango fruit. HWT at 53 °C for 4 min suppressed of decay on pepper fruit compared with control and HWT at 45 °C for 15 min [7].

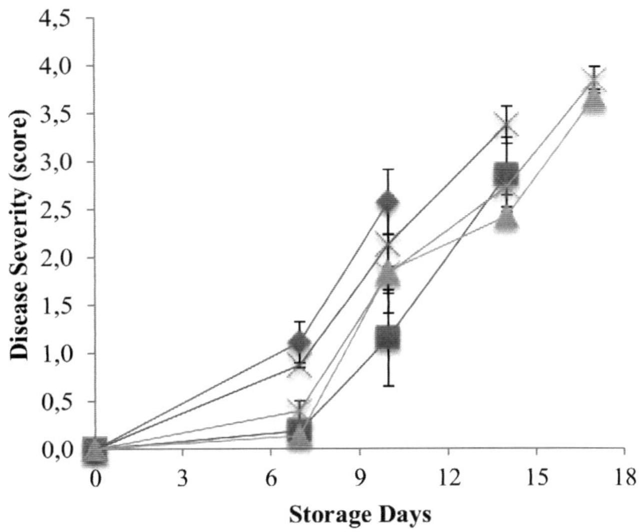

Figure 1. Disease severity of chili treated with UV-C irradiation for 15 min and 20 min, HWT at 45 °C for 15 min and at 53 °C for 4 min, and untreated (control) during storage at 25 °C packed with LDPE (0.08 mm thick). Vertical bars represent standard error of the means (SE) for triplicate samples.

3.2 Quality of Chili

Weight loss of chili increased significantly during period of storage at 25 °C for all treatments. Significant increase was shown on untreated chili (control) compared with treated chilis (Figure 2). Among treated chilis, treatment with UV-C for 20 min showed significant suppresses the increasing of the weight loss. Weight loss of mango increased gradually during period of storage treated with UV-C and HWT, however, significant different was not found among them [6].

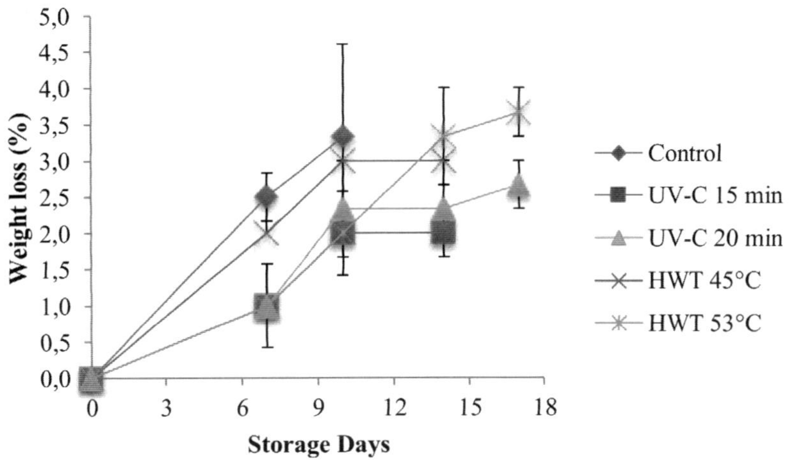

Figure 2. Weight loss of chili treated with UV-C irradiation for 15 min and 20 min, HWT at 45 °C for 15 min and at 53 °C for 4 min, and untreated (control) during storage at 25 °C packed with LDPE (0.05 mm thick). Vertical bars represent standard error of the means (SE) for triplicate samples.

IOP Conf. Series: Earth and Environmental Science **327** (2019) 012007 doi:10.1088/1755-1315/327/1/012007

Skin colour as hue value of chili decreased during period of storage at 25 °C for all treatments (Figure 3), however, significant difference in hue value was not found among them.

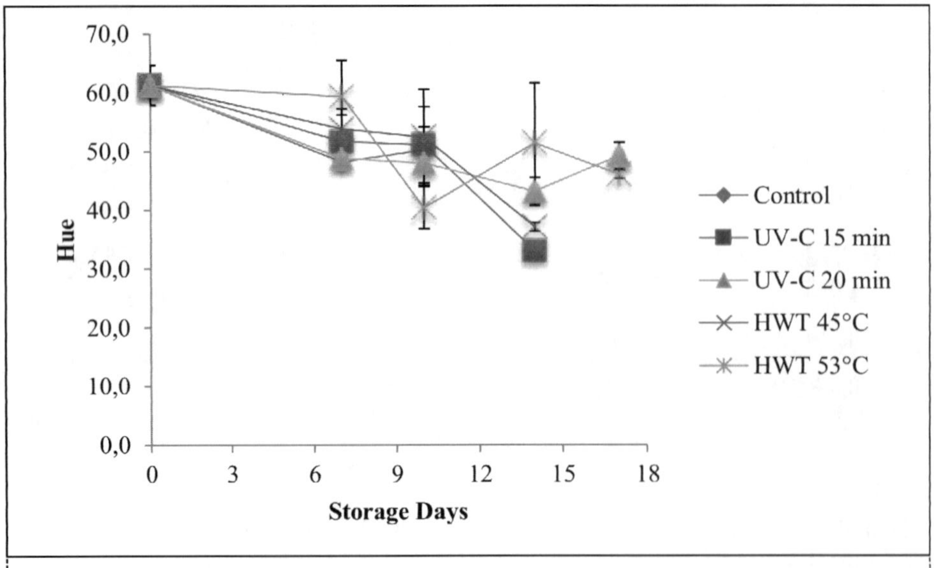

Figure 3. Hue value of chili treated with UV-C irradiation for 15 min and 20 min, HWT at 45 °C for 15 min and at 53 °C for 4 min, and untreated (control) during storage at 25 °C packed with LDPE (0.08 mm thick). Vertical bars represent standard error of the means (SE) for triplicate samples.

Firmness of chilli decreased significantly during period of storage at 25 °C for all treatments. Rapid decrease in firmness was shown on untreated chili compared with treated chilli. However, significant difference was not found among them.

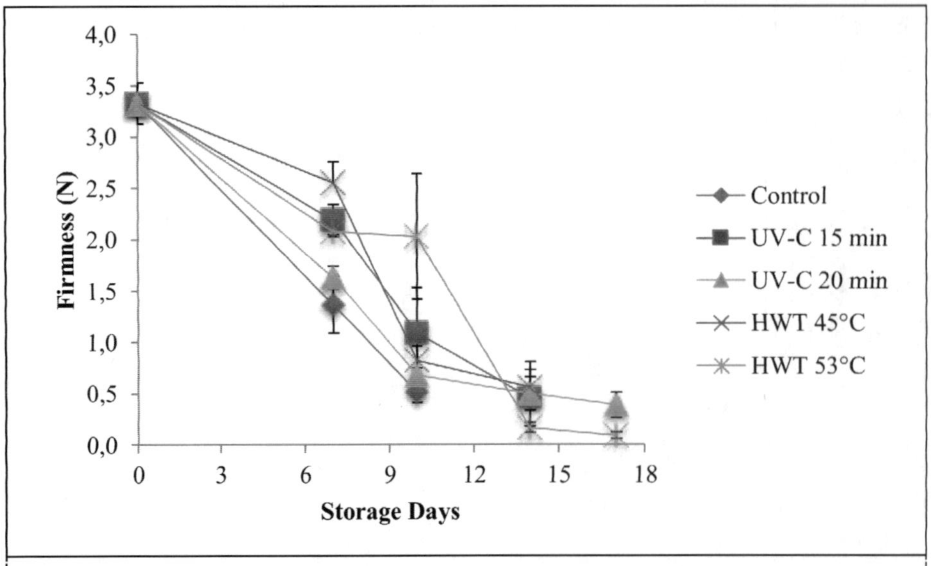

Figure 4. Firmness of chili treated with UV-C irradiation for 15 min and 20 min, HWT at 45°C for 15 min and at 53°C for 4 min, and untreated (control) during storage at 25 °C packed with LDPE (0.08 mm thick). Vertical bars represent standard error of the means (SE) for triplicate samples.

Titrable acidity increased for chili treated with UV-C for 15 min, HWT at 45 °C for 15, and HWT 53 °C for 4 min on day 7, after, the value decreased gradualy until the end of storage. While for chili treated with UV-C for 20 min, the titrable acidity was stable relatively during period of storage, however, significant difference was not found among treatments tested. UV-C radition more effective to maintain the titrable acidity compared with HWT [6].

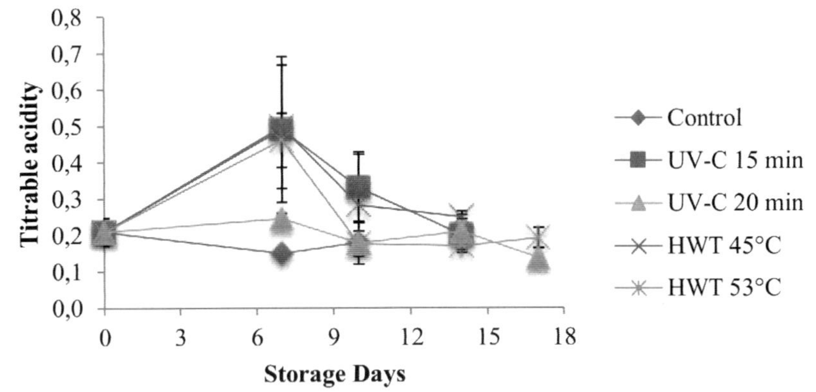

Figure 5. Titrable acidity of chili treated with UV-C irradiation for 15 min and 20 min, HWT at 45 °C for 15 min and at 53 °C for 4 min, and untreated (control) during storage at 25 °C packed with LDPE (0.08 mm thick). Vertical bars represent standard error of the means (SE) for triplicate samples.

pH values of chili decreased below 7 showing on untreated and HWT chilis, while chiles treated with UV-C showed on increases of pH above 7 during period of storage, and significant diference in pH values was found among all treatments. After day 7, pH value decreased significantly on untreated and treated chili with HWT 45 °C for 15 min, while chili treated with HWT at 53 °C for 4 min decreased gradualy until day 14 of storage. For chili radiated with UV-C, the pH value was maintained above 7 for both the treatements until the day 14 and 17 for UV-C for 15 min and 20 min, respectively.

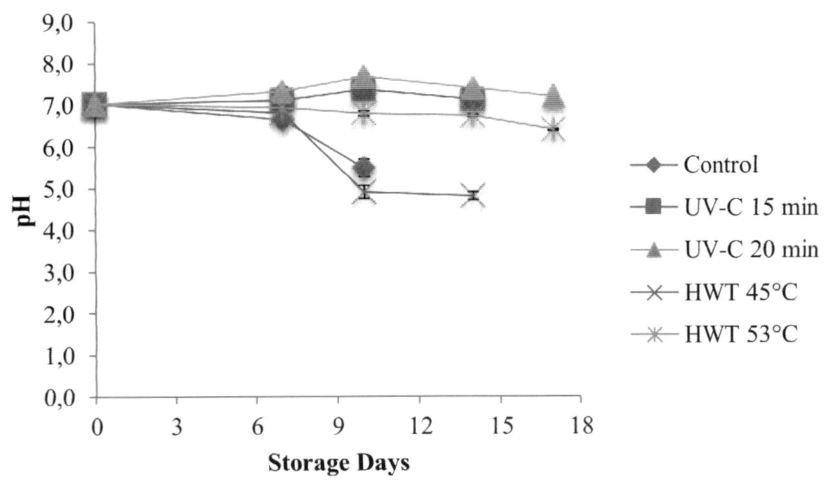

Figure 6. pH of chili treated with UV-C irradiation for 15 min and 20 min, HWT at 45 °C for 15 min and at 53 °C for 4 min, and untreated (control) during storage at 25 °C packed with LDPE (0.08 mm thick). Vertical bars represent standard error of the means (SE) for triplicate samples.

3.3 MDA Content

MDA content increased on untreated chili and radiated UV-C chili on day 7, and then this value decreased until the end of storage, while on HWT chili, MDA content decreased and then increased gradually. However, significant difference in MDA content was not found among treatements.

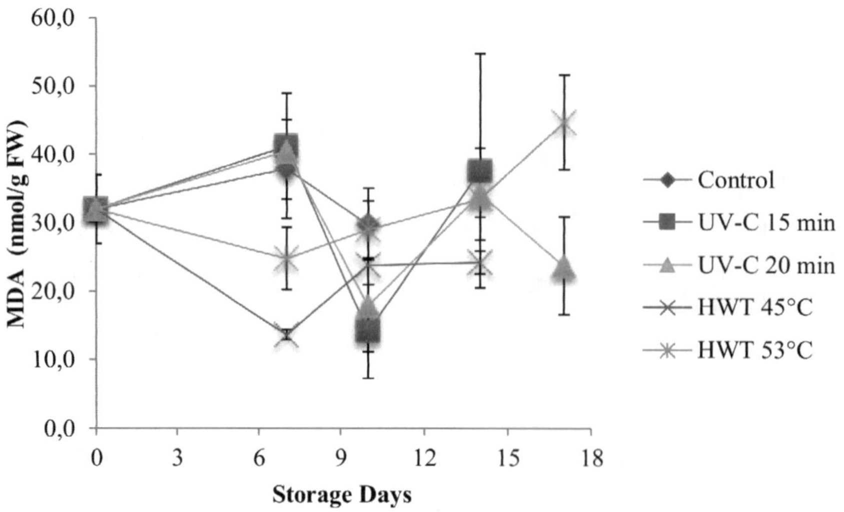

Figure 7. MDA content of chili treated with UV-C irradiation for 15 min and 20 min, HWT at 45 °C for 15 min and at 53 °C for 4 min, and untreated (control) during storage at 25 °C packed with LDPE (0.08 mm thick). Vertical bars represent standard error of the means (SE) for triplicate samples.

3.4 Total Microbes

The total microbes increased during period of storage at 25 °C for all treatments tested. However, significant difference in total microbes was not found among treatments. Nevertheless, chilli treated with UV-C at 53 °C for 4 min showed a lower value of microbes compared with other treatments until the end of storage.

Table 1. Total of microbes (log CFU/g sample) of chili treated with UV-C irradiation for 15 min and 20 min, HWT at 45 °C for 15 min and at 53 °C for 4 min, and untreated (control) during storage at 25 °C and packed with LDPE (0.08 mm thick).

Storage Days	Total of microbial (Log CFU/g sample)				
	Control	UV-C 15 min	UV-C 20 min	HWT 45 °C	HWT 53 °C
0	1.5a	1.5a	1.5a	1.5a	1.5a
7	1.5a	1.6a	1.5a	1.7a	1.6a
10	1.8a	1.7a	1.7a	1.7a	1.7a
14		2.2a	1.8a	cnc	1.6a
17			1.8a		1.7a

4. Conclusion

In this study, the individual effect of UV-C radition and HWT on harvested quality of chili were investigated. From the results, chili treated with UV-C for 20 min and with HWT at 53 °C for 4 min prolonged the shelf life of chili compared with others treatments. This study suggests that combination UV-C irradiated with HWT under low temperature storage might be more effective to maintain the harvested quality of chili.

Acknowledgement
The authors would like to thank Institute for Research and Community Service Andalas University for the financial support, to The United Graduate School of Agricultural Science Gifu University for laboratory support of this research work.

References

[1] Chanchaichaovivat A., Panijpan B, Ruenwongsa P 2007 *Screening and Identification of Yeast Strains from Fruits and Vegetables: Potential for Biological ConTrol of Postharvest Chilli Anthracnose (Colletotrichum Capsici)*. Biol. Control **42** 326–335

[2] Chanchaichaovivat A, Panijpan, B, Ruenwongsa P 2008 *Putative Modes of Action of Pichia Guilliermondii Strain R13 in Controlling Chilli Antracnose After Harvest.* Biol. Control **47** 207–215

[3] Parry D W 1990 *Plant Pathology Agriculture* (Australia: Cambridge University Press)

[4] Dedej S, Delaplane K S, and Scherm H 2004 *Effectiveness of Honey Bees in Delivering the Biocontrol Agent Bacillus Subtilis to Blueberry Flowers to Suppress Mummy Berry Disease*. Biol. Control **31** 422–427

[5] Pal K K and Gardener B M 2006 *Biological Control of Plant Pathogens* The Plant Health Instructor.

[6] Sripong K, Jitareerat P, Tsuyumu S, Uthairatanakij A, Srilaong V, Wongs-Aree C, Ma G, Zhang L, and Kato M 2015 *Combined Treatment with Hot Water and UV-C Elicits Disease Resistance Against Anthracnose and Improves the Quality of Harvested Mangoes* Crop Protection **77** 1–8

[7] Gonza ́lez-Aguilar G A, Gayosso L, Cruz R, Fortiz J, Ba ́ez R, and Wang C Y 2000 *Polyamines Induced by Hot Water Treatments Reduce Chilling Injury and Decay in Pepper Fruit* Postharvest Biol. Technol. **18** 19–26

[8] Liamnimitr N, Thammawong M, Techavuthiporn C, Fahmy K, Suzuki T, and Nakano K 2018 *Optimization of Bulk Modified Atmosphere Packaging for Long-Term Storage of 'Fuyu' Persimmon Fruit* Postharvest Biol.Technol. **135** 1-7

Mapping of Irrigation Service Capacity by Remote Sensing in Batang Tampo Irrigation System-Lintau Buo

A Asmuti[1]

[1] Department of Agricultural Engineering, Andalas University, Padang, Indonesia

Corresponding author's e-mail address: asmuti@ae.unand.ac.id

Abstract. Batang Tampo Irrigation System (DI Tampo) area 1,437 ha, has 65 irrigation service are in Sub Basin Batang Tampo. Main river in Sub Basin is 25 km and a part of Sub Basin Kuantan-Inderagiri. This area lied on hilly terrain 350-2262 m above sea level. Normaly, cropping intensity of rice farms is 2 to 2,5. But in dry period in downstream area it has 1 or failed because of water stress or dried. Even The rainfield area is not planted in the period of dry season. The area of rice field along the river or irrigation system show that it has better cropping intensity. It should have to know how is irrigation service capacity of Batang Tampo Irrigation System. Technology of Remote Sensing it can be used to evaluate of the capacity of irrigation sevice by using of information data satelite, like Digital Elevation Model (DEM) for topografi data, image, land level and Enhanced Vegetation Index (EVI) and rain (Herdianto dkk., 2010). This data usely combine with Geographic Information System (GIS) for spacial information data. This study aims to map the ability of DI Tampo service by utilizing remote sensing technology with a vegetation index as its main parameter, upstream - downstream variations, variations otf the position rice farms land level (Hilly or Valley) or terrain land level, and distance of rice fields from irrigation water sources. The method to identification of the capacity irrigation service is EVI by remote sensing and survey by the farmers to verivication data. Hydrology data is collected by the authority. Rice field area is identified by GIS and topography data by DEM. EVI value, which is a representation of photosynthetic activity and canopy density of paddy cropped or cropping intensity on paddy field each year. Average EVI data is collect for five years (2012 – 2017, excepted 2015). The result of this research shows: Irrigation service capacity influences by the topography of rice farms, mainly in the upstream. In upstream, some area of rice field was not cropped of paddy (different crop) because of water scarsity and different land level of rice farm (Hilly and river). In the downstream Irrigation service capacity influences by the distance of paddy field from the water resource.

Keywords: Irrigation, EVI, irrigation capacity

1. Introduction

Tampo Irrigation Area covers 65 irrigation systems from upstream to downstream with a total area is 1,437 ha with the main water resources of Batang Tampo, which is located in the Batang Tampo sub watershed with an area of around 26,450 ha. Thus the area of DI Tampo is 5.43% of the area of Sub watershed of Batang Tampo. Under normal conditions, the rice fields in Batang Tampo area have planting intensity 5 times in two years and some of them twice in a year. However, in certain conditions, part of the rice fields that near the downstream and in certain locations are only able to plant once in a year and sometimes crop failure due to lack of water and drought. With the discharge

IOP Conf. Series: Earth and Environmental Science **327** (2019) 012008 doi:10.1088/1755-1315/327/1/012008

of dry season around 300 liters/second, Batang Tampo water flow is unable to fulfill the needs of irrigation water for DI Tampo as a whole.

When the dry season begins to occur, the areas along the river and irrigation networks show higher planting index values than areas far from the network [8]. With the length of the main river reaching 25 km at an altitude between 350-2262 m above sea level, the form of Sub watershed of Batang Tampo is wavy and variations in steep altitude in the upstream section, ramps in the middle and somewhat flat in the lower part. The rice fields in sub watershed of Tampo also have height variation because they are along the main and undulating river paths according to the Batang Tampo cross section, so most of the paddy fields are near the upsteram and the middle are tiered, and slightly sloping and flat in the lower reaches.

With such shape of the watershed, the rice fields in the upstream and along the river flow have greater chance of getting water flow, especially in the dry season. The undulating shape of the land also indicates that there is lower chance of getting the water in higher areas on the ridge (hilly/guguak), water must be flowed from a place that is relatively further compared to rice fields that are in lower place in the valley (Lakuak) to raise water level. Thus Guguak rice fields require relatively larger water because some will be lost on the way. Another problem is the method of taking water and the amount of water for irrigation from a part of the irrigation area has not been measured properly. There are indications of users in the upstream section or those adjacent to water sources or irrigation channels using the water excessively.

From the description of this problem, it should to know: What is the ability of Batang Tampo Irrigation system to be able to serve rice fields along the flow of Batang Tampo river according to variations of: upstream-downstream, the distance from the channel irrigation and the location of rice fields according to variations of guguak-lakuak.

Technology that has been used to evaluate the performance of an irrigation system is remote sensing, which is using information provided by satellites such as the Digital Elevation Model (DEM) for topographic data, image, surface thermal (surface temperature), plant index (vegetation index), and rainfall [9]. The uses of satellite data can be combined with the Geographic Information System (GIS) to describe spatial information about the data that obtained. In the field of irrigation, GIS is used to visualize and explain the results of evaluations of irrigation performance and map theoretical irrigation water needs [5].

This study aims to map the ability of DI Batang Tampo service by utilizing remote sensing technology with a vegetation index as its main parameter, knowing the production, cropping intensity and according to upstream-downstream and variations in the distance of rice fields from irrigation water sources, variations in the position of rice fields and how farmers should handle the water limitations. The study was conducted with a five year times series data analysis of the vegetation index (2012–2017 excepted 2015) satellite MODIS-USGS (MODERATE resolution of the Imaging Spectroradiometer-United States Geological Survey). The vegetation index that used is NDVI and EVI which calculates and describes the condition of plants from a combination of electromagnetic waves in the red, near infrared, and blue spectrum derived from plant leaves. Observations of satellite data were verified in the field to see the accuracy of satellite data compared to conditions in the field. Field data also to know production, cropping intensity and irrigation and rice farms management. The research output will provide a map of irrigation service capabilities in Batang Tampo Irrigation System according to the variation of upstream-downstream rice fields, distance from channels or water sources and variations according to the height of the water source guguak-lakuak. The irrigation service capacity will indicate by: EVI value as reflection of cropping intensity.

2. Research Methodology

The research will be conducted at DI Tampo which is geographically located in the Sub watershed of Batang Tampo area and administratively is located in the North Lintau District and Lintau Buo District, Tanah Datar District, West Sumatra. Sub watershed of Batang Tampo is part of the Kuantan-Indragiri River which empties into the east coast of Sumatra Island. The shape of the Batang Tampo

Sub watershed extends from north to south, is at altitude between 350-2262 m above sea level, bumpy and variations in the height of the DAS steep in the upstream, ramps in the middle and somewhat sloping and flat in the lower reaches with the length of the main river Tampo Rod reaches 25 km.

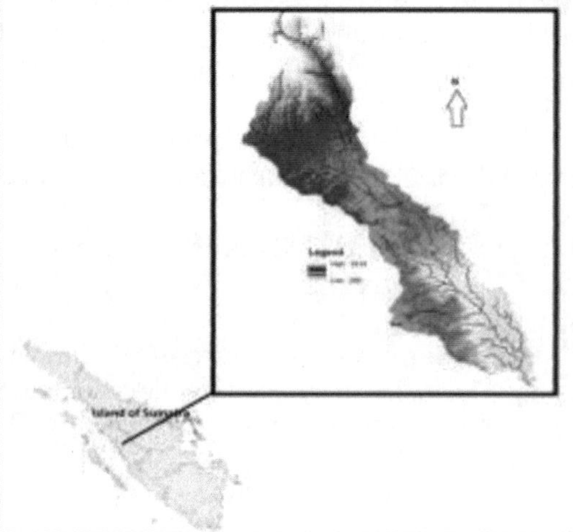

Figure 1. Sub Das Batang Tampo

Identification of geographic conditions and topography in Batang Tampo is done using primary and secondary data. Primary data is collected using the results of monitoring of satellite data using Information provided by the Digital Elevation Model (DEM) for ASTER GDEM topographic data with a resolution of

30 m which will provide topographic data in Batang Tampo and Batang Tampo Sub watershed. The elevation of benchmark points is measured by GPS Mapping, the results of this activity: a). Divide the Batang Tampo Sub watershed area into three land slope groups, which are: 1). Upstream with land slope> 15%, 2). Middle section with slope between 8% to 15% and, 3). Downstream with a small slope less than 8%. The identified areas are confirmed by maps of irrigation areas in the relevant agencies such as the agricultural service and the Public Works and Public Housing Agency (Dinas PUPR). The output of this activity is the Batang Tampo Sub Das map according to the upstream-middle-downstream region. b). Determine the location or path of the irrigation network. Based on the map the irrigation network is calculated and measured the distance of the nearest and farthest paddy fields. c). from topographic maps, you can find locations or grooves of the guguak rice fields and other rice fields. d). Based on the results of the identification of the three conditions above (a, b and c) the location and number of samples are measured cropping intensity and productivity of farmers' rice fields according to upstream-downstream, distance from the channel irrigation to the rice fields and the location of the guguak-lakuak.

The number and size of the sample will be determined based on the results of field observations taking into account the diversity of field conditions. The results of this mapping are: a.) Map of Irrigation Area according to topographic classification: Upper, Middle and Lower. b) Map of irrigation networks, C-Topographic map of rice field location by height. The results of the vegetation index analysis will show the highest to lowest vegetation index on DI Batang Tampo. The output of the vegetation index value is a map of the vegetation index IN Batang Tampo which will provide an overview of the level of planting intensity. The results of the mapping will show the locations of rice fields with a certain level of cropping intensity which are described in different colors that can be clustered. Planting intensity is one parameter that shows the level of irrigation services in that location. The clustering results are validated and verified in the upstream-downstream field, far from the source of the irrigation water and from the location of the guguak lakuak and will be classified into irrigation water: Good, lacking and poor and not available at all. Based on the results of regional identification and identification of water-deficient rice fields, a map of the water shortage area is obtained and the area according to upstream-downstream, distance from the channel and location of the guguak lakuak. Furthermore, based on the above categories, data validation and verification is carried out on the map with the field on selected samples to obtain data: Irrigation service capacity , cropping intensity and rice productivity.

3. Results and Discussion

Vegetation index analysis will show the highest to lowest vegetation index on DI Batang Tampo. Vegetation index value is a map of the vegetation index in Batang Tampo which will provide an overview of the level of planting intensity. Normaly : 0-1. On Batang Tampo Irrigation system EVI in rice farm is : 0.24-0.62. The results of the mapping will show the locations of rice fields with a certain level of cropping intensity which are described in different colors that can be clustered. By verication data EVI is clastered: < 0.20 (red) 0.21–0.4 (ligh green) 0.41–0.60 (green), 0.61–0.80 (ligh blue) and > 0.81 (blue).

Planting intensity is: 1–3, one parameter that shows the level of irrigation services in that location. The clustering results are validated and verified in the upstream-downstream field, far from the source of the irrigation water and from the location of the guguak–lakuak.

Based on the results of regional identification and identification of water-deficient rice fields, a map of the water shortage area is obtained and the area according to upstream-downstream, distance from the channel and location of the guguak lakuak. Furthermore, based on the above categories, data validation and verification is carried out on the map with the field on selected samples to obtain data: Irrigation service capacity, cropping intensity and rice productivity.

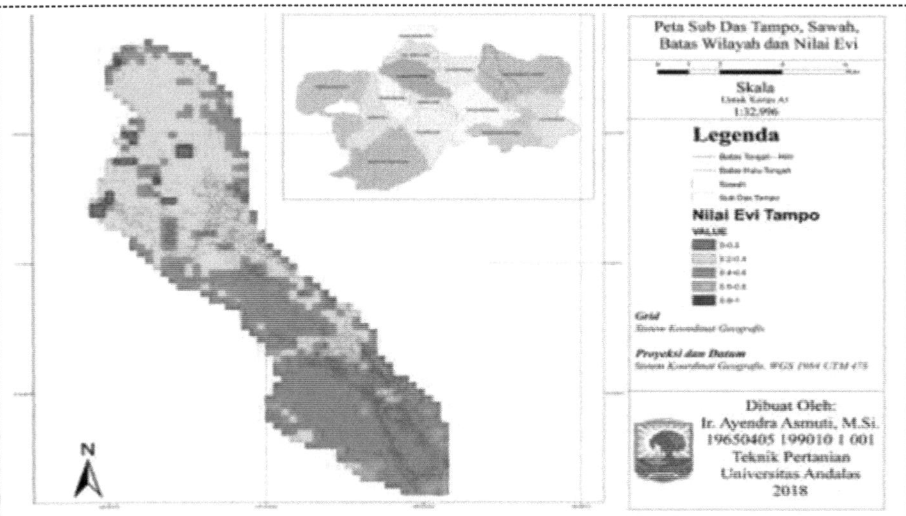

Figure 2. EVI Value Cluster

Table 1. EVI Cluster rice field

E V I	Color	Cropping Intensity	%	< 300 m from the river/canal	> 300 m from the river/canal	Guguak	Lakuak
0–0.2	Red						
0.21–0.40	Ligh green	1 – 2.0	73	25% by sample	75% by sample	67%	33%
0.41-0.60	Green	2.0 – 2.5	25	73% by sample	25%	33%	65%
0.61- 0.80	Ligh blue	2.5 – 3.0	2	2% by sample	-	-	2%

> 0.81	Blue

4. Conclusion
The analysis irrigation service capacity by remote sensing can be done for first step to identification irrigation system capabilities. The EVI index on rice field is 0.24 – 0.62. Furthermore, for accuracy measurenment it have more land actifities to verivication data. The water scarsity in rice field in Tampo Irrigation System influences of topography and rice field level from the river or water resource.

References
[1] Asmuti A 1995 *Analisis Keragaan Teknis dan Ekonomi Beberapa Taraf Teknis Irigasi Gravitasi Padi Sawah (Studi Kasus Irigasi di Sumatera Barat)* (Bogor : Tesis S2 IPB Bogor)
[2] BWS V 2009 *Studi Pengembangan Kawasan Batang Tampo di Kabupaten Tanah Datar* (Sumatera Barat: Laporan Akhir Balai Wilayah Sungai V Sumatera)
[3] Burman N D, P R Nixon, J L Wriht and W O Pruitt 1980 Water Requirement
[4] Dastane N G 1974 *Effective Rainfal in Irrigated Agriculture* (Rome: Irrigation and Drainage Paper FAO) 62p
[5] Doorenboos and J Pruit 1993 *Irrigation Water Requirement* (Rome: FAO)
[6] El-Magd I A and Tanton T W 2005 Remote sensing and GIS for estimation of irrigation crop water demand *International Journal of Remote Sensing* **26** *(10)* 2359-2370
[7] Fischer G, Tubiello F N, van Velthuizen H, and Wiberg D A 2007 Climate Change Impacts on Irrigation Water Requirements: Effects of Mitigation 1990–2080 *Technological Forecasting and Social Change* **74** 1083–1107
[8] Herdianto R, Elvi R S, and I Agus 2017 *Dampak El Nino Tahun 2015 terhadap Stres Tanaman di DAS/Daerah Irigasi Tampo* (Jambi: Kongres Rapat Anggota Tahunan dan Seminar Nasional INACID Jambi-Indonesia, 10-11 March 2017)
[9] Herdianto R, Paik K, Coles N, and Smettem K J 2013 Transitional Responses of Vegetation Activities to Temperature Variations: Insights Obtained from a Forested Catchment in Korea *Journal of Hydrology* **484** 86-95
[10] Herdianto R, Paik K, Coles N, and Smettem K J 2010 *Coupling Vegetation Index and Terrainreference.*

Study Relationship of Rice Chlorophyll Content Index (CCI) Value with Rice Prediction Yield Production on Rice Cultivation in West Sumatera

F Hidayah[1*], Santosa[2], R E Putri[2]

[1] Masters Program in Engineering Of Agricultural ,Technology Of Agricultural, Andalas University, Padang
[2] Program in Engineering Of Agricultural ,Technology Of Agricultural, Andalas University, Padang

Corresponding author's e-mail address: fitrihidayah222@gmail.com

Abstract. Yield mapping is conducted to inform farmers about the variability of production that is used as an illustration of the yield of rice plants in a particular rice field area, so that it can be a reference for farmers in increasing their agricultural yields for the next planting season and determining the point area of land that produces the highest production. Through a yield mapping, the production results in an area of land can be predicted by looking at the correlation value and mapping between the values of Chlorophyll Content Index (CCI) of rice plants at different ages and yields at each point of observation area. Measurement of CCI values was carried out on rice plants at the age (25, 40, 60, and 70 DAP). Measurement of CCI values was carried out at 20 observation points in the Banda Langik rice field area located on the Sungai Bangek, Lubuk Minturun, West Sumatra. CCI values and yields of rice plants obtained on each field are then processed into maps of CCI values and yield mapping using GIS (Geographic Information System) applications with the kriging method. Based on research that has been carried out during one planting season shows that the CCI value has a relationship with the yield of rice plants. This is evidenced by the correlation values generated (r) 25 DAP (r = 0.815), 40 DAP (r = 0.877), 60 DAP (r = 0.810) and 70 DAP (r = 0.917). Based on the mapping of CCI values and mapping of production results, it was shown that production results in a land area could be predicted based on an overview of the CCI values of rice plants at different ages.
Keywords: Chlorophyll Content Indeks (CCI), Kriging, Yield Mapping, Spatial Variability

1. Introduction

The rice plant (*Oryza sativa* L.) is a food commodity that is used as a staple food for the Indonesian people. At present, rice plants in Indonesia have decreased productivity [10]. There are various ways that can be used to prevent a productivity decline, one of them is the development of pre-existing technologies to help increasing the productivity of rice plants.

The plants chlorophyll value is closely related to plant health. [12] explains the chlorophyll value obtained from the measurement of SPAD values on plants that have a close relationship with the health of plants determined based on yields of harvest on rice plants obtained through the method of cropping (Crop Cutting Test (CCT)). The fertile and enough nutrient plants will look green on the leaves and indicate sufficient Nitrogen (N) content, on the contrary, if the nutrient content is adequately fulfilled, the productivity of the plant will rise [4]. By knowing the SPAD value contained in plants will also provide information about the N content contained in the leaves, and the number of N in the leaves correlated with the chlorophyll value in the leaves of rice plants [7].

The number of N values contained in rice leaves varies at each stage of growth. In increasing the productivity, farmers face some obstacles such as crop development in the area of right time and difficulties in monitoring crop development quickly in an area of land observed, therefore a calculation technology is needed to increase the predicted rice productivity based on the index value of chlorophyll content and yields on rice plants, which can cover a wide, reliable area, with fast time and accurate results [2].

The development of methods to predict crop yields in Indonesia based on spatial variability of plant fertility has not been widely carried out and to observe the development of rice plants cannot be conducted quickly, be proper and cover a large area of land if measurements use only chlorophyll meters. For this reason, measurements of greenness were carried out on plants of different ages that had been determined for harvest time and mapping of production results that could predict yields and describe yields on rice plants, based on the Chlorophyll Content Index (CCI) obtained from reading Chlorophyll meter CCM-200 Plus on leaves and yields from rice plants, and looked at the relationship between yields with the value of Chlorophyll Content Index (CCI) from CCM-200 Plus readings at different ages of rice plants.

This mapping of production is known as yield mapping. Yield mapping is a part of the concept of precission farming, which is a principle of integrated and sustainable agricultural technology [6]. Through yield mapping, it is expected to provide information to farmers regarding the variability of production results that will be used as an illustration of the yield of rice plants in a particular rice field area, so that it can be a reference for farmers in increasing their agricultural yields for the next planting season and determining the area points that produces the highest production.

The purpose of this study was to evaluate the relationship between the amounts of Chlorophyll Content Index (CCI) content and yield on rice plants which was seen based on the variability of the yield (yield) in the field plot obtained through the Crop Cutting Test (CCT) method and through Mapping the production of rice is expected to be a determination of the results of production predictions and can be a picture of land for sustainable agriculture in the future.

2. Materials and Methods

2.1 Tools and Materials

The tool used in this study consisted of CCM-200 plus chlorophyll meter, stationery (pencils and books), measuring instruments (meters and scales), a set of data processing equipment consisting of laptops and USB, software for data processing and data analysis consists of Microsoft Word, Excel, Statistical Product and Service Solutions (SPSS), GIS and SAS.

In this study rice plants with an area of 0.5 ha were used as research samples. For the data sample, 5 clumps of rice were collected in each plot of rice fields at each of the designated observation points.

2.2 Method of Collecting Data

The collection of data samples in this study was carried out in two ways: Grid Sampling and Crop Cutting Test (CCT).

2.2.1 Grid Sampling Point. Grid sampling point is the process of data collection carried out to describe the condition of the land based on the fertility level of the plant by dividing plant samples into several parts. The trick is to divide the paddy fields into several groups and then divided them into plots. Furthermore, the data collection process is carried out directly in the field to get information directly from the plants, which are used as sample data to be processed. For 0.5 ha of land divided into 20 grids, then determine the observation points of each plot. For the sampling grid process can be seen in Figure 1.

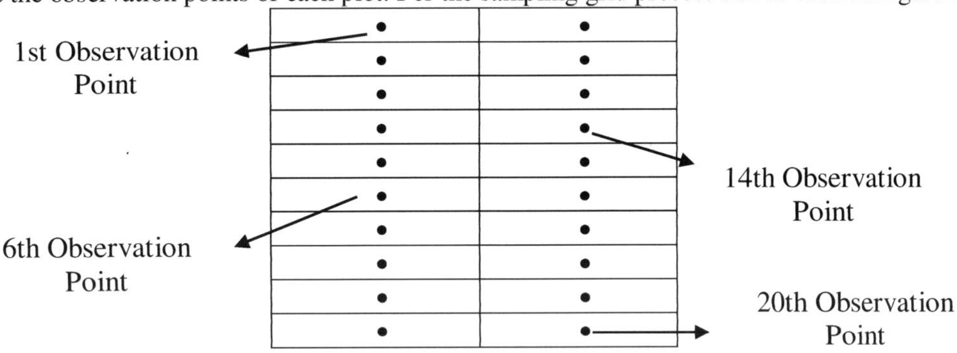

Figure 1.Grid Sampling Point

Figure 1 is a method of dividing land into several plots which are used as observation points. Giving observation points on research land is done sequentially to facilitate the process of data collection, observation and management.

2.2.2 Tile Method or Crop Cutting Test (CCT). Crop Cutting Test (CCT) is a method of taking and collecting plant data samples that is conducted directly in the field used at harvest time for each plot of land that is used as a research sample (Putri et al., 2016). The steps to do CCT are: (1) Determining and marking the plot of the rice fields that are used as experiments and determine the sizes of the selected fields, (2) harvesting by CCT, (3) threshing crops from harvested crops from the CCT area, (4) winnowing and weighing rice or wet grain obtained from CCT rice fields.

2.3 Data Collection of Chlorophyll Content on Plant Leaves Using CCM-200 plus Chlorophyll Meter

Observation of the amount of Chlorophyll Content Index (CCI) of leaves was carried out in accordance with the specified observation points on the land. Then taking data on chlorophyll values in plants is done and collected according to the observation points that have been selected to be used as data samples. The process of reading the CCI status in the leaves was carried out in four stages of growing plants, namely when the rice was 25, 40, 60, and 70 days after planting (DAP). The first data collection was carried out when the rice plants were 25 DAP, the data was collected on the 25th day of the DAP because it was still a vegetative phase on rice plants. In this phase is the phase of plant development that will be the determinant of rice productivity, then the determination of data collection time on the 40th, 60th, and 70th days after planting because it is a generative (reproductive) phase in rice plants [1].

Plants used as data samples were examined at each stage of growth in each plot of rice fields, each of which consisted of 20 observation points. In this study 5 clumps, 5 stems, 5 leaves and 3 reading points of CCI values on the leaves of rice plants were taken using CCM-200 plus chlorophyll meter. Collecting and retrieving data on CCI values in leaves using corophyll meter CCM-200 plus randomly. Then the CCI value obtained from each reading on each plant sample at each point is then taken the average value. The average value of CCM-200 plus chlorophyll readings is used as a data sample. The measurement of the chlorophyll value contained in the leaves will be collected between 10:00 - 15:00 noon. When measuring sunlight it must be relatively bright to avoid noise in the data. Measurements are taken at the very top of the leaf, the topmost leaf before the flag leaves out and leaves the flag after leaving, because this is an area commonly used in the practice of collecting data on the greenish value of leaves [12]. The part of the leaf that will be measured for its CCI value is located at the end, middle and base of the rice leaves. The CCI measurement techniques carried out during the field are:

1. Measurements are conducted at each sample point that has been selected
2. Sampling data according to the midpoint of each map.
3. Selection plants randomly and making measurement readings three times on each leaf selected, so that the data taken is accurate.

2.4 Location and Point of Each Coordinate Map Rice Fields
The process of taking the coordinate points at each point that has been determined as an observation point on the research land is done using GPSmap 62sc GARMIN to determine the coordinates of each plot of rice fields from each field.

No Point	Coordinat	No Point	Coordinat
1	00° 49' 52.0" 100° 21' 18.1"	11	00° 49' 51.8" 100° 21' 18.9"
2	00° 49' 51.7" 100° 21' 18.1"	12	00° 49' 51.6" 100° 21' 18.9"
3	00° 49' 51.4" 100° 21' 18.0"	13	00° 49' 51.3" 100° 21' 18.8"
4	00° 49' 51.2" 100° 21' 18.0"	14	00° 49' 50.9" 100° 21' 18.7"
5	00° 49' 50.9" 100° 21' 17.8"	15	00° 49' 50.7" 100° 21' 18.6"
6	00° 49' 50.5" 100° 21' 17.8"	16	00° 49' 50.4" 100° 21' 18.6"
7	00° 49' 50.2" 100° 21' 17.8"	17	00° 49' 50.1" 100° 21' 18.5"
8	00° 49' 49.9" 100° 21' 17.7"	18	00° 49' 49.9" 100° 21' 18.4"
9	00° 49' 49.7" 100° 21' 17.6"	19	00° 49' 49.5" 100° 21' 18.3"
10	00° 49' 49.4" 100° 21' 17.5"	20	00° 49' 49.3" 100° 21' 18.3"

2.5 GIS Data Management
At this data management stage, the process of collecting all data obtained from the measurement results is carried out, then the data is processed into raw data before the map making process is done using a GIS (Geographic Information System) application. The data processed in the GIS is spatial data, namely the data of CCI values contained in rice leaves which are measured using CCM-200 plus chlorophyll meter and harvest data on rice plants.

The method used for processing the map of the CCI value of rice plants is by the kriging method. Kriging is a geostatistical interpolation technique that considers both distance and variation between data from sample points when performing data retrieval can estimate values in unknown areas. Estimations that has been discovered by this method use a linear weighted combination of data values at all points to be predicted [3]. This kriging method wasselected due to this method can predict, which is continuous from a group of data samples. Data retrieval is done using interpolation techniques (estimating values at certain location points). The stages of map making according to (PT. WMT-II 2012) are:

1. Collection and collection of CCI value data on leaves
2. Data processing
3. Determination of CCI leaf values based on the age of rice plants
4. Preparation
5. Data management by kriging method
6. Analyzing test results
7. Visual analysis resulting from the interpolation of the greenness of rice plants
8. Making a map of the model
9. Visual analysis of model maps
10. Decision making
11. Making the results of the analysis of the results of prediction based on the map of greenness on the leaves of rice plants

2.6 Data Correlation Test

Correlation test on the data was conducted to see the strong relationship between the two data, namely the data of CCI values and the yield of rice plants on a land area. Regression analysis is basically the study of the dependence of the dependent variable (bound) with one or more independent variables (explanatory variables or independent variables), with the aim of estimating and / or predicting the average population or the average value of the dependent variable based on the value of the independent variable known [7]. Correlation analysis aims to measure the strength of linear associations between two variables. For the correlation test data in this study will be tested using the SPSS program. The strong relationship between the data will be expressed in the form of numbers between 0-1, if the number generated by the correlation of data is 0, it indicates that there is no relationship between the data and if it shows the number 1, it shows the perfect relationship between the data. If the smaller the correlation coefficient, the greater the error to make predictions. Data from the CCI value mapping and production results are said to be interconnected if the data shows more than 0.6 and less than 1. As a reference to see the strong relationship between leaf CCI data and harvest yield, it can be seen in Table 1.

Table 1. General Guidelines for Determining Correlations

r	Relationship Criteria
0	There is no correlation
0-0,5	Weak correlation
0,5-0,8	Medium correlation
0,8-1	Strong / tight correlation
1	Perfect correlation

3. Result and Discussion

3.1 Description of Research Sites

This research was carried out in the rice field area of Sungai Bangek Village, LubukMinturun, Koto Tangah District, Padang City. The land used is agricultural land from the Banda Langik farmer group, the name of the farmer group comes from the name of the rice field area. The condition of the research land has a clay sand soil type, based on the measurement of the texture triangle the soil type is a sample of soil A, and has a good irrigation, so it is good for rice cultivation. Geographically on the rice field map, BandaLangik is located at the coordinate point (00 ° 49 '52.2 "SL - 100 ° 21' 17.8" EL). The rice seedlings used in the study were IR 42 rice and the land consisted of 20 observation points in the form of plots / tiles, the size of each plot of land was 25x10 m.

3.2 Relationship of Chlorophyll Content Index (CCI) with Production Results on Rice Crops

Data correlation test between CCI value and production results in this study was conducted to find out that the CCI value has a relationship with the production results. Through the data correlation test, the relationship and the influence of CCI values on different ages and production results will be seen. The data that will be tested is the CCI value data from the measurement results obtained from the field at different ages, ie at the age of 25, 40, 60 and 70 DAP plants and the yield of rice plants at each observation point.

Table 2. Correlation of CCI of Rice Crops with Production Results at the ages of 25, 40, 60 and 70 DAP

	Production Result (kg/m^2)	25 DAP	40 DAP	60 DAP	70 DAP
Production Result (kg/m^2)	1.000				
25 DAP	.815	1.000			
40 DAP	.877	.775	1.000		
60 DAP	.810	.808	.866	1.000	
70 DAP	.917	.832	.903	.880	1.000

In Table 2 shows that the CCI value has a relationship with production results, can be seen from the correlation values of each at different ages, namely at the age of 25, 40, 60, and 70 DAP. At the age of 25 DAPshows the correlation value is r = 0.815, based on the correlation value produced at the age of 25 DAP, the CCI value has a relationship with the production results, but the CCI value has not had a major effect. CCI values have a relationship with production because in this phase the CCI value affects the *peranakan*process, growth, leaf growth and growth in rice plants.

At the age of 40 DAP the correlation value between CCI values and production results has increased that is r = 0.877, based on this correlation value shows the CCI value has a strong relationship to production results compared to the age of 25 DAP. In this phase the CCI value affects the production, because the CCI value can determine the health of rice plants during growth. Plant health affects the number of tillers that will be produced in each clump. CCI values in rice plants have increased during fertilizing. Fertilization is done to meet the nutritional needs of plants. This increase in CCI value in rice cultivation at age 40 DAP shows that fertilization can increase the CCI value in plants and chlorophyll values affect the growth process in plants because according to [12] explains that the value of SPAD has a close relationship with plant health which can be determined based on results harvest production. The first fertilization was carried out at the vegetative period, which was at the age of 23. Theincreasing in CCI value in the paddy cropping age of 40 DAP showed an increasing in the N value so that it increased the

CCI value in plants, and this also indicated that fertilization in the vegetative period was very influential in supporting growth.

At the age of 60 DAP the correlation between the value of CCI and production results decreased. At the age of 60 DAP the correlation value between the CCI value and the yield of r = 0.810 based on the correlation value showed that the relationship between CCI value and yield at the age of 60 DAP was lower than when the rice plants were 25 and 40 DAP. Based on observations that have been conducted in the field, this is caused by, at the age of 60 DAP the CCI value in plants is low. The decline in CCI values that occurred at the age of 60 DAP because fertilization had not been carried out. The second fertilization on the land was done at the age of rice 63 DAP, thus causing a decrease in CCI values on rice plants, if it was correlated between the correlation value with fertilization showed that there was an influence of N value on the development of chlorophyll values in rice plants so that the correlation between CCI values and yield at the age of 60 DAP decreased.

Correlation value between CCI value and yield at 70 DAP is r = 0.917. Based on the correlation value shows that at 70 DAP the CCI value has a strong relationship with production results, seen from the correlation value produced. At the age of 70 DAP the chlorophyll value greatly determines the yield of rice because at the age of 70 DAP the panicle of the rice plant has been released.

At the age of 70 the required DAP nutrients must be fulfilled as needed by the plant because it is a generative (reproductive) period of rice plants. Plant growth phase which is suspected to have a close relationship with the productivity of rice plants is a plant in the early generative phase (pinnacle initiation) when rice plants are being produced [11].

Based on observations at each stage of growth of rice plants at each age of different plants showed that the relationship most related between the value of chlorophyll and production compared with other ages is when rice is 70 DAP. According to [12] explained in his research that at the age of 70 DAP the value of chlorophyll in rice plants was higher than the age of other rice plants. In this study showed that the highest CCI value on plants and the most influential on production results was at the age of 70 DAP, therefore at the age of 70 DAP it was very important to determine the yield of rice for this periodget susficient nutrients and must get farmer's attention.

3.3. Mapping of Chlorophyll Content Index (CCI) and Production Results on Rice Crops

The mapping of chlorophyll values and yields of rice plants in this study was carried out to see the overall distribution of data on the land. The CCI value mapping function and production results are used to predict the yield of rice plants based on the CCI value level predicted through the colors contained on the map. Mapping on the CCI value of rice plants is based on the age of different rice plants, then at each point observations were made which areas produced the highest yields predicted through maps of chlorophyll values in rice plants.

This mapping will also function as a means of detecting nutritional needs in rice plants which are predicted through the level of CCI values of rice plants projected by the map. To see the data distribution of CCI values and yields of rice plants on the map along with how their relationship at the ages of 25, 40, 60 and 70 DAP can be seen in the map below which was made with the GIS program using the Krigging method.

1. Chlorophyll Content Index (CCI) Value Map on Land Age 25, 40, 60 and 70 DAP

On the land map, it will be seen how the data distribution of CCI values of rice plants at each observation point and on the map can also be seen how the increase in CCI values based on different ages in Figure 2.

IOP Conf. Series: Earth and Environmental Science **327** (2019) 012009 doi:10.1088/1755-1315/327/1/012009

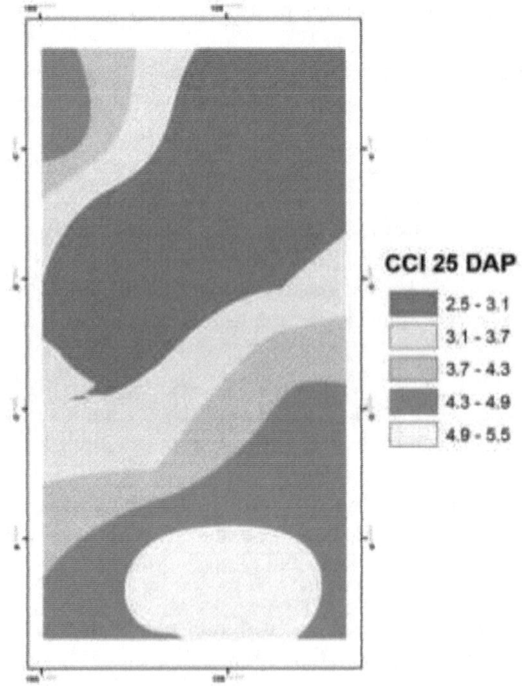

(a)

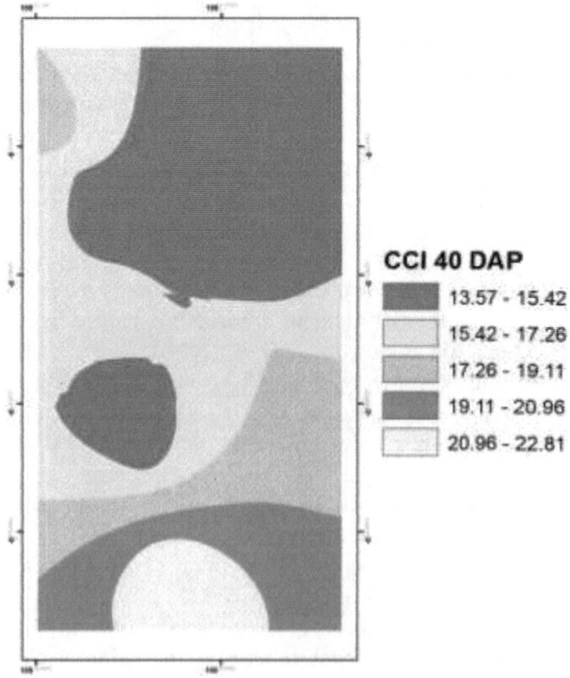

(b)

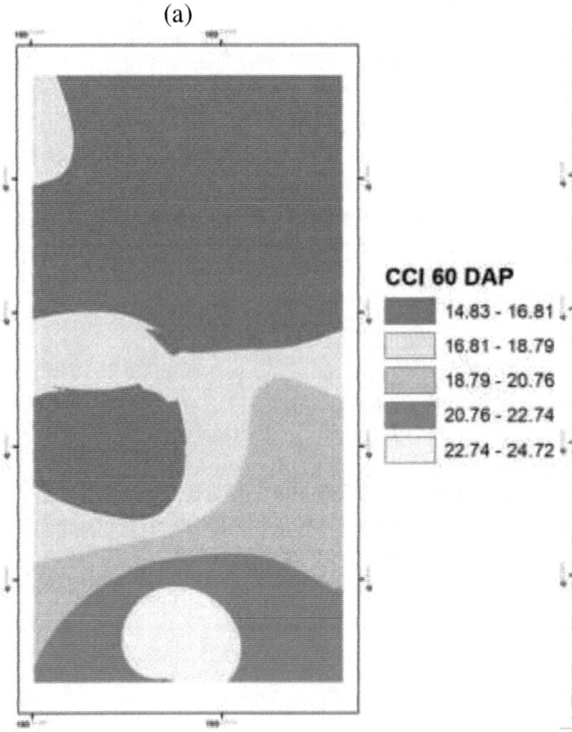

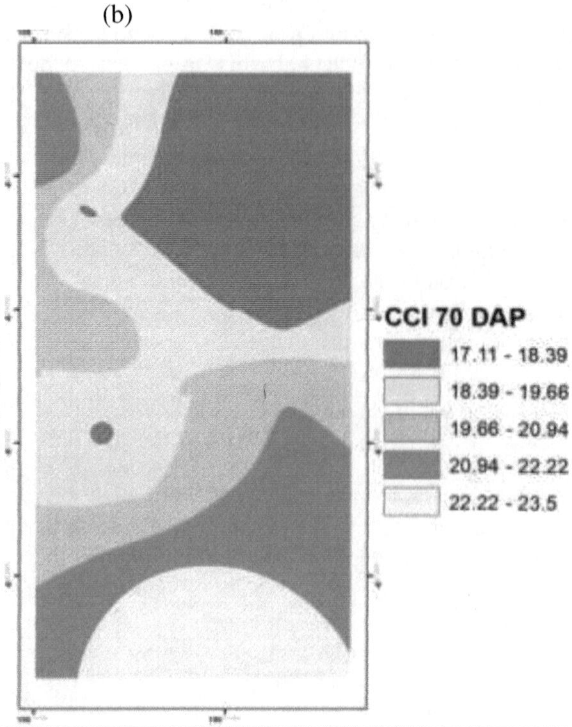

(c) (d)

Figure 2. Map of CCI Value aged 20, 40, 60, and 70 DAP

Based on Figure 2 in the CCI value mapping it can be seen that the CCI value in rice plants changes at different ages. In Figure 2a can be seen mapping the CCI value of rice plants at the age of 25 DAP. In Figure 2b is mapping the CCI value of rice plants at the age 40 DAP. In Figure 2c is a mapping of CCI values at the age of 60 DAP and Figure 2d shows a mapping of chlorophyll values at the age 70 DAP. In the CCI land value mapping map legend can be used as a benchmark for detecting CCI values at each observation point.

Based on the colors in the map legend, it can be determined that, green indicates the CCI value (very low), yellow (low), orange (medium), purple (high) and white (very high). In Figure 2a can be seen the map of rice plants at the age of 25 DAP, based on the legend map shows that at the age of 25 DAP the CCI value is very low, namely (2.5 – 3.1) low (3.1 – 3.7) medium (3.7 – 4.3), high (4.3 – 4.9) and very high (4.9 – 5.5) In Figure 2b at the age of 40 DAP, based on the legend of the map shows that the CCI value is very low, namely (13.57 – 15.42) low (15 – 17.26) medium (17.26 – 19.11) high (19.11 – 20.96) and very high (20.96 – 22.81). In Figure 2c it can be seen the map of rice plants at the age of 60 DAP, based on the legend the map shows that at the age of 60 DAP a very low value (14.83 – 16.81) is low (16.81 – 18.79) medium (18.79 – 20.76) high (20.76 – 22.74) and very high (22.74 – 24.72) and in Figure 2d can be seen the map of rice plants at the age of 70 DAP, based on legend shows that at 70 DAP the value was very low, namely (17.11 – 18.39) low (18.39 – 19.66) medium (19.66 – 20.94), high (20.94 – 22.22) and very high (22.22 – 23.5).

Based on the CCI value level map can be seen which observation points have low CCI values and those that require nutrition. For areas that have very low value, fertilization should be carried out to meet the needs of plant nutrients to support plant growth in the area of the land.

2. Map of Production Results

Mapping of production results on the land is done to see the production results at each observation point to be used as an illustration for the production of subsequent crops. This production map can be used as a benchmark to increase production in the following season (Figure 3).

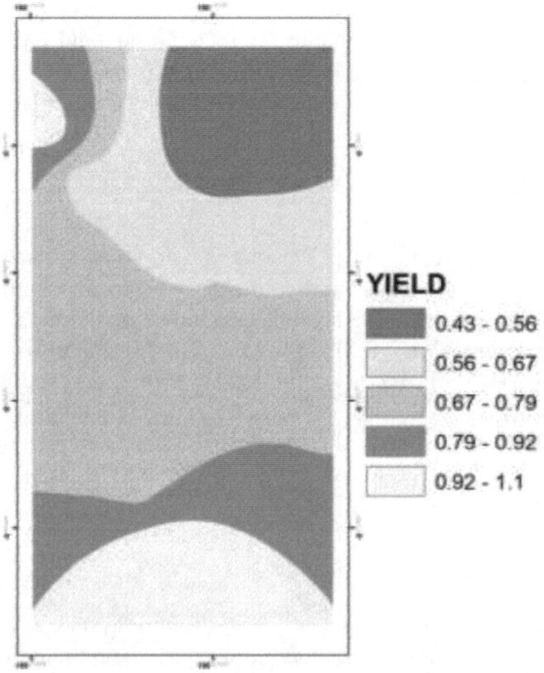

Figure 3. Production Results Map

Based on the colors in the map legend Figure 3 can be seen green indicates the area of the plant that has a CCI value (very low) yellow (low) orange (medium) purple (high) white (very high). From the legend the map on land 1 shows very low values, namely (0.43 – 0.56), low (0.56 – 0.67) medium (0.67 – 0.79), high (0.79 – 0.92) and very high (0.92 – 1.1). Based on the legend on the map for observation points that produce the least production can be used as an illustration to determine which observation points that require additional nutrients to support plant growth on the land, to increase production yields on rice plants.

CONCLUSION

From the observations that have been conducted it can be concluded that the CCI value of plants can be used as an illustration to see the health and conditions of plant growth. Based on the research that has been carried out during one planting season shows that there is indeed an influence of CCI values on the yield of rice plants obtained through the CCI value mapping measurement does not need to be done one by one against the leaves of rice plants, because the nondestructive method used in the process of measuring the CCI value carried out, can cover a large area and also through CCI value mapping we can predict the nutritional needs of plants to increase production in areas of land that require adequate nutrition.The evaluation results that have been carried out during one planting season shows that CCI values have a relationship with production results on rice plants seen from the yield produced at each observation point.

REFRENCES

[1][IRRI] International Rice Research institute. 2002. Rice Almanac. *Source Book for the Most Important Economic Activity on Earth.* International Rice Research institute. Metro Manilla, Philippines. 257 p.

[2]Barret, E., & Curtis, L. F. 1982. *Introduction to Enviromental Remote Sensing* (2nd Edition ed). London: Chapman and Hall.

[3]Bohling, G. 2005. *Introduction ToGeostatistics and Variogram Analysis*. Kansas Geology Survey. Diaksespada 22 Februari 2017 dandidownloadmelaluihttp://people.ku.edu/~gbohling/cpe940/variogram.pdf.

[4]Cen, H., Shao, Y., Song, H., and He Y.,. 2006. *Non-destructive Estimation of Rape Nitrogen Status Using SPAD Chlorophyll Meter*. ICSP 2006

[5]Cressie, N. A. C. 1993. *Statistics for Spatial Data*. Wiley Series in Probability and Mathematical Statistics: Applied Probability and Statistics. New York: John Wiley & Sons, Inc.

[6]Darmawan, S. 2007. *Precision Farming*SebagaiSistemPendukungKeputusan. *http://ppibaraki.wordpress.com/2007/11/29/51/* [Diunggahpadatanggal 20 Juni 2016]

[7]Gholizadeth A., Amin M.S.M., Anuar A.R. and Aimrun W. 2009. Evaluation of SPAD Chlorophyll Meter in Two Different Rice Growth Stages and its Temporal Variability. European Journal of Scientific Research. Vol. 37: 591-598.

[8]Gujarati, N.D. 2003. *Basic Econometrics*. 4th ed. New York: McGraw-Hill Companies, Inc.

[9]Hayati, F.D. 2012. PengujianTeknikInterpolasiSediaanTegakandanBiomassaBerbasis IHMB padaHutanLahanKering PT. TrisetiaIntiga, KabupatenLamandau, Kalimantan Tengah. [skripsi]. Bogor: InstitutPertanian Bogor.

[10]Pingali P.L, M. Hossain, and R.V. Gerpacio. 1997. *Asian Rice Bowls: The Returning Crisis? International Rice Research Institute (IRRI)*. Manila.

[11]Prasad, R. 2006. Fertilizer Nitrogen: Requirements and Management. *In*: Global Aspects of Food Production. P: 199-226 M.S. Swaminathan and S.K. Sinha (*Eds.*). Tycooly International Riverton, New Jersey-United States.

[12]Putri, R. E., Yahya, A., Adam, N. M., & Aziz, S. A. (2016). Variability of Rice Yield With Respect To Crop Health. JurnalTeknologi, 78(1-2), 79-85.

Energy Consumption on Tillage Operation in Low Land Paddy Cultivation

M I A Lubis[1], R E Putri[2], Andasuryani[2] Santosa[2], F Arlius[2], and A Hasan[2]

[1] Magister Student of Biosystem and Bioprocess Engineering, Andalas University, Padang, Indonesia
[2] Department of Agricultural Engineering, Andalas University, Padang, Indonesia

Corresponding author's e-mail address: m.iqbalabdilubis.ua@gmail.com

Abstract. Energy consumption is important to know for efficient on using energy. This study is determined using energy for tillage operation in low land paddy cultivation in Padang Pariaman, West Sumatera Province. The aims of this study is to identify any energy expenditure which used for tillage operation. The data were collected in five paddy field which was located at $0°40'41.63'' - 0°40'37.15''$ latitude and $100°16'39.64'' - 100°16'46.38''$ longitude and average areas is 792.4 meters-square. The energy expenditure were collected on tillage operation which are machinery, fuel, and human energy. All of the energy expenditure will be calculating by manual calculation (conversion factor of energy), exception human energy it also calculating by realtime measurement equipment (*Garmin Forerunner 35*). The biggest consumption energy occured at fuel energy and the lowest one is machinery energy. The fuel energy were 270.6195 and 314.6556 MJ/ha for the average of fuel energy on first and secondary tillage operation, respectively. For first and secondary tillage operation, the average energy of machine were 6.6683 and 9.2414 MJ/ha, respectively. Human energy shown the average energy expenditure on first and secondary tillage operation were 19.4979 and 24.5826 MJ/ha by equipment, while the energy calculation uses conversion factor were 14.8358 and 19.4979 MJ/ha.
Keywords: Energy Consumption; Tillage Operations; Low Land Paddy Cultivation

1. Introduction

Tillage is an important operation which doing in agricultural cultivation, for example in paddy cultivation. Tillage is an operation to prepare the land, which the output of the operation is soil puddled in minimum depth 25 cm [1]. The other functionn of this operation is weeding area, water leveling, restructure soil, and to increase soil fertility for the plant. Tillage operation usually done by the farmer at least twice in one planting season [2]. Tillage is an operation that include in paddy cultivation to maintaining the soil fertility, like: to make soil surface roughness, aerate the soil, increasing porosity, and to stimulate decompositioning [3]. Which is this operation concluded taht tillage operation as one factor to optimizing the plant's growth [4].

Talking about tillage as one of the agricultural process, it can not be merged of energy. Agricultural process is one of the most consumer in terms of input energy, such as machinery, human power, fertilizer, electrical and others [5]. Recent dtudy was summarized which is on tillage operation at least need three inputs energy (machinery, fuel, and labour) [6].

The aims of this study are to reported energy distribution on every energy resources, to identify the most energy expenditure, and to compare the human energy which solved beetwen by conversion factor calculation and realtime measurement (by using garmin forerunner 35 and HRM-strap).

2. Methods

This study has done for 2 months, started on August to September. This study has done at five paddy fields area is 792,4 m.sq, where located at $100°16'39.64'' - 100°16'46.38''$ longitude and $0°40'41.63'' - 0°40'37.15''$ latitude (Figure 1). This study used experimental methods. The tools that used in this study are hand tractor-Yanmar TF 75, garmin *forerunner* 35, heart rate monitor strap (HRM-strap), stopwatch, measuring glass, GPS, and laptop. While the materials used are paddy fields and fuel.

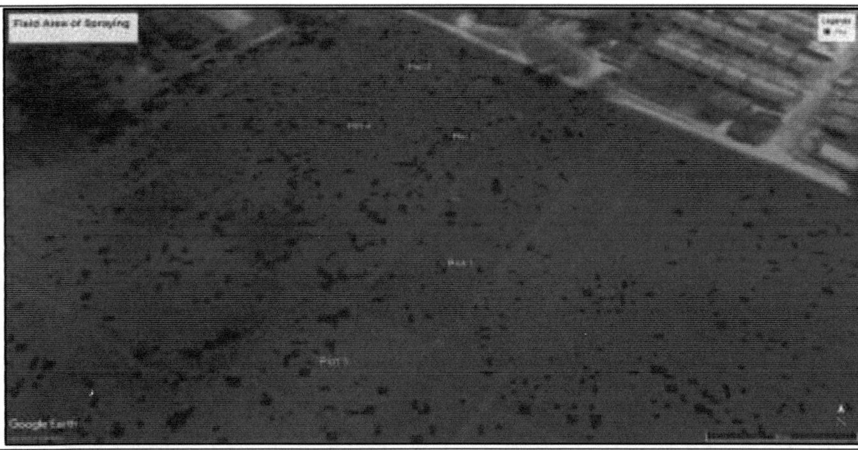

Figure 1. Paddy Fields

Tillage operation has been done for two times. The intervals between first and secondary tillage is 13 days. Some parameter that analyzed are machinery energy, fuel energy, and labour energy. Several formula that used are:

2.1 Machinery Energy

In agricultural practices, any several machines that used. Those are hand tractor, tractor 4WD, transplanter, sprayer, rotary tiller, and eachothers. The machine that used in tillage operation on paddy field is hand tractor. Every working of machine will be give energy expenditure. To know how much the machinery energy, it can be used this formula [6]:

$$ME = \frac{Cf_m \, x \, m}{FC \times E_t} \qquad (1)$$

Where ME is machinery energy ($MJ.ha^{-1}$), Cf_m is conversion factor for machine ($MJ.kg^{-1}$), m is mass of the machine (kg), FC is field capacity that should be gotten by dividing paddy field areas with working time ($ha.h^{-1}$), and E_t is economic life time of machine (h).

2.2 Fuel Energy

Hand tractor is not be able separate from fuel. It's because the power source of engine came from fuel. Fuel that used by machine also having energy. The fuel energy could be know by using this formula [6]:

$$FuE = \frac{Fu_c \times Cf_{fu}}{A} \qquad (2)$$

Where FuE is fuel energy ($MJ.ha^{-1}$), Fu_c is total of fuel consumption (L), Cf_{fu} is conversion factor for fuel ($MJ.L^{-1}$), and A is total areas (ha).

IOP Conf. Series: Earth and Environmental Science **327** (2019) 012010 doi:10.1088/1755-1315/327/1/012010

2.3 Labour Energy

Machines as tool which can not separate from agricultural process, certainly need labour to operate them. This study using two manner for calculating the labour energy. Those are manual calculating (equation 3) and realtime calculating (equation 4). For manual calculating it can be solved by using this formula [6]:

$$LE_m = \frac{n \times t \times Cf_l}{A} \quad (3)$$

Where LE_m is labour energy by manual calculation ($MJ.ha^{-1}$), n is number of labour (no dimensional), t is working time (h), Cf_l is conversion factor for labour ($MJ.h^{-1}$), and A is total areas (ha).

Manual calculating is different with realtime calculating. Realtime calculating do by wearing the realtime equipments to labour (Figure 2). Where is the tools on left side (HRM-strap) is use on the chest of operator and the other side is use on operator's hand as like as watch (garmin *forerunner* 35). This tools should be connected to each other, where the HRM-strap should be recorded the operator's heart rate that should be read on garmin *forerunner* 35. The results of garmin *forerunner* are heart rate, total working time, distance, and calories (Figure 3). By this results we can used to calculate the labour energy with this formula:

$$LE_r = \frac{Cal \times (4{,}1868 \times 10^{-3})}{A} \quad (4)$$

Where LE_r is labour energy by realtime calculation ($MJ.ha^{-1}$), Cal is total calorie of labour (Calories), $4{,}1868 \times 10^{-3}$ is conversion factor of calorie (MJ/Calories), and A is total areas (ha).

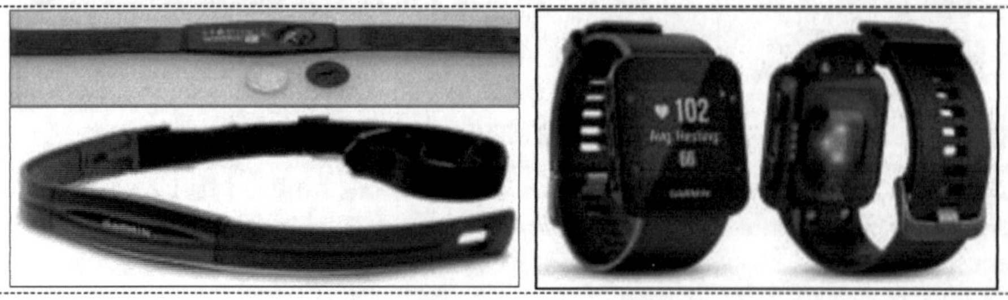

Figure 2. Realtime Equipments

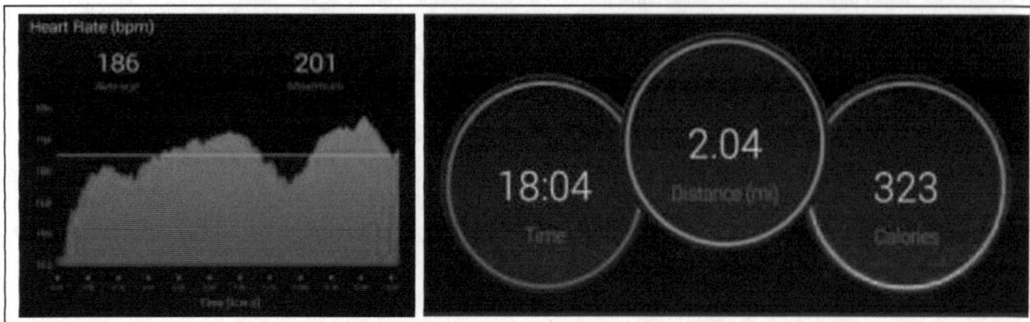

Figure 3. Results of The Equipment

3. Results and Discussion

3.1 Energy Distribution

This study discussed about tillage operation which had been conducted on August to September on 5 paddy fields (Figure 1). Tillage operation shown in this study at least need three input energy. Energy of this study was distributed as first and secondary tillage operation that shown in Figure 4a and 4b, respectively.

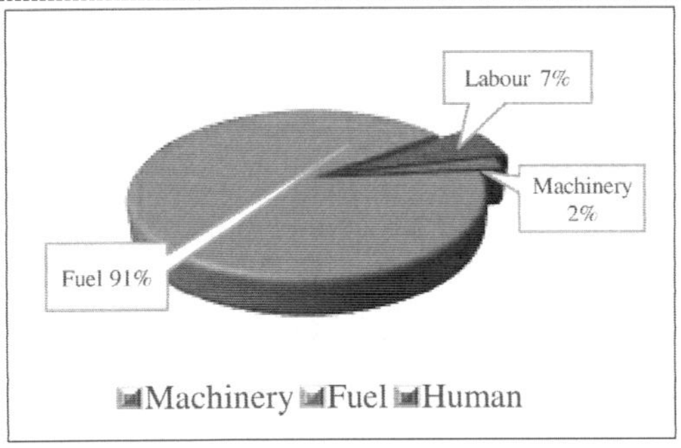

Figure 4. (a) Distribution Energy: First Tillage

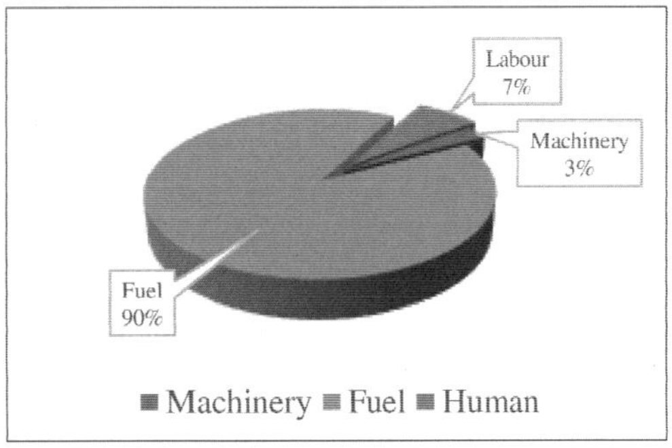

Figure 5. (b) Distribution Energy: Secondary Tillage

First and secondary tillage operation were described that labour energy both of them is equals, but in machinery and fuel have differences. In machinery energy, first tillage shown lower 1% than the secondary tillage. It caused by working time in secondary tillage more length than primary tillage. It seems like recent study that reported pudling activity (secondary tillage) is longer than first tillage [7]. It's because in secondary tillage need levelling paddy field area, which the levelling activity is not enough in one operation and at least two until four operation until the levelling land and soil puddling is good and ready for planting. For fuel energy, the highest is also on secondary tillage operation. It was affected by working time.

3.2 The Most Energy Expenditure

On tillage operation the most energy expenditure occured in fuel energy, exactly in the secondary tillage (1573.2779 MJ.ha-1). The smallest energy expenditure occured in machinery energy on first tillage operation (33.3414 MJ.ha-1). If we analyzed in each of input energy resources, both of the first and secondary tillage the highest is fuel energy, then human, and the smallest one is machinery. But, between first and secondary tillage operation, the highest energy occured in the secondary tillage operation for each input energy resources. Completely displayed in Table 1.

Table 1. Energy Expenditure

Paddy Fields	Energy Expenditure (MJ/ha)					
	Machinery		Fuel		Human	
	First Tillage	Secondary Tillage	First Tillage	Secondary Tillage	First Tillage	Secondary Tillage
1	6.0035	8.3228	308.1433	346.6612	17.7291	21.0022
2	3.8823	7.2281	198.9968	255.1241	13.9108	21.9050
3	9.4285	8.5213	276.6347	249.4247	24.4063	20.6329
4	6.2989	9.6869	254.5013	322.7822	19.2857	24.5054
5	7.7282	12.4480	314.8214	399.2857	22.1574	34.8673
Total	**33.3414**	**46.2071**	**1353.0975**	**1573.2779**	**97.4893**	**122.9129**

3.3 Human Energy

In this study, measurement of human energy has been done in two ways. They are realtime measurement (equipment) and conversion factor calculating. The first and secondary tillage operation shown human energy which calculating by conversion factor and equipment were 74.1791 and 97.4893 MJ.ha-1 then 102.8032 and 122.9129 MJ.ha-1, respectively (Table 2).

Table 2. Measurement of Human Energy

Paddy Fields	Energy Activities Measuring			
	First Tillage		Secondary Tillage	
	Conversion Factor	Equipment	Conversion Factor	Equipment
1	13.3567	17.7291	18.5168	21.0022
2	8.6374	13.9108	16.0814	21.9050
3	20.9769	24.4063	18.9585	20.6329
4	14.0140	19.2857	21.5518	24.5054
5	17.1940	22.1574	27.6947	34.8673
Total	**74.1791**	**97.4893**	**102.8032**	**122.9129**

Figure 5 shown R.square for firt and secondary tillage operation 0.9835 and 0.8799, respectively. It means, 98.35% conversion factor calculating affects equipment calculating on first tillage and 87.99% conversion factor calculating affects equipment calculating on secondary tillage. By using formula (CORREL) in Excel apps. we can identified how much relation between calculating by conversion factor and equipment. Correlation between conversion factor and equipment measuring were 99.17 and 93.80%, respectively for first and secondary tillage operation. It means that human energy which calculating by conversion factor and equipment has high correlation.

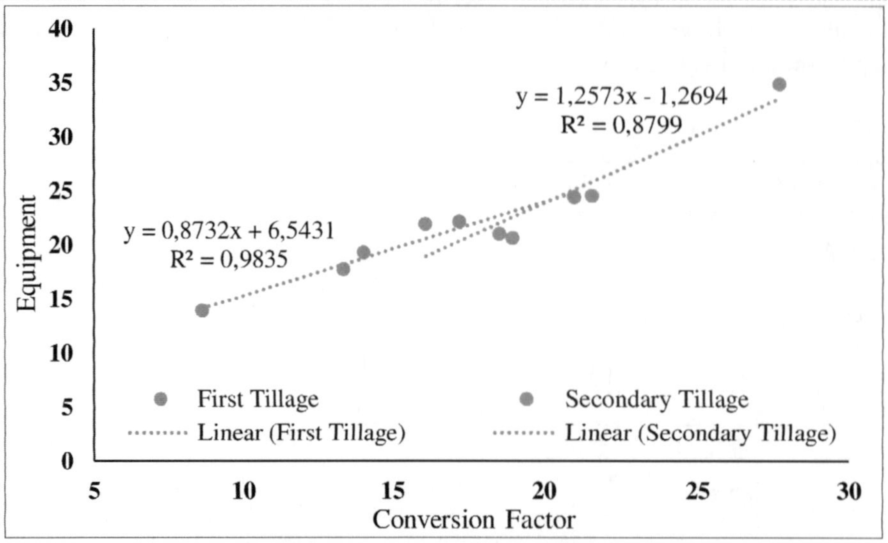

Figure 6. Measurement Relation of Human Energy

4. Conclusion

Tillage is one of important operation that needed to do in paddy cultivation. In this operation thr energy was distribute in three kind of energy, they are fuel, machinery, and human energy. Fuel spent energy as the bigest with 91% and 90%, then human energy with 7% and 7%, and the smallest is machinery energy as much as 2% and 3 %, they are respectively for first and secondary tillage. Secondary tillage was spent energy expenditure more than primary tillage. The averages of human on first and secondary tillage operation were 19.4979 and 24.5826 MJ/ha by equipment, while the energy calculation uses conversion factor were 14.8358 and 19.4979 MJ/ha.

Acknowledgements
The authors would like to thank Andalas University Grand number 95/UN.16.17/PP.PGB/LPPM/2018 for providing the laboratory facilities and financial project of this project.

References

[1] Agricultural Ministry 2016 *Technical Instructions Paddy Cultivation Jajar Legowo Super.* (Jakarta: Agricultural Research and Developement Agency-Indonesia) 58p
[2] Suastika I W, Basaruddin N, and Tumarlan T 1997 *Paddy Field Cultivation on Tidal Land* (Jakarta: Agricultural Research and Developement Agency-Indonesia) 18p
[3] Yule D F and Willcocks T J 1996 *Tillage and Cultural Practices* (*Developments in Soil Science* vol 24) chapter 8 pp 261–302
[4] [IRRI] International Rice Research Institute 2018 Step by Step Production *Rice Knowledge Bank of IRRI* (2018) 26
[5] Ozkan B, Ceylan R F, and Kizilay H 2011 Comparison of Energy Inputs in Glasshouse Double Crop (Fall and Summer Crops) Tomato Production *J. Renewable Energy* **36** 1639–1644
[6] Muazu A, Yahya A, and Ishak W I W 2015 Energy Audit for Sustainable Wetland Paddy Cultivation in Malaysia *J. Energy* **87** 182–191
[7] Muazu A 2015 *Modelling For Energy Optimization in Wetland Paddy Production in Malaysia.* (Malaysia: Universiti Putra Malaysia) 373p

IOP Conf. Series: Earth and Environmental Science **327** (2019) 012011 doi:10.1088/1755-1315/327/1/012011

Pyrolysis and Characterization of Liquid Smoke from Cacao Pod Husks

R E Putri[1(1)], A Kasim[2], Emriadi[3], and A Asben[2]

[1]Postgraduated Doctoral Program, Andalas University, Padang
[2]Faculty of Agricultural Technology, Andalas University, Padang
[3]Faculty of Mathematics and Natural Science, Andalas University, Padang

Corresponding author:rahmistppmedan@gmail.com

Abstract. Cacao pod husks were treated by the pyrolysis process to produce liquid smoke. Cacao pod husks can be utilized to produced their useful chemicals by pyrolysis. The reactor of pyrolysis used is made from cylindrical stell reactor with a small capacity, rust resistant and portable. A sample was pyrolyzed for 5 hours. The parameters measured to determine the quality of liquid smoke are the yield of pyrolysis products, pH, the concentration of acetic acid and phenol concentration. The quality of liquid smoke also tested by looking at the chemical components of liquid smoke by using gas chromatography-mass spectroscopy (GC-MS). The pyrolysis process produced products in the form of liquid smoke, charcoal, tar, and gas. The yield of pyrolysis products are 14.77; 40.43; 1.67; and 43.13%, respectively. Liquid smoke obtained has a pH of 3.2 levels, acetic acid 9.306 ppm, and phenol levels 1.688 ppm. GC-MS measurement result showed that methanol, acetic acid, and acetol are the highest concentration of chemical obtained in liquid smoke.

1. Introduction

These Liquid smoke is the process of vapors condensation from the combustion of raw materials which containing lignin, cellulose, hemicellulose and other carbon compounds. Liquid smoke is known to be useful as a preservative and latex coagulation material. This is because of the functional properties of liquid smoke such as antifungals, antibacterials, and antioxidants. Sources of liquid smoke producing can come from coconut shells [1], palm oil waste [2], waste wood-based composites [3], and the waste of fruit and cocoa leaf skin [4]. In this study, Indonesian waste of cacao is used, which the cacao production ranks is third highest in the world [5].

Cacao is an industrially important plant because it produces cocoa beans which are the basic ingredients in making chocolate as a popular food in the world. However, in addition to the production of cocoa beans, are also produced the waste of cacao pod husks. According to [6] every 1 ton of dry cacao production produced 10 tons of wet cacao husks. This cacao pod husks cause odors that are not good for the environment. Cacao pod husks waste is only used as a substitute for firewood, animal feed, or as compost. To increase economic value and reduce the risk of environmental impact, technology is needed to process them.

Cacao pods husks potentially to be used as liquid smoke because they contain lignocellulose. The chemical composition of cacao pods husks, are lignin 60.67%, holocellulose 36.47%, alpha cellulose 17.57% and hemicellulose 18.90% [4]. While the research from [6] published that the composition of cacao pods is cellulose 35%, hemicellulose 10.8-11%, lignin 14.6%, pectin 6.1%, crude fiber 22.6-32.5%, crude protein 5.9-7.7% and ash 9.1-10.1%. Pyrolysis can be used to decompose organic compounds from biomass to produce various products such as liquid smoke, biochar, tar, and gas. Compared to the combustion process pyrolysis requires low temperatures and results in low air pollution emissions [7].

In this study, cacao pod husks were converted by pyrolysis. The pyrolysis process uses a reactor made from cylindrical stell reactor. The pyrolysis equipment used is applicable and can scale up to a larger scale or industrial scale.

2. Material and Methods

2.1. Materials

Cacao pod husks used comes from the cacao plantations of West Sumatra, Indonesia. The cacao pods were separated from beans, and then sun-dried for a week. Before being fed into the pyrolyzer, the cacao pod was reduced in size about 3 cm x 4 cm. The water content of the cacao pod was analyzed using Moisture Meter.

2.2. Pyrolysis process of the cacao pod husks [8]

The pyrolysis of cacao pod husks was carried out in cylindrical stell reactor. The length and diameter of the pyrolyzer were 38 cm and 28 cm, respectively. 3 kg of cacao pod husks were placed in the pyrolyzer. The sample is pyrolyzed for 5 hours until there is no more liquid smoke dripping into the container.

2.3 pH measurement of liquid smoke

Measuring the pH of liquid smoke using a pen type pH meter, pH-009 (I) before measuring, the pH meter is calibrated first.

2.4 Measurement of total phenol

Total phenol levels were determined by the Follin-Ciocalteu method. Standard gallic acid curves are made with a variation of 4 concentrations; 250; 500; 1000; and 1250 ppm and absorbance measured at $\lambda = 765$ nm. The sample measurement procedure was carried out by inserting 1 mL of sample and 1 mL of Folin-Ciocalteu reagent into a 10 mL flask. The mixture is then shaken for five minutes After that, 7,5 % Na_2CO_3 is added aligned with distilled water to a volume of 10 mL. The solution was incubated for 1 hours at room temperature. Absorbance mesured at $\lambda = 765$ nm using a Shimadzu UV 1800 spectrophotometer. Total phenol levels were determined by the Follin-Ciocalteu method. Standard gallic acid curves are made with a variation of 4 concentrations; 250; 500; 1000; and 1250 ppm and absorbance measured at $\lambda = 765$ nm. The sample measurement procedure was carried out by inserting 1 mL of sample and 1 mL of Folin-Ciocalteu reagent into a 10 mL flask. The mixture is then shaken for five minutes After that, 7,5 % Na_2CO_3 is added aligned with distilled water to a volume of 10 mL. The solution was incubated for 1 hours at room temperature. Absorbance mesured at $\lambda = 765$ nm using a Shimadzu UV 1800 spectrophotometer.

2.5 Measurement of total acid by titration [9]

As much as 2 mL of liquid smoke was added to aquadest until the volume was 100 mL. Then add phenolphthalein indicator 3 drops. Titrate with 0.1N NaOH until the solution change colour to pink. Record the volume of NaOH used for titration. Calculate the acid content in liquid smoke with the formula:

$$\text{Acid Levels} = \frac{\text{mL titrant x N.NaOH x Acetic Acid } (g/mol)}{\text{Liquid Smoke } (mL)} \tag{1}$$

Information:

mL titrant = NaOH volume for titration

N NaOH = Normality of solution
Molecular Weight of acetic acid = 60 g/mol

2.6 Analysis of Liquid Smoke Chemical Content
Liquid smoke prepared from pyrolysis was qualified by Gas Chromatography-Mass Spectrometry (GC-MS). GC-MS brand Shimadzu QP 2010 plus with the operational condition of injection temperature of 210 ° C, the column temperature of 100 ° C, the temperature rise of 10°C/min, helium carrier gas, the pressure of 111 KPa, Rtx_5MS column type (30m x 0.25mm).

3. Result and Discussion

3.1. Pyrolysis Process of Cacao Pod Husks

The raw material is put into a pyrolyzer with the amount of material filled as much as 2/3 of the height of the pyrolyzer. The material in the pyrolyzer is heated so that the phase changes to vapor/gas. Steam/gas that has come out of the material will flow to the top of the pyrolyzer and exit through the pipe then flow into the condenser for the condensation process. Liquid smoke that has changed phase into a liquid phase goes out through the condenser and enters the storage tube. Charcoal as residue will remain in the pyrolyzer.

The products of this pyrolysis process are liquid smoke, tar, charcoal, and gas. The yield of pyrolysis products is shown in Table 1. The yield of liquid smoke from cacao pod husks is 14.77%. The biggest result from pyrolysis is charcoal with yield 40.43%. While from the results of measurements of water content and cacao pod husks density, the results were 14.34% and 0.8 g/mL respectively. This is caused by the water content in the raw materials which only sun-drying, not oven drying before pyrolysis. The yield of liquid smoke produced by the pyrolysis process will be affected by the raw materials used, including water content and density of raw materials.

Pyrolysis at high and long temperatures will cause the formation of liquid smoke to decrease because the temperature in the cooling water increases so that the smoke produced does not condense completely. The condensation process in this study does not use continuously water flowed on the cooling system so the condensation process does not take place perfectly. This condition is in accordance with [10] that liquid smoke by wood material processes can be produced maximally if the condensation process happens perfectly. The percentage of yield also depends on the pyrolysis temperature used and characterization of raw material. Besides climate, season, the age of the plant, type of the plant, raw materials and the way of burning also influence the liquid smoke yield on pyrolysis process [11].

Table 1. Yield of pyrolysis products from cacao pod husks

Biomass	The weight of Material (kg)	Temp Max (°C)	Pyrolysis Products			
			Liquid Smoke (%)	Ter (%)	Charcoal (%)	Gas (%)
Cacao pod huks	3	147	14.77	1.67	40.43	43.13

3.2. Characterization of Liquid Smoke

3.2.1. Acidity (pH)
Liquid smoke products from the pyrolysis process contain acidic compounds. pH value is one of the quality parameters of liquid smoke. Measurement of pH value in the liquid smoke is to determine the level of pyrolysis decomposition of raw materials. The measurement results of the pH value of the liquid smoke from cacao pod husks are 3.2. The pH value obtained from the pyrolysis process at 147°C is higher than the liquid smoke of rubber fruit shells which pyrolyzed at 300°C is between 1.96-2.65 [12]. High temperatures will cause the compounds in the rubber shell to break down into acidic compounds [13]. This low pH value indicates that liquid smoke quality is high, especially in terms of

as a food preservative. Low pH value has an effect on long-lasting value and storage power of the smoke product [14].

3.2.2. *Total Acetic Acid*

From the measurement of the liquid smoke's pH value, it can be seen that the liquid smoke of cacao pod husks contains acetic acid compounds. The level of acetic acid from the liquid smoke in this study was 9.306 ppm. This acidic liquid smoke was the major contribution to food preservation activity that effects its storage time and flavor. The quality of liquid smoke influenced by the composition of chemical compounds in liquid smoke. The chemical compounds in liquid smoke influenced by pyrolysis conditions and type of raw material [15]. According to [16], bio-oil compounds contain hydroxyl (−OH) or carboxylic (−COOH) groups.

3.2.3. *Phenol Levels*

In the liquid smoke of cacao pods husks, contain phenol groups which derived from lignin degradation. Identification of phenol in liquid smoke is expected to represent the quality criteria so that the target user is more appropriate. Determination of phenol in this study using UV: 1800 spectrophotometer at 725 nm. Total phenol content obtained from the liquid smoke of cacao pods husks is 1.688 ppm. The phenol content obtained was lower than the phenol content of the 2.5098 ppm rubber fruit shell which was pyrolyzed at 300°C [12].

The temperature of cacao pods husks pyrolysis in the study was only 147°C. This low temperature causes lignin to not decompose completely. The quality of liquid smoke produced depends on the characteristics of the raw material and the temperature achieved during the pyrolysis process [15]. According to [13], the decomposition of chemical components in wood gradually decomposes hemicellulose starting at 200°C, at 240°C decomposition of cellulose into pyrolignite solution, CO, and CO_2 ; at a temperature of 240-400°C the decomposition process of cellulose and lignin occurs into pyrolignite solution, CO, CH_4, H_2 gas and more tar and at temperatures above 400°C aromatic layer formation occurs.

3.2.4. *Components of Liquid Smoke*

The components in liquid smoke were assumed to consist of thermally degraded products of biomass components such as cellulose, hemicellulose, and lignin. The components of liquid smoke chemistry from cacao pod husks were analyzed using GC-MS. The chromatogram can be seen in Figure 1.

Figure 1 shows that liquid smoke from pyrolysis of cacao pod husks shows the separation of chemical components through the chromatogram peaks that appear. The peaks began at a retention time of 1.359 to 7.524 minutes and based on system data identified as many as 24 compounds in the liquid smoke of cacao pod husks. These compounds are thought to have names and concentrations as listed in Table 2.

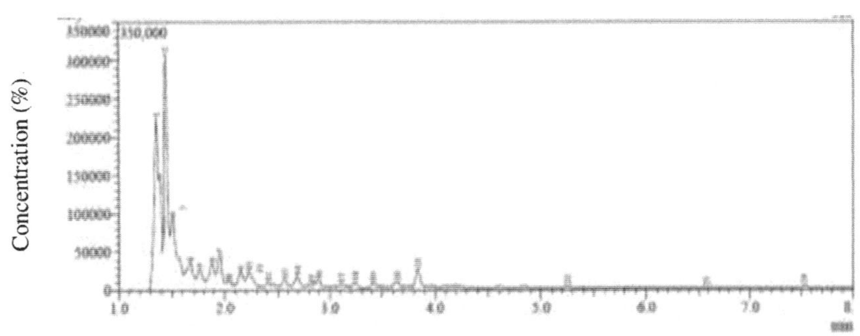

Figure 1. Liquid Smoke Chromatogram of Cacao Pod Husks

Table 2. The chemical content of liquid smoke from cacao pods husks

No.Peak	Retention Time (minute)	Compound	Concentration (%)
1	1.359	Methanol	25.78
2	1.445	Acetic Acid	20.57
3	1.511	Acetol	10.15
4	1.764	Cyclopentanone	3.14
5	1.879	2-Cyclopentene-1-one	3.70
6	1.953	Furfuryl alcohol	4.43
7	2.040	3-Methylpyridine	1.21
8	2.148	Butyrolactone	2.42
9	2.227	2-Methyl-2-cyclopentenone	2.60
10	2.414	3,4-Dimethylpyridine	0.83
11	2.568	3-Methyl-2-cyclopentenone	1.37
12	2.690	Benzene sulfonic acid	1.52
13	2.825	2,3,3-Trimethyl-1,4-pentadiene	0.45
14	2.896	Tetrahydrifurfuryl alcohol	0.77
15	3.110	Corylon	0.35
16	3.242	2,3-Dimethyl-2-cyclopentene-1-one	0.63
17	3.441	o-Cresol	1.49
18	3.838	phenol	1.97
19	5.250	2-Methoxy-4-methylphenol	0.23
20	6.574	p-Ethylquaiacol	0.08
21	7.524	2,6-Dimethoxyphenol	0.39

From GC-MS spectra data, the dispersed liquid smoke of cacao pod husks consist of 24 components. The highest compounds in liquid smoke was Methanol 25.78%, Acetic Acid 20.57%, and Acetol 12.40%. From 24 compounds identified in the liquid smoke of cocoa pods, there was 1 carboxylic acid group compound, 4 alcoholic compounds, 7 ketone group compounds, 4 phenol group compounds, 2 pyridine compounds, 1 benzene compund, 1 corylon, 1 butyrolactone and 3 unidentified compunds. This result is the decomposition of lignocellulose from cocoa pod husks which consists of cellulose, hemicellulose, lignin, and extractive substances. The decomposition of hemicellulose biomass was acetic acid, methanol, furfural, aldehydes, ketones, and o-acetyl groups [17].

Pyrolysis products from lignin are derivatives of phenol, guaiacol, and syringol [16]. This result is not much different from the material obtained by Mansur et al., (2014), that the chemical composition of liquid smoke from cacao pod husks is carboxylic acids, ketones, phenolic compounds, aldehydes, furans, benzenediols, heterocyclic aromatics, and heavy components. Meanwhile, the persistence of phenolics in liquid smoke from coconut shell from 2.48% to 23.40% at each varied temperatures [1]. The most chemical content in liquid smoke from the shell, away, and the length of palm oil is acetic acid and phenol [2].

4. Conclusion

Cacao pod husks have the potential to be developed as liquid smoke containing useful chemicals. The pyrolysis process produced products in the form of liquid smoke, charcoal, tar, and gas. The yield of pyrolysis products are 14.77; 40.43; 1.67; and 43.13%, respectively. Liquid smoke has a pH of 3.2 levels, acetic acid 9.306 ppm, and phenol levels 1.688 ppm. GC-MS measurement result showed that methanol, acetic acid, and acetol are the highest concentration of chemical obtained in liquid smoke. To get a quality pyrolysis product is influenced by the operational conditions of pyrolysis. So that in this study there needs to be further developed in the heating process. The optimum temperature of pyrolysis is achieved to degrade lignocellulose biomass if condensation process happens perfectly.

5. References

[1] S. Yuniningsih and S. P. A. Anggraini, 2013,"*Characterization of Liquid Smoke from Coconut Shell to Be Applicated as Safe Food Preservatives for Human Health*," vol. 3, no. 2, pp. 1–5,

[2] A. G. Haji,2013"*Komponen Kimia Asap Cair Hasil Pirolisis Limbah Padat Kelapa Sawit*," *J. Rekayasa Kim. dan Lingkung.*, vol. 9, no. 3.

[3] Y. Feng, S. Chen, and J. Mu,2010. "*Characterization of Products from Pyrolysis of Waste Wood-based Composites Containing Urea-formaldehyde Resins*," pp. 185–189.

[4] M. Wijaya, 2014,*Pemanfaatan limbah kakao sebagai bahan baku produk pangan*. Jurusan Kimia FMIPA Univ.Negeri Makasar.

[5] Departemen Perindustrian, 2007,"Gambaran Sekilas Industri Kakao," [Online]. Available: https://kemenperin.go.id/290/Paket-Informasi-Kakao.

[6] D. Mansur, T. Tago, and H. Abimanyu, 2014,"*Conversion of Cacao Pod Husks by Pyrolysis and Catalytic Reaction to Produce Useful Chemicals,*" *Biomass and Bioenergy*, vol. 66, no. July, pp. 275–285.

[7] D. Czajczyńska *et al.*,2017, "Potential of pyrolysis processes in the waste management sector," *Therm. Sci. Eng. Prog.*, vol. 3, pp. 171–197.

[8] R. E. Putri, A. Kasim, Emriadi, and A. Asben, 2018,"Characterization Of Liquid Smoke From Coconut Shell Based Pyrolysis Process," in *International Conference on Advance and Scientific Innovation.*

[9] Prasetyowati, M. Hermanto, and S. Farizy, 2014,"Pembuatan Asap Cair Dari Cangkang Buah Karet SEbagai Koagulan Lateks," *J. Tek. Kim.*, vol. 20, no. 4, pp. 14–21.

[10] A. Demirbas, 2005,"Pyrolysis of ground beech wood in irregular heating rate conditions," vol. 73, pp. 39–43.

[11] I. K. Budaraga, Y. Marlida, and U. Bulanin, 2016,"Liquid Smoke Production Quality from Raw Materials Variation and Different Pyrolysis Temperature," vol. 6, no. 3, pp. 306–315, .

[12] Prasetyowati, A. P. Novianty, and M. R. Haryuni, 2014,"Pembuatan Asap Cair dari Limbah Kulit Singkong (Manihot Esculenta L. SKIN) untuk Bahan Pengawet Kayu," *Tek. Kim.*, vol. 20, no. 1, pp. 64–75.

[13] C. E. Byrne and D. C. Nagle, 1997,"Wood monoliths-characterization," *Carbon N. Y.*, vol. 35, no. 2, pp. 267–273.

[14] Haji, Abdul Gani, Z. A. Mas'ud, B. W. Lay, and S. H. Sutjaho,2006, "Karakterisasi Asap Cair Hasil Pirolisis Sampah Organik Padat," *J. Agroindustrial Technol.*, vol. 16, no. 3, pp. 1–9, .

[15] T. Nakai, S. N. Kartal, T. Hata, and Y. Imamura, 2007,"Chemical characterization of pyrolysis liquids of wood-based composites and evaluation of their bio-efficiency," vol. 42, pp. 1236–1241.

[16] E. A. Smith, S. Park, A. T. Klein, and Y. J. Lee, 2012,"Bio-oil Analysis Using Negative Electrospray Ionization: Comparative Study of High-Resolution Mass Spectrometers and Phenolic versus Sugaric Components," .

[17] D. Fengel and Gerd Wegener, 1984,*Wood: Chemistry, Ultrastructure, Reaction.* New York, .

IOP Conf. Series: Earth and Environmental Science **327** (2019) 012012 doi:10.1088/1755-1315/327/1/012012

Quality of Leather Using Vegetable Tannins Extract of *Acacia Mangium* Bark from Waste of Industrial Plantation

S Mutiar[1], A. kasim[2], Emriadi[3], A Asben[2]

[1] Postgraduate Doctoral Program, Andalas University, Padang, 25163, Indonesia
[2] Faculty of Agriculture, Andalas University, Padang, 25163, Indonesia
[3] Faculty of Math and Science, Andalas University, Padang, 25163, Indonesia

Corresponding author's e-mail address: srimutiar@unidha.ac.id

Abstract. The purpose of this study was to determine the characteristics of leather using vegetable tanning ingredients from the bark of *Acacia mangium Willd.* Tanning agent ingredients are provided in the form of bark powder and extract. The first phase of the preparation of the bark into powder and the second stage is the extraction of tannin from *A.mangium* bark. Extraction using water at 100^0C for 30 minutes then evaporated so that tannin extract was obtained. Then the second stage is carried out the tanning process using both materials. Data analysis uses paired two sample for means T-test. The results of statistical analysis showed significant differences in the characteristics of the leather produced. Chemical analysis of leather tanned using *A.mangium* bark obtained moisture content of 16.76%, ash of 3.42%, tannin bound 18.99% and degree of tanning 38.21% while chemical analysis of leather tanned using tanneries *A.mangium* bark extract which is moisture content 16.03%, ash 4.18%, bound tannin 32.41% and 84.82% tanning degree. The physical properties of leather using *A.mangium* bark powder were tensile strength 493.84 kg / cm2, elongation 34.94%, zwik strength / lastability 6.66 mm (leather cracked). Compared to the characteristics of leather using tanner *A.mangium* bark extract, tensile strength was 254.32 kg / cm2, elongation 44.39%, zwik / lastability strength 8.63 mm (leather not cracked).

1. Introduction

Timber production from Industrial Plantation Forests leaves waste in the form of untapped bark. One type of plant developed is, A. mangium Willd. The potential of this bark may still be used as a tannin preparation [1]. Some studies say tannins are widely found in plants such as stems, roots, leaves, fruit, and skin. According to [2], tannins are the most important substances in plants, found in the wood, bark, fruit leaves, and roots, which occur naturally. and is used as a leather tanning material [3].

A. mangium Willd. which is often referred to as mangium is one type of plant that is widely developed in the industrial plantation forest program. According to FAO, as much as 67% of the total mangium forest in the world is in Indonesia. Harvesters can be carried out at the age of 5-8 years used for pulp making in the paper industry [4]. The utilization of mangium wood has so far experienced a broader spectrum, both for fiber wood, woodworking, and energy wood (fuel and charcoal). Various studies have been conducted to support the expansion of the use of mangium wood in the form of whole wood, particles, fibers or wood derivatives. In addition to the wood, it has also been investigated mangium bark extraction as a tanner. Some studies use acacia plants as tanneries [5, 6, 7, 8]. Extraction aims to attract secondary metabolites with the help of solvents [9]. Tannins can be dissolved in water, alcohol, acetone, but insoluble in benzene, chloroform, and organic solvents from petroleum ether [7]. That tannin is a water-soluble compound [10].

The purpose of this study was to determine the characteristics of the leather by using vegetable tanning ingredients from powder and tannin extract from A.mangium bark. The quality of the leather with A.mangium bark is observed, and likewise, the leather with tannin agent of extract from A.mangium bark. Observations in the form of chemical analysis and permitting physical properties of leather.

2. Materials and Methods

The ingredients used are aquades, goat skin and *A.mangium* bark. The bark obtained from PT.Arara Abadi. The bark used is a tree that has been ripe for harvest.

2.1 Research Equipment

The equipment used was an analytic frame *Kern*, centrifuge *Hettich 320R*, cabinet dryer, disc mill, vacuum rotary evaporator R-1001 VN, 70 mesh sieve, spectrophotometer *Shimadzu UV 1800*, soxhlet, universal testing machine CY-6730, colorflex Ez hunter and calipers.

2.2 Preparation of Material tanning agent of A.mangium of Bark

A.mangium bark material is cut to 0.5 - 1cm in size and dried. Then it was put into a disc mill and filtered with a 70 mesh filter. The material that passes the filter is partly prepared for the extraction process.

2.2.1 *Tannin Extraction.* Tannin extraction was carried out using distilled water. Comparison of ingredients with solvents 1:20. The extraction process was carried out by the heating temperature of 100^0C for 30 minutes. The time is calculated after the temperature is reached. Then the mixture is filtered using filter paper to obtain a filtrate. The filtrate is concentrated using a vacuum rotary evaporator so that it is obtained tannin extract. Then, the extract yield was calculated.

2.2.2 *Tanning Process.* Goatskin tanning uses skin tanning modification methods from the Yogyakarta Balai Besar Kulit Karet dan Plastik (BBKKP), namely: The skin used is goatskin. Skin preparation for tanning which consists of soaking until the skin is obtained. The skin is continued with tanning using *A.mangium* bark (not extract) and tannin extract from *A.mangium* bark. Then the skin is overnight for 24 hours. After that, the skin is washed to remove residual tanning substances that are still attached, then drying by means of opposition. The leather is prepared for testing. Testing the characteristics of leather consists of chemical analysis and physical properties. Chemical analysis was carried out based on the testing method on SNI-0234: 2009. Meanwhile, measurement of physical properties of leather refers to ISO 3376: 2011.

2.3 Data analysis

Statistical data analysis was performed to compare the two results of the leather analysis. The process of data analysis is done using paired two sample for means t-Test.

3. Discussion

The results of the analysis of *A.mangium* bark powder can be seen in Table 1. *A.mangium* bark powder contains tannin at 24.80% and 25.59% in water-soluble ingredients, 2.73% ash content and pH value equal to 6. Tannin content of a large vegetable tanning ingredient of 10% can be used as a tanner [11]. Based on the chemical analysis carried out, *A.mangium* bark powder can be used as a tanner.

Table 1. Analysis of *A.mangium* bark

Components	Persentase
Moisture content	10.76 ± 1.50
Water soluble	25.59 ± 4.41
Tannin	24.80 ± 1.31
Ash	2.73 ± 0.53
pH	6 ± 0,0

3.1 A. mangium Tannin Extract

The extract obtained was calculated as yield, namely the percentage comparison of the extract results with the material used. The yield of tannin extract produced can be seen in Table 2. The yield of tannin extract obtained was 33.63%. In addition to tannin content there is also an analysis of 78.51% water-soluble material, 12.10% moisture content, 3.71% ash and pH 6. This tannin extract is used as a vegetable tanner. Whereas the tannin content of the extract is 48.57%. The use of extracts on plant tanneries allows these substances to penetrate rapidly into the skin because the extracted substances have small molecules and active binding capacity on the skin molecules. This molecule and its binding capacity can be enlarged by changing the density and pH so that it can increase the tanning power of the skin. The pH of the solution affects the solubility of a material [12].

Table 2. Analysis of extracts from the *A.mangium* bark

Components	Persentase
Moisture content	12.10 ± 2.95
Tannin	48.57 ± 0.51
Ash	3.71 ± 0.51
pH	6 ± 0,0

Chemical analysis of leather Table 3 shows the tanned chemical analysis data using bark powder and *A.mangium* bark tannin extract.

Table 3. The results of chemical analysis of leather with different tanning material forms from the *A.mangium* bark

Parameters	*Form of materials*		Description
	Powder of bark (%)	Extract tannin (%)	
Moisture content	16,76 ± 1,61	16,03 ± 0,92	ns
Fat	3,55 ± 0,14	3,56 ± 0,18	*
Ash	3,42 ± 2,75	4,18 ± 1,71	*
Water soluble	7,54 ± 0,74	5,58 ± 0,67	*
Raw skin content	49,71 ± 6,13	38,20 ± 3,87	*
Bonded tannin	18,99 ± 9,84	32,41 ± 9,24	*
The degree of tanning	38,21 ± 7,55	84,82 ± 7,36	*
Ph	7 ± 0,0	7 ± 0,0	ns

Description: * (T count> T table = significant), ns (T count <T table = non significant)

Based on statistical analysis there is a significant difference between the leather chemical properties produced. Tanning with tannin extracts from A. mangium bark showed bonded tannins and a higher degree of tanning compared to the use of bark powder directly. Where tannins are bound and tanning levels of leather are 32.41% and 84.82%. When compared with leather tanned with bark powder the results are lower where the tannin content is 18.99% and the tanning degree is 38.21%. Judging from the bound tannin value and the degree of tanning the tanning extract tanning material is better than the use of *A. Mangium* wood powder due to different tannin levels. The extraction process can have an effect on tannin polymers which can form intramolecular bonds in the intermolecular cross-ring environment with skin collagen. Likewise, the mechanism of vegetable tanning forms the amount of hydrogen bonds between several phenolic hydroxyl groups from tannins with carboxyl and amine compounds in skin collagen.

3.2 The Physical Properties of Leather

The results of the measurement of the physical properties of leather with tanning powder with tannin extract from *A.mangium* bark can be seen in Table 4.

Table 4. Results of measurement of physical properties of leather

Parameters	Bark powder	Extract of bark	Description
Tensile strength (kg/cm^2)	493,84 ± 31,59	254,32 ± 11,26	*
Elongation (%)	34,94 ± 1,32	44,39 ± 6,91	ns
lastibility (mm)	6,66 ± 0,46 (cracked nerf)	8,63 ± 0,39 (nerf doesn't crack)	*
Thickness (mm)	0,9 ± 0,1	0,9 ± 0,1	ns
Color	Light Brown	Brown	*

Description: * (T count> T table = significant), ns (T count <T table = non significant)

Based on the results of the statistical analysis the measurement of tensile strength of tanned skin tanned with bark powder and tannin extract showed significantly different results. Tensile strength using bark powder results in a higher tensile strength compared to the tanned skin using tannin extract. However, it shows that there is a significant effect on leather elongation. Leather with tannin extract produced greater skin elongation 44.39% compared to tanned with bark powder which was 34.94%. Vegetable tanning material directly has limited properties of leather which are rather stiff compared to various treatments such as extracts and tanning combinations [13].

A significant difference is also found in the measurement of zwik skin strength, where the skin is tanned by using *A.mangium* bark powder which produces cracked nerf, while the leather using tannin extract from the bark is not cracked. The tanning process has a strong effect on the physical properties of the skin compared to skin that has not been tanned. In addition to strengthening physical properties, it can also increase the thickness of the skin due to the bonds that occur in the skin resulting in thicker and fuller skin [12]. Leather with vegetable tanners is generally light brown or reddish according to the color of the tanner.

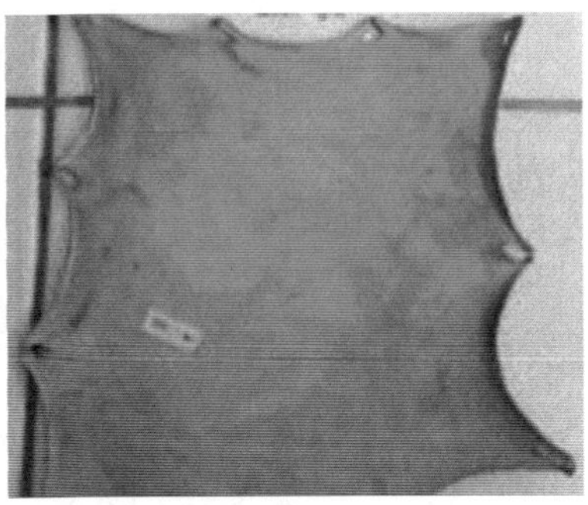

Figure 1. (a) Leather with tanned *A.mangium* bark powder

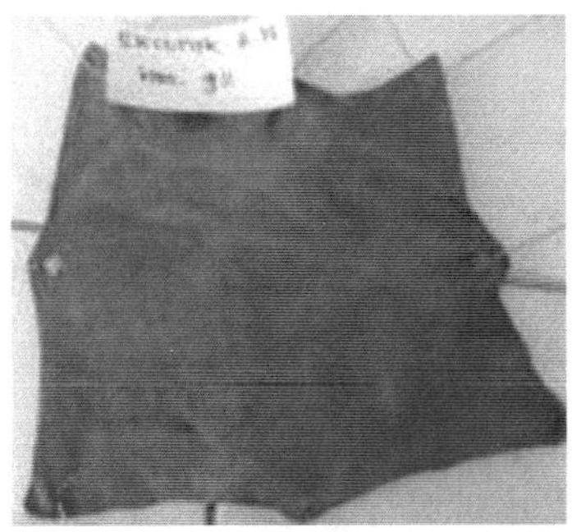

Figure 1. (b) Leather with tannin extract ingredients from the *A.mangium* bark

4. Conclusion

The results of research that has been done can be concluded that the use of tannin extracts from *A.mangium* bark produces a leather that is better than the use of bark powder. The bound tannin value and tanning degree using tannin extracting tanneries were higher than leather tanned with bark powder. Based on statistical analysis there are significant differences in the characteristics of leather with tanneries and extracts from *A.mangium* bark. Based on the measurement of physical properties of leather showed results that were not significantly different from the tensile strength of tanned skin. The tanned leather with bark powder produces leather nerf cracked while tanning ingredients with extracts produce Leather nerf that is not cracked.

Acknowledgment

The author's acknowledgment the Ministry of Research, Technology and Higher Education and Lembaga Pengelolaan Dana Pendidikan (LPDP) through the BUDI-DN for funding and technical support.

Reference

[1] Zakir, M. J., Ramalingam, S., Balasubramanian, P., Rathinam, A., Sreeram, K. J., Rao, J. R., & Nair, B. U. (2015). Innovative material from paper and pulp industry for leather processing. *Journal of Cleaner Production*, *104*, 436–444. https://doi.org/10.1016/j.jclepro.2015.05.001.

[2] Mekonnen, D., Habtamu, S., Sathiyamoorthy, M., & Selvi, V. (2013). Preparation of eco-friendly leather by process modifications to make pollution free tanneries. *Journal of Engineering Computers & Applied Sciences*, *2*(5), 17–22.

[3] Vieira, M. C., Lelis, R. C. C., da Silva, B. C., & de Lima Oliveira, G. (2011). Tannin extraction from the bark of Pinus oocarpa var. oocarpa with sodium carbonate and sodium bisulfite. Floresta e Ambiente, 18(1), 1-8.

[4] Notohadiningrat, T. (2006). Hutan tanaman industri dalam tataguna sumberdaya lahan. *Ilmu Tanah Universitas Gadjah Mada*, 1–9.

[5] Liao, J., Qu, B., Liu, D., Zheng, & Naiqin, &. (2015). Ultrasonics Sonochemistry New method to enhance the extraction yield of rutin from Sophora japonica using a novel ultrasonic extraction system by determining optimum ultrasonic frequency. *Ultrasonic Sonochemistry*, *27*, 110–116. https://doi.org/10.1016/j.ultsonch.2015.05.005.

[6] Haroun, M., Khirstova, P., & Covington, T. (2013). Analysis of Commercial Vegetable Tannin materials and Related Polyphenols of Selected Acacia Species in Sudan. *Journal of Forest Product & Product*, *2*(1), 21–28.

[7] Rhazi, N., Hannache, H., Oumam, M., Sesbou, A., Charrier, B., Pizzi, A., & Charrier-El Bouhtoury, F. (2014). Green extraction process of tannins obtained from Moroccan Acacia mollissima barks by microwave: Modeling and optimization of the process using the response surface methodology RSM. *Arabian Journal of Chemistry*. https://doi.org/10.1016/j.arabjc.2015.04.032.

[8] Hussain, G., Abass, N., Shabir, G., Athar, M., Saeed, A., Saleem, R., … Khan, M. A. (2017). New acid dyes and their metal complexes based on substituted phenols for leather: Synthesis, characterization and optical studies. *Journal of Applied Research and Technology*, *15*(4), 346–355. https://doi.org/10.1016/j.jart.2017.03.002.

[9] Mukhriani. (2014). Ekstraksi, pemisahan senyawa, dan identifikasi senyawa aktif. *Jurnal Kesehatan*, *VII*, 361–367.

[10] Missio, A. L., Tischer, B., dos Santos, P. S. B., Codevilla, C., de Menezes, C. R., Barin, J. S., Tondi, G. (2017). Analytical characterization of purified mimosa (Acacia mearnsii) industrial tannin extract: Single and sequential fractionation. *Separation and Purification Technology*, *186*, 218–225. https://doi.org/10.1016/j.seppur.2017.06.010.

[11] Mahdi, H., Palmina, K., & Glavtch, I. (2006). Characterization of Acacia niloticaas an indigenous tanning material of Sudan. *Journal of Tropical Forest Science*, *18*(3), 181–187.

[12] Kasim, A., Novia, D., Mutiar, S., & Pinem, J. (2013). Karakterisasi kulit kambing pada persiapan penyamak dengan gambir dan sifat kulit tersamak yang dihasilkan. Majalah Kulit, Karet dan Plastik. Vol. 29. No. 1.
[13] Hassan, E. A., & Ibrahim, M. T. (2014). Ebtesam A. Hassan 1*; Mohamed T. Ibrahim 1 and Sally K. A. 2. *Journal of Agricultural and Veterinary Sciences*, *15*(1), 87–94.

IOP Conf. Series: Earth and Environmental Science **327** (2019) 012013 doi:10.1088/1755-1315/327/1/012013

The Effects of Various Way of Processing Black Glutinous Rice (*Oryza sativa L. Processing Var Glutinosa*) on Digestibility and Energy Value of the Products

Rini[1], R Yenrina[2], T Anggraini[2], N E Chania[2]

[1] Faculty of Agricultural Technology, Andalas University
[2] Faculty of Agricultural Technology, Andalas University

Corresponding author's e-mail address: rinibahar59@yahoo.com

Abstract. The aim of this research is to study about the nutritional value, digestibility, and energy value from various ways of black glutinous rice processing. This research used explorative design using five processing methods and three replications. The results showed that the highest water content was found in boiled black glutinous rice (75.61%), the highest ash content in roasted black glutinous rice (1.19%), the highest fat content in puffing black glutinous rice (10.51%), the highest protein content in puffingblack glutinous rice (11.09%), the highest starch digestibility in roasted black glutinous rice (80.12%), the highest amylose content and the lowest amylopectin in roastedblack glutinous rice (10.32% and 89,67% respectively), the highest starch digestibility in puffingblack glutinous rice(75.61%), the highest antioxidant activity in raw black glutinous rice (60.75%), and the highest energy content (408 kcal) in puffing black glutinous rice.
Keywords: Black Glutinous Rice, Digestibility, Nutritional Value, Processing

1. Introduction

Black glutinous rice (*Oryza sativa L Var. Glutinosa*) is widely available in Indonesia with estimated total production around 42,000 tons per year. Glutinous rice (or sticky rice) has characteristics such as not transparent, has a distinctive smell, and almost all of the starch is amylopectin. Glutinous rice is usually processed into simple foods such as steamed black glutinous rice or boiled black glutinous rice. Glutinous rice is almost completely dominated by amylopectin thus the very sticky feature. Winarno (2004) stated that glutinous rice has a high starch content consisting in 1-2% amylose and 98-99% amylopectin. The higher the amylopectin level, the stickier the rice became.

Starch digestibility means the level of the easiness of the starch to be hydrolyzed and become simpler substance (Mercier, 1988). Indrasari, Wibowo, and Darajat (2008) reckoned that measuring starch digestibility could be done by using α-amylase enzyme. This enzyme has the ability to breakdown starch into simple sugars through hydrolysis process. According to Rimbawan and Siagian (2004), the higher the starch digestibility is the easier the starch to breakdown into glucose which could lead to the increased level of blood glucose. The increase level of blood glucose will also cause the increase need of insulin to convert glucose into energy.

Processing the glutinous rice to become a variety of processed products could affect the nutritional value, starch digestibility, and energy value of the products. The methods that could be use for processing the glutinous rice are roasting, steaming, puffing, and boiling. Roasting is a cooking method which utilized heat without using oil. This method is the one that is mostly developed. This method is done to improve the flavor of the food. Steaming is a cooking method which utilize liquid or steam (moist cooking method). Basically, the food is cooked by the steam, therefore the food should not be in direct contact with the boiled water underneath the filter. Puffing is method of food processing where the food is expanded in volume as a result of temperature and pressure involvement which results in changes in the structure of the food material. Boiling is a cooking method using boiled water as a heat transfer medium. Harris and Karmas (1989) stated that the factors that affects the loss of nutrients during the boiling process are the material surface area, concentration of dissolved substances in water, and the action of stirring the water. There is still no further information about the nutritional value, starch digestibility, and energy value produced by the variety of processed black glutinous rice as yet.

2. Materials and Methods

2.1 Ingredients and Tools

The ingredient used in this research was black glutinous rice obtained from Pasar Raya Padang, West Sumatera. The chemical substances used are diethyl ether, ethanol 95%, NaOH 1N, K_2SO_4, HgO, H_2SO_4, H_3BO_3, $Na_2SO_2O_3$ 0.1 N, starch solution 0.5%, HCl 0.02 N, HCl 30%, cotton wool, aluminium foil, filter paper, iodine solution, acetic acid, boiling solvent, boiling stone, distilled water, methyl orange indicator, MM-MB indicator, phenolphthalein 1% indicator, phosphate buffer solution 0.1 M, luff schoorl solution, KI 30% solution, α-amylase enzyme solution, dinitrosalicylate acid solution, pure amylose, standard maltose, and pure starch.

Tools used were pots, stoves, frying pan, ignition plates, hot plates, furnaces, cup saucer, pipette dropper, water bath, desiccator, spectrophotometer, test tube, volumetric flask 250 ml, soxhlet complete with the condenser, oven, filter paper, cotton, centrifuge, spoon, steam bath or electric heater, Kjeldahl heater connected to a steam absorber through an aspirator, Kjeldahl flask, burette, Erlenmeyer 500 ml, fat flask, vortex, bomb calorimeter, and analytical scales.

2.2 The Research Design

This research used explorative design consists of three replications. This research uses steaming, roasting, boiling, puffing and control as methods of processing.

2.3 Implementation of Research

2.3.1 Boiled Black Glutinous Rice. Boiling the black glutinous rice is done by following these steps:

1. 750 ml of water is boiled (100°C) in a pan within ± 10 minutes over medium heat
2. Put 200gr of black glutinous rice into the boiling water for about ± 15 minutes
3. Remove the cooked black glutinous rice from the boiling water and drained the excess water
4. Boiled black glutinous rice is ready to be analyzed

2.3.2. Steamed Black Glutinous Rice. Steaming the balck glutinous rice is done by following these steps:

1. Clean wash the black glutinous rice and let it soak for about 2 hours
2. Boil (100°C) 1 L of water in the steam pan for about ± 10 minutes over medium heat
3. Put 200 gr of soaked black glutinous rice into the steaming pan for about ± 15 minutes
4. Remove the cooked black glutinous rice from the boiling water and drained the excess water
5. Steamed black glutinous rice is ready to be analyzed

2.3.3. Roasted Black Glutinous Rice. Roasting black glutinous rice is done by following these steps:

1. Soak 200 gr of black glutinous rice and drain the excess water
2. Mash the soaked black glutinous rice to size 75 mesh
3. Put the mashed black glutinous rice into the pan and roast it in medium heat
4. During the roasting process, stir the black glutinous rice evenly
5. The roasting process is carried out at 100-120°C in temperature for 30 minutes
6. Roasted black glutinous rice is ready to be analyzed

2.3.4 Puffing Black Glutinous Rice. Puffing black glutinous rice is done by following these steps:

1. Put 5 gr margarine into a pre-heated pan (60°C) and wait until it melts
2. Wash clean 10 gr of black glutinous rice (water content 13.5-14%), drained the excess water, and put it in the pan from step 1.
3. Close the pan with a lid and shake the pan once in a while in order to make all of the black glutinous rice exposed to margarine evenly
4. This process will make the black glutinous rice give out a popping sound. Leave it for ± 5 minutes (90-110°C) until the popping sound stops.
5. Puffing black glutinous rice is ready to be analyzed

2.4 Observation

In this study, chemical observations were carried out, i.e., observation of the water content (Gravimetric), ash content, fat content, protein content, starch digestibility, amylose content, amylopectin content, energy value, and antioxidant activity.

3. Results and Discussion

3.1 Water Content (Gravimetric)

Based on the results of this study, the water content of various processed black glutinous rice products could be seen in Table 1.

Table 1. Water Content of Various Processed Black Glutinous Rice (BGR)

Treatment	Water Content(%) ± SD
Control BGR	12.53 ± 0.2738
Boiled BGR	73.26 ± 0.7673
Steamed BGR	67.37 ± 0.8129
Roasted BGR	10.27 ± 0.5175
Puffing BGR	7.77 ± 0.4221

SD: Standard Deviation

The water content of the control black glutinous rice and the four processed black glutinous rice ranges from 7.77-73.26 %. The water content of the control black glutinous rice (12.53%) is higher than the ones from the Directorate of Nutrition, Health Departement of Indonesia (2001) which was 12%. This difference is caused by the difference of variety and the place of growth. The water content from the roasting process and the puffing process are decreased compared to the control.

The highest water content is found in the boiled black glutinous rice (73.26%) due to the involvement of water in the process. On the other hand, steamed black glutinous rice was processed using hot steam (moist heat), which basically used convection heat transfer, thus resulted in lower water content (67.37%) compared to boiled glutinous rice.The lowest water content is found in roasted black glutinous rice and puffing black glutinous rice (10.27% and 7.77% respectively) since the processing was done in dry environment. The roasted black glutinous rice was processed by mashing it and roasting it in high temperature which made the water content evaporates. This founding is in line with the research by Ravi et al., (2015) which stated that the roasting proses would result in water evaporation and decrease the water content after the process is done.

3.2 Ash Content

From the results of this study, the ash content of various processed black glutinous rice can be seen in Table 2.

Table 2. Ash Content of Various Processed Black Glutinous Rice (BGR)

Treatment	Ash Content(%) ± SD
Control BGR	0.59 ± 0.3436
Boiled BGR	0.26 ± 0.1131

Steamed BGR	0.39 ± 0.0002
Roasted BGR	1.19 ± 0.0023
Puffing BGR	0.79 ± 0.0036

SD: Standard Deviation

The ash content of the control black glutinous rice and the four processed black glutinous rice ranges from 0.26-1.19%. The highest ash content is found ini roasted black glutinous rice (1.19 %), higher compared to research by Refdi and Prima (2017) which is 0.68 %. This shows that the roasting process resulted in the increase in ash content. This happens due to the interaction between the metal equipments that were used during the process. This founding is in line with the research by Faria et al., (2012) which stated that the possible interaction between the metal equipments couldn't be ignored. The ash content of puffing black glutinous rice is 0.79%, which shows an increase compred to the control. This was caused by the puffing processing that used margarine. The existence of sodium in the margarine will cause the increasing ash content since sodium is one of the main salt group (Deman,1997).

Boiled and steamed black glutinous rice shows a decrease value of ash content compared to roast and puffing black glutinous rice. This was due to the dissolving of salt mineral in water during the boiling process, which was in accordance with Chen et al., (1999), whom stated that the treatment of soaking in water could causes the dissolution of some minerals, water-soluble vitamins, albumin, and sugar. According to, Sediaoetama (1993), the act of processing food could decrease the mineral content of the food due to the damage done by pH, oxygen, light, and heat or any combination of said components during the process. In accordance with Santoso et al., (2006) that stated that the chemical structure of minerals contained in food could change due to the act of processing or the interaction with other materials, thus resulting in the increase or decrease level of mineral solubility depending on the process.

3.3 Fat Content

From the results of this study, the fat content of various processed black glutinous rice can be seen in Table 3.

Table 3. Fat Content of Various Processed Black Glutinous Rice (BGR)

Treatment	Fat Content (%) ± SD
Control BGR	2.19 ± 0.3302
Boiled BGR	0.49 ± 0.4088
Steamed BGR	0.35 ± 0.0400
Roasted BGR	1.92 ± 0.3308
Puffing BGR	10.51 ± 0.4765

SD: Standard Deviation

The fat content of the control black glutinous rice and the four processed black glutinous rice ranges from 0.35-10.51%. The highest fat content is found in puffed black glutinous rice (10.51%) while the lowest in steamed black glutinous rice (0.35%). The increase of fat content in the puffed black glutinous rice is due to the addition of margarine that contains fat during the process.

Boiling and steaming process resulted in fat content of 0.49% and 0.35% respectively, which are lower compared to the control (2.19%). The decrease in fat content is due to the treatment of washing and soaking during the process resulting in the loss of the epidermis of the black glutinous rice that contains fat. This also causes the high water content in boiled and steamed black glutinous rice. Mastuti (2008) stated that the high water content found in end products is usually the cause of the low fat content. Besides, the water content in the final product it is also influenced by the hydrolysis process. Ketaren (1986) stated that fat hydrolysis will form free fatty acids and glycerol, where the higher the temperature used, the higher the level of fat damage.

3.4 Protein Content

From the results of this study, the protein content of various processed black glutinous rice can be seen in Table 4.

Table 4. Protein Content of Various Processed Black Glutinous Rice (BGR)

Treatment	Protein Content (%) ± SD
Control BGR	9.45 ± 0.7663
Boiled BGR	2.95 ± 0.3669
Steamed BGR	2.55 ± 0.3689
Roasted BGR	8.33 ± 0.2010
Puffing BGR	9.04 ± 0.0556

SD: Standard Deviation

The protein content of the control black glutinous rice and the four processed black glutinous rice ranges from 2.55-9.45%. The control black glutinous rice protein content is 9.45%, which is higher compared to the data from the Nutrition Directorate, Health Department of Indonesia, (2001) which is 6.7% and research by Zulaikah, (2002) which is 8.0%. The highest protein content is found in the control black glutinous rice (9.45%) and the lowest in steamed black glutinous rice (2.55%). This was due to the water content, where the lower the water content, the higher the protein content found. According to Riansyah et al., (2013), the decrease of water content in a food material will cause the increase of protein content. The use of heat in the process could reduce the water content which resulted in the increase of protein content. The drier the food, the higher the protein content found, which is in accordance with Albert's (2013) whom stated that by decreasing the water content, the food would have higher content of protein, carbohydrate, fat, and other minerals. According to Buckle, Edwars, Fleet and Wotton (1987), the protein content is inversely proportional to the water content. The higher the water content, the lower the protein content found and vice versa.

3.5 Starch Content

From the results of this study, the starch content of various processed black glutinous rice can be seen in Table 5.

Table 5. Starch Content of Various Processed Black Glutinous Rice (BGR)

Treatments	Starch Content (%) ± SD
Control BGR	71.75 ± 0.1371
Boiled BGR	31.59 ± 0.1640
Steamed BGR	54.83 ± 0.6967
Roasted BGR	80.12 ± 0.1569
Puffing BGR	67.36 ± 0.1569

SD: Standard Deviation

The starch content of the control black glutinous rice and the four processed black glutinous rice ranges from 31.59-80.12%. The highest starch content is found in roasted black glutinous rice (80.12%), which is allegedly caused by the mashing treatment and the absence interaction with water during the process resulted in the increase of starch content. The starch content in boiled black glutinous rice was decreased (40.16%), which is in accordance with Rauf (2015), which stated that starch does not dissolve in cold water but in hot water instead where the starch granules became dissolved. Decrease in starch content was also found in puffing black glutinous rice (67.36%). In steamed black glutinous rice, the starch content also decreased significantly (16.92%), which was due to the use of hot steam during the process, resulted in the decrease of starch content.

3.6 Amylose and Amylopectin

From the results of this study, the amylose content of various processed black glutinous rice can be seen in Table 6.

Table 6. Amylose content of Various Processed Black Glutinous Rice (BGR)

Treatment	Amylose Content (%) ± SD

Control BGR	10.12 ± 0.5446
Boiled BGR	6.80 ± 0.6364
Steamed BGR	8.28 ± 0.1790
Roasted BGR	10.32 ± 0.6013
Puffing BGR	9.41 ± 0.0184

SD: Standard Deviation

The amylose content of the control black glutinous rice and the four processed black glutinous rice ranges from 6.8-10.32%. The highest amylose content was found in boiled black glutinous rice (10.32%), while the lowest was found in roasted black glutinous rice (6.80%), which was allegedly due to the soaking and steaming treatment that used heat. Rauf (2015), stated that starch granules dissolved in hot water. Amylose is easier to become a water soluble component compared to amylopectin when heated. This is due to the molecular weight of both types of starch. Amylose has a smaller molecule compared to amylopectin, while the higher the solubility of the starch, the greater the swelling became. Since amylose has a greater solubility in hot water compared to amylopectin, amylose has a greater swelling power.

The decrease in amylose content is due to the increase in temperature resulted in the decrease of amylose constituent component in starch. The increase of starch solubility due to the increase heating of the starch suspension is caused by the amylose depolymerization. High temperature could cause the depolymerization of starch molecules (Yuliasih, Irawadi, Sailah, Pranamuda, Setyowati and Sunarti, 2007). This resulted in a simpler amylose molecule, which has a shorter chain thus easily dissolves in water. Amylose is a starch component that has a straight chain and is water soluble. (Ben, Zulianis, and Halim, 2007). The amylopectin content of various processed black glutinous rice can be seen in Table 7.

Table 7. Amylopectin content of Various Processed Black Glutinous Rice (BGR)

Treatment	Amylopectin Content(%) ± SD
Control BGR	89.87 ± 0.5446
Boiled BGR	93.19 ± 0.6364
Steamed BGR	91.71 ± 0.1790
Roasted BGR	89.67 ± 0.6013
Puffing BGR	90.58 ± 0.0184

SD: Standard Deviation

The amylopectin content of the control black glutinous rice and the four processed black glutinous rice ranges from 89.67-93.19%, while amylopectin content in control black glutinous rice or without processing is 89.87%. Haryadi (2008) stated that the low amylose content (0-2) and the high amylopectin content (98-99%) found in glutinous rice are responsible for the sticky and shiny characteristics of the glutinous rice.

3.7 Starch Digestibility

Based on the results of this study, the starch digestibility of various processed black glutinous rice can be seen in Table 8.

Table 8. Starch digestibility of Various Processed Black Glutinous Rice (BGR)

Treatment	Starch Digestibility (%) ± SD
Control BGR	39.29 ± 0.1171
Boiled BGR	75.61 ± 0.6925
Steamed BGR	62.54 ± 0.2177
Roasted BGR	43.32 ± 0.1738
Puffing BGR	40.69 ± 0.3059

SD: Standard Deviation

The starch digestibility of the control black glutinous rice and the four processed black glutinous rice ranges from 39.29-75.61%. The lowest starch digestibility is found in black glutinous rice without processing or control (39.29%). Increased starch digestibility is found in boiled and steamed black glutinous rice (75.61% and 62.54% respectively). This is allegedly caused by the long amount of time spent during the process that used water (Thornburn, Brand and Truswell, 1986 in Amalia, Rimbawan, and Dewi, 2011). Cooking process could cause the different levels of starch digestibility. The starch gelatinization process caused the changes in the structure of amylose from crystalline into amorphous, thus the increase in starch digestibility (Harlampu, 2000).

The factor that affects the starch digestibility is the amylose content. The result is in accordance with the literature that stated the higher the amylose content, the lower the starch digestibility. This is supported by Frei, Siddhuraju, and Becker (2003) whom stated that amylose content is one of the factors that affect the starch digestibility. Amylose is a glucose polymer that has a non-branching structure with a more crystal like feature and a more extensive hydrogen bond. The hydrogen bond in amylose is also stronger compared to amylopectin, therefore is harder to be hydrolyzed by digestive enzymes (Behall and Hallfrisch, 2002). This non-branching structure strongly tied the amylose and making it hard to digest (Rimbawan and Siagian, 2004). Other factors that affects starch digestibility is the presence of anti-nutrition and anti-amylase (food fiber and tannin), and the chemical structure of the starch (starch resistance).

In correlation to the level of food fiber, which is part of a food that cannot be digested by digestive enzymes, it could increase the food mix viscosity inside the intestine thus inhibits the food and enzyme interaction (starch). According to Winarno (2004), the fiber content in black glutinous rice is 1.3%. Therefore, the other factor that affects the starch digestibility is food fiber. Food that contains a high level of fiber will be tougher to digest or has a lower digestibility.

3.8 Antioxidant Activity

Based on the results of this study, the antioxidant activity of various processed black glutinous rice can be seen in table 9.

Table 9. Antioxidant activity of Various Processed Black Glutinous Rice (BGR)

Treatments	Antioxidant activity (%) ± SD
Control BGR	60.75 ± 0.244
Boiled BGR	14.05 ± 0.904
Steamed BGR	5.66 ± 0.417
Roasted BGR	39.39 ± 0.36 4
Puffing BGR	46.73 ± 0.315

SD: Standard Deviation

Antioxidant activity of control and puffing black glutinous rice is measured in the concentration of 1000 ppm while boiled, steamed, and roasted black glutinous rice was measured in the concentration of 100 ppm. Antioxidant activity of the control black glutinous rice and the four processed black glutinous rice ranges from 5.66% - 60.75%. The highest antioxidant activity is found in control black glutinous rice (60.75%) and the lowest was in steamed black glutinous rice (5.66%). Antioxidant activity found in boiled and steamed black glutinous rice was significantly decreased (14.05% and 5.66% respectively) compared to the control. This is allegedly caused by the use of water as the heat transfer medium during the process. Furthermore, the act of soaking the black glutinous rice caused the antioxidant to dissolve during the process of steaming.

On the other hand, antioxidant activity in roasted and puffing black glutinous rice slightly decreased (39.39% and 46.73% respectively) compared to control. This is allegedly caused by the short amount of time spent during the process. This founding is in accordance with the research by Suhartatik (2013), which stated that the heating process also affects the antioxidant activity of the food. The longer heating process could damage the antioxidant activity. According to Gordon, et al., (2001) antioxidant activity is affected by factors such as lipid content, antioxidant concentration, temperature, oxygen pressure, and the general chemical component of the food i.e., protein and water.

Anggarini et al., (2015) stated that antioxidant activity in black glutinous rice ranges from 30%-55% measured in the concentration of 1000 ppm. One of the antioxidant found in black glutinous rice is anthocyanin and it could be damage by high temperature. According to Nugraheni (2014), high temperature could induce damage towards the logarithmic pigment with heating time at constant temperature. Anthocyanin extracted from the black glutinous rice was damaged after processing at 80-100°C. This damage happened because anthocyanin sustain an increase in oxidation reaction at high temperature (Hou, et al., 2013). This is also in line with the research by Suhartatik (2013), the higher the heating temperature, the more anthocyanin is damaged. Similarly, the longer the heating process the more number of anthocyanin was degraded.

3.9 Total Energy

Based on the results of this study, the total energy of various processed black glutinous rice can be seen in Table 10.

Table 10. Total Energy of Various Processed Black Glutinous Rice (BGR)

Treatment	Total Energy(kcal) ± SD
Control **BGR**	351 ± 0.1331
Boiled BGR	142 ± 0.2769
Steamed BGR	233 ± 0.0987
Roasted BGR	379 ± 0.1263
Puffing BGR	408 ± 0.1197

SD: Standard Deviation.

Total energy of the control black glutinous rice and the four processed black glutinous rice ranges from 351-408 kkal. The highest total energy is found in puffing black glutinous rice (408 kkal), which was caused by the use of margarine during the process. The higher fat content it has the higher energy value it has since margarine contains 80% of fat and contains energy 9 kkal per gram (Winarno, 2004). Total energy found in control black glutinous rice is 351 kkal, which is lower compared to the research by Soeharto (2004) that stated the total energy was 356 kkal.

The lowest total energy is found in boiled boiled black glutinous rice (142 kkal). The use of water as the heat transfer medium caused the fat to be taken out, thus the higher water content found in boiled and steamed black glutinous rice. According to Mastuti (2008), the high water content found in the end productusually resulted in a low fat content.

Fat is one of the main content found in food and also the main energy source. According to Andarwulan et al., (2011), fat has an important role in the increase calorie of food due to the higher total energy it has compared to other nutrition substances. Every burning process of 1 gram carbohydrate and protein will produce 4 kkal energy, while the burning process of 1 gram fat will produce 9 kkal (Apriyantono, 1989).

4. Conclusion

The results of this research shows that the highest water content is found in boiled black glutinous rice (73.26%), the highest ash content is found in roasted black glutinous rice (1.19%), the highest fat content is found in puffing black glutinous rice (10.51%), the highest protein content is found in puffing black glutinous rice (11.09%), the highest starch content is found in roasted black glutinous rice (80.12%), the highest amylose and the lowest amylopectin content is found in roasted black glutinous rice (10.32% and 89.67% respectively),the highest digestibility is found in boiled black glutinous rice (75.61%), the highest antioxidant activity is found in control black glutinous rice (60.75%), and the highest total energy is found in puffing black glutinous rice (408 kkal).

Based on these findings, we can conclude that the best way of processing the black glutinous rice is by puffing it, which resulted in high fat content and high total energy produced. Meanwhile, the best digestibility is found in boiled black glutinous rice.

REFERENCES

[1] Albert R Reo 2013 *Quality of Red Snapper Processed with Differences in Garand Concentration and Duration of Drying. Tropical Fisheries and Marine Journal* (online), Vol IX-1, (http: ejournal.unsrat.ac.id) [accessed November 8, 2018]
[2] Andarwulan, N K, Feri and H Dian 2011 *Food Analysis* (Jakarta:Dian Rakyat) pp. 66-194.
[3] Anggraini T, Novelina, Limber U and R. Amelia 2015 Antioxidant Activities from Several Red, Black and White Cultivars from West Sumatra, Indonesia *Pakistan Nutrition Journal* **2** 112 - 117
[4] Amalia S N, Rimbawan and M Dewi 2011 Glycemic Index Value of Several Types of Sweet Corn Processing (*Zea mays saccharata* Sturt) *Nutrition and Food Journal* **6** (1) 36-41
[5] Apriyantono A, D Fardiaz, N L Puspitasari, Sedarnawati and S Budiyanto 1989 *Guide to the Food Analysis Laboratory* (Bogor: IPB)
[6] Behall K M and J Hallfrisch 2002 Plasma Glucose and Insulin Reduction After Consumption of Breads Varying in Amylose Content *European Journal of Clinical Nutrition* **56** (9) 913-920.
[7] Ben E S, Zulianis and A Halim. 2007 Preliminary Study of Separation of Amylose and Amylopectin of Cassava Starch by Butanol-Water Fractionation *Journal of Pharmaceutical Science and Technology* **12** (1) 1-11
[8] Buckle R A, Edwards, G H Fleet and M Wotton. 1987 *Food Science* (Jakarta: Press UI)
[9] Chen K N and Chen M J 1999 *Statistical Optimization: Response Surface Methodology. Inside: Erdogdu F* (Ed). Optimization in Food Engineering (Florida: CRC Press)
[10] Nutrition Directorate of the Indonesian Ministry of Health 2001 *Nutrient content of Black Glutinous Rice* (Jakarta: Bhrata)
[11] Deman J M 1997 *Food Chemistry* Second Edition Edited by Kosasih Panmawinata (Bandung: Bandung Institute of Technology) pp 103-112
[12] Faria Simone Aparecida dos Santos Conceicao, Bassinello P Z, Penteado Marilene de Vuono C P 2012 Nutritional Composition of Rice Bran Submitted to Different Stabilization Procedures *Brazilian Journal of Pharmaceutical Science* **48** (4) 651-657.
[13] Frei M P, P Siddhuraju and K Becker 2003 Studies on the in Vitro Starch Digestibility and The Glycemic Index of Six Different Indigenous Rice Cultivars from The Philippines *Food Chemistry* **83** (3) 395-402
[14] Gordon M H, J Pokorny, N Yanishlieve and M Gordon 2001 *Antioxidants in Food* (New York: CRC Press)
[15] Harlampu S G 2000 Resistant Starch a Review of the Physical Properties and Biological Impact of RS3 *Carbohydrate Polymers* **42** (3) 285-292
[16] Harris R S and E Karmas 1989 *Evaluasi Gizi pada Pengolahan Bahan Pangan* (Bandung ITB – Press) 729p
[17] Haryadi 2008 *Teknologi Pengolahan Beras* (Yogyakarta: Gajah Mada University Press) 239p
[18] Indrasari S D, P Wibowo and A A Daradjat 2008 *Kandungan Mineral Beras Varietas Unggul Baru.* [Disampaikan pada Seminar Nasional Padi, Sukamandi, 23-24 Juli 2008] (dalam proses publikasi)
[19] Ketaren S 1986 *Minyak dan Lemak Pangan* (Jakarta: Universitas Indonesia Press)
[20] Mastuti R 2008 Pengaruh Suhu dan Lama Waktu Menggoreng Terhadap Kualitas Fisik dan Kimia Daging Kambing Restukturisasi *Jurnal Ilmu dan Teknologi Hasil Ternak* **3** (2) 23-31
[21] Mercier C and P Colonna 1988 *Starch and enzymes: Innovations in the Products, Process and Uses*. (Chimic: Biofutur) pp55-60
[22] Nugraheni M 2014 *Pewarna Alami* (Yogyakarta: Graha Ilmu) 180p
[23] Rauf R 2015 *Kimia Pangan* (Yogyakarta: Andi)
[24] Ravi U, L Menon, M Hepzibah and S Saha 2015 Flour Pretreatment on The Quality Characteristics and Oil Absorption of Traditional South Indian Festive Snack *Indian Journal of Traditional Knowledge* **1** (1) 139-143
[25] Refdi C W and P Y Fajri 2017 Komposisi Gizi dan Pati Tepung Beras Rendang dari Beberapa Sentra Produksi di Kota Payakumbuh Sumatera Barat *Jurnal Teknologi Pertanian Andalas 21* (1) 42-43
[26] Rimbawan and S Albiner 2004 *Indeks Glikemik Pangan* (Bogor: Penebar Swadaya) 144p
[27] Sediaoetama A D 1993 *Ilmu Gizi Jilid II* (Jakarta: Dian Rakyat)

[28] Santoso J, G Satako, Y S Yumiko, S Takeshi 2006 Mineral Content of Indonesian Seaweed Solubility Affected by Basic Cooking *Journal of Food Science and Technology* **12** (1) 59-66
[29] Suhartatik N 2013 Stabilitas Ekstrak Antosianin Beras Ketan (*Oryza sativa* glutinosa) Hitam selama Proses Pemanasan dan Penyimpanan *Agritech* **33** 384–390
[30] Winarno F G 2004 *Kimia Pangan dan Gizi* (Jakarta: Gramedia Main Library) 251p
[31] Yenrina R 2015 *Metode Analisi Bahan Pangan dan Komponen Bioaktif* (Padang: Andalas University Press) 159p
[32] Yuliasih I, T T Irawadi, I Sailah, H Pranamuda, K Setyowati, and TC Sunarti 2007 Pengaruh Proses Fraksinasi Pati Sagu Terhadap Karakteristik Fraksi Amilosanya *Jurnal Teknologi Industri Pertanian* **17** (1) 29-36
[33] Zulaikah S 2002 *Ilmu Bahan Makanan I* (Surakarta: Fakultas Ilmu Kedokteran Universitas Muhammdiyah Surakarta)

IOP Conf. Series: Earth and Environmental Science **327** (2019) 012014 doi:10.1088/1755-1315/327/1/012014

Genetic Diversity Based on RAPD Marker of Ramie Plants (Boehmeria nivea [L.] Gaud) In West Sumatera

R Mayerni[1], Yusniwati[2], D Yulfa[2], and S R O S Chan[2]

[1,2] Faculty of Agriculture ,Andalas University, Limau Manis Padang, 25163 Indonesia

Corresponding author's e-mail address: renimayerniagr@unand.ac.id

Abstract. This research has been conducted at the Experimental Farm of Agricultural Faculty of Andalas University, two districts in West Sumatera (Lima Puluh Kota and Agam) and the Laboratory of the Department of Agronomy and Horticulture Agricultural Faculty, Bogor. The purpose of this study was to determine the genetic diversity of the ramie plants collection Experimental Farm, Agricultural Faculty of Andalas University and exploration at two districts in West Sumatera. This research was conducted by molecular characterization, using RAPD markers *(Random Amplified Polymorphic DNA*) with 5 primers (OPF 4, OPF5, OPH7, OPX2, and OPX 17). The samples used were 4 clones of ramie plants (Padang 3, Ramindo, Matur and Situjuah) with 3 replications. Fragments were sized with reference to the size markers and scored (present/absent). The similarity coefficients calculated and dendograms produced using NTSYS software (version 2.02). From the results of the study the results of the similarity level coefficient were 0.56-0.78, and formed 2 large groups (I, II). M1 and M2 clones are in group I, while in groups II (IIa and IIb). Group IIa (R1, R2, R3, P3, M1) and Group IIb (S1, S2, S3, P1).

1. Introduction

Ramie plants (*Boehmeria nivea* (L.) Gaud.) is one of the annual plants that have many benefits. Ramie plants as raw material in the textile industry [8], pulp [5], paper [25], land conservation [2], as compost material [19], traditional medicine [9], and animal feed [11].

Fiber from ramie plants as a raw material for the textile industry, because produce high-quality fibers [21]. In addition, its fiber has a smooth, long and strong texture [14,15,26]. Rami bast fibre clearly shows extra xilary fibre that is found in the cortex around floem [17].

According from [16,4,22], ramie plants are easy to grow in the tropics and will high produce if planted in the lowlands to highlands (10 - 1500 m above sea level). In West Sumatra forest, ramie plants are still found wildly. Knowledge of genetic diversity is very important in plant breeding activities to determine steps to improve the quality and quantity of ramie plants.

And to know genetic diversity can used molecular characterization observations, information about molecular character is needed as a differentiator between plant accession and estimation of distance or genetic diversity in kinship analysis so that it can be used as basic material for improving ramie plants.

The further the genetic distance between parents in one species, the greater the chance to produce new varieties [24]. [12] state that to improve the genetic properties of wild ramie germplasm better through molecular characterization.

RAPD technique (*Random Amplified Polymorphism* DNA) is one of the PCR-based molecular methods that can be used to identify individuals at the DNA level [6,7]. This method is also relatively cheaper and popular [10] and ability to quickly detect polymorphisms at a number of band analyzed [1].

This can be seen in the results of research by [13] that of the 37 ramie accessions using 31 primers with the RAPD technique which included 29 wild accessions and 8 commercial accessions with polymorphism levels of 95.5%. As well as based on research by [18], out of 5 clones of ramie plants using 10 primers showed a level of polymorphism of 93%. [21] also added that the high level of polymorphism with molecular markers in ramie plants would be very useful in analyzing population structure and effective in managing plant genetic resources.

2. Material and Methods

2.1. Materials

Fresh leaf samples collected from 4 ramie clones (Ramindo 1, Padang 3, Matur and Situjuah), CTAB extraction buffer (Tris-HCl (1 M, pH 8), NaCl (1.4 M), EDTA (0.5 M, pH 8)) was used for DNA extractions. DNA Taq Polymerase was used for amplification reactions.

2.2. Methods

Sample from leaf segments 3-5 (15-20) grams. Leaf samples were sterilized using 70% alcohol and washed using distilled water. DNA Isolation was performed using the CTAB method. amplification DNA was performed using PCR. Reaction mixtures contained: Taq polymerase reaction mixture 6.25 μL, DDH_2O 4.25 μL, ramie DNA 1 μL and primer 1 μL. Primers used were either from [3], OPX 17, OPX 02, OPH 07, OPF 04, OPF 05 and were purchased from QIAGEN. PCR conditions were as follows: initial denaturation at 94 ^{0}C for 5 min followed by 45 cycles at 94 ^{0}C, 1.5 min at 37 ^{0}C, 1 minute at 72 ^{0}C and a final extension reaction at 72 ^{0}C for 5 minutes. PCR products were analyzed by agarose gel electrophoresis following staining with ethidium bromide

2.3. Data Analysis

Fragments were sized with reference to the size markers and scored (present / absent). The similarity coefficients calculated and dendograms produced using the *Unweighted Pair Group Method with the Arithmetic Average* (UPGMA) and the NTSYSpc software (version 2.02) [20].

3. Result and Discussion

Identification by morphological characters is still influenced by the environment. So as to support the results of the analysis, kinship analysis can be carried out based on molecular character using the RAPD technique. This analysis was carried out on 4 clones of ramie plants using 5 primers [18]. The primers used are OPX 17, OPH 07, OPF 04, OPH 05 and OPX 02. Of the 5 RAPD primers used there are differences in the produced of bands (Table 1).

Table 1. Profile of bands

Primary Code	Monomorphic	Polymorphic	Number of Band	Polymorphism(%)
OPX 17	2	9	11	81
OPH 07	0	13	13	100
OPF 04	0	10	10	100
OPH 05	0	9	9	100
OPX 02	0	11	11	100
Total	2	42	54	96.2

It can be seen that the total bands produced from PCR amplification in 4 clones of ramie plants is 54 bands. From the samples tested each one primer produces a range of 9-13 polymorphic ribbon PCR fragments, so that it can be said that the RAPD technique can select polymorphism in a fast time. From the results of testing using 5 primers showed a high value of polymorphism which is indicated by a total of polymorphic bands of 42 bands and a level of polymorphism of 96.2%. The high value of the resulting polymorphism shows the high diversity in the samples tested. In addition, the results of PCR amplification using 5 primers only produced 1 monomorphic band in the use of OPX 17 primers, whereas for the other 4 primers it did not produce monomorphic bands. To visualize DNA bands formed from the results of amplification and electrophoresis (Figure 1-5).

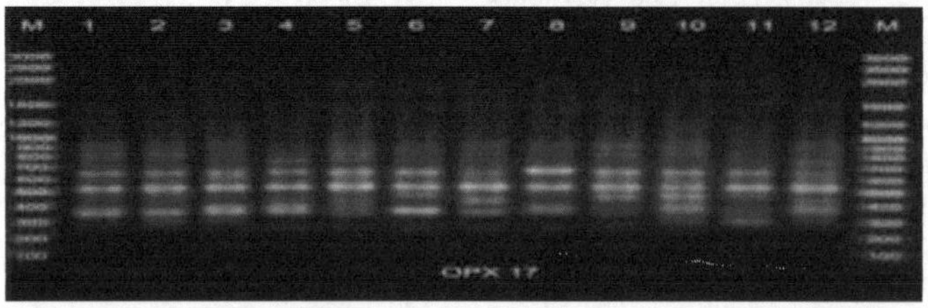

Figure 1. Visualization of Gel Electrophoresis Using OPX 17

Description: Column 1-12 is a clone, (M = DNA *Ladder*), 1,2,3 (Ramindo 1), 4,5, 6 (Matur), 7,8,9 (Situjuah), P1, P2, P3 (Padang 3)

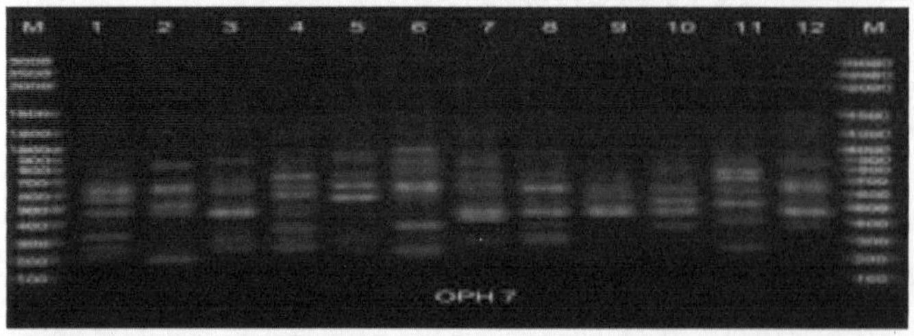

Figure 2. Visualization of Gel Electrophoresis Using OPH 07

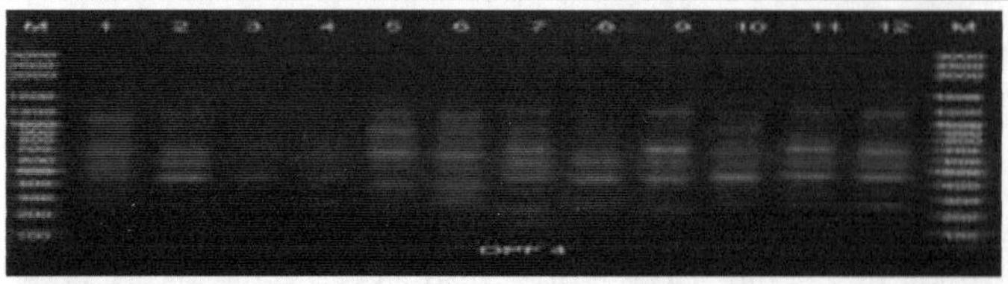

Figure 3. Visualization of Gel Electrophoresis Using OPF 04

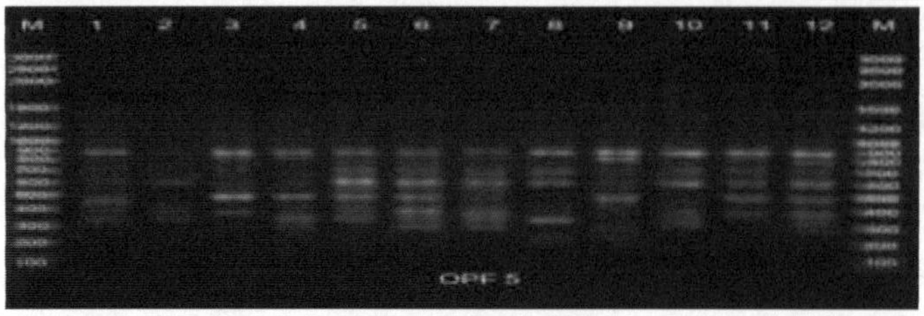

Figure 4. Visualization of Gel Electrophoresis Using OPF 05

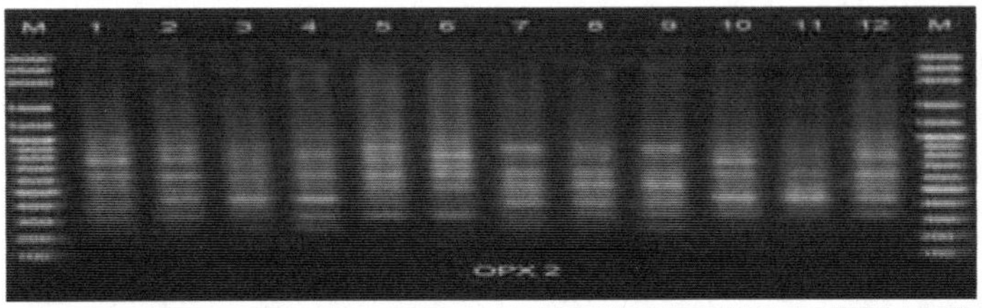

Figure 5. Visualization of gel electrophoresis using OPX 02

RAPD technique as a molecular marker in this study shows very high polymorphism. From 5 primers used, all produced polymorphisms> 50% (Table 1.). The number of bands produced ranged from 9-13 polymorphic bands, with the most bands produced by OPH 07 and the least by OPH 05. Furthermore the percentage of polymorphism ranged from 81-100%. The lowest percentage of polymorphism (81%) is produced by OPX 17 primer while the other 4 primers produce 100% polymorphism.

The ability of the 5 primers used in this study to detect polymorphisms has also been reported from [3], it was reported that the results of analysis of 5 clones using 10 RAPD primers with the 5 best primers of which were tested in the study produced a total band, 101 bands with a total of 94 polymorphic bands and the percentage 93%.

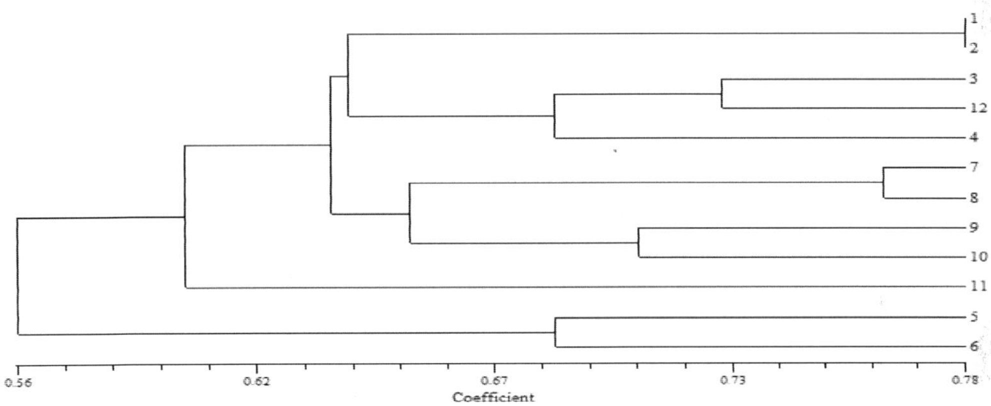

Figure 6. Dendogram of 4 Clones Rami Plants
(1, 2,3 = Ramindo, 4,5,6 = Matur, 7,8,9 = Situjuah and 10, 11, 12 = Padang 3)

Similarity analisis of ramie plants produces a dendogram with a coefficient of 0.56-0.78 or 56-78%, the overall character is molecularly united at a coefficient of 0.56, there are 2 main groups namely groups I and II. Group I consists of clones 5 and 6. Group II into 2 groups again IIA (1,2,3,12,4), IIB (7,8,9,10) and IIC (only 11). The closest similar clones are 1 and 2 (Ramindo replicates 1 and 2). Ramindo, Situjuah, Padang 3 and Matur clones (1 replication) are included in one group while Matur clones (2 replications) form their own groups.

[23] stated that the greater the value of the similarity coefficient, the greater the level of similarity between plants compared. Conversely the smaller the number of similarity coefficients, the smaller the level of similarity of these plants. That is, the greater the number of similarities, the closer the level of kinship and vice versa if the smaller the number of similarities, the further the relationship. Implications in plant breeding, with the existence of a long-standing family relationship, high heterotic, high recombinant values are very important for assembling hybrid seeds or plants. High genetic diversity is more desirable because closely related individuals will have a close genetic distance, whereas if they are distant, they will have a far genetic distance.

4. Conclusion

The results of molecular analysis using RAPD markers showed that the high level of diversity produced by the primer with a percentage of polymorphism was 96.2%. As for the similarity analysis, the coefficient value is 56-78% which indicates that between clones have close kinship.

References

[1] Azrai, M. 2005. Pemanfaatan Markah Molekuler dalam Proses Seleksi Pemuliaan Tanaman. *Jurnal Agro Biogen* **1** (1): 26-37.

[2] Banerjee, P., D. P. Ray., S. Debnath., dan P. K. Biswas. 2015. Quality Attributes of North-East Himalayan Soil for Cultivation of Ramie (*Boehmeria nivea* L. Gaudich.) *International Journal of Agriculture, Environment and* Biotechnology **8** (4): 879-884.

[3] Chan, S. R. O. S. 2018. Identifikasi Tanaman Rami (*Boehmeria nivea* (L.) Gaud.) Koleksi Kebun Percobaan Fakultas Pertanian Universitas Andalas. Tesis. Program Pascasarjana Universitas Andalas. Padang.

[4] Dahlan, D. 2011. Buku Ajar: Mata Kuliah Budidaya Tanaman Industri. Universitas Hasannudin. Makassar.

[5] Dewi, T. K., Dandy., dan W. Akbar Pengaruh Konsentrasi NaOH, Temperatur Pemasakan, dan Lama Pemasakan Pada Pembuatan Pulp dari Batang Rami dengan Proses Soda. *Jurnal Teknik Kimia* **17 (2)**: 68-74.

[6] Dwiatmini, K., N. A. Mattjik., H. Aswidinnoor., dan N.L. T. Matius. 2003. Analisis Pengelompokan dan Hubungan Kekerabatan Spesies Anggrek Berdasarkan Kunci Determinasi Fenotipik dan Marka Molekuler RAPD.*Jurnal Hortikultura* **13**(1):16-27

[7] Eltanti, F. 2015. Karakteristik Morfologi dan Molekuler 18 Genotipe Cabai Hias (*Capsicum* spp.).Skripsi. Fakultas Pertanian Institut Pertanian Bogor. Bogor.

[8] Fan, P., F. He., Y. Yang., M. Ao., J. Ouyang., Y. Liu., dan L. Yu. 2015. *In-Situ* Microbial Degumming Technology with *Bacillus* sp. HG-28 for Industrial Production Of Ramie Fibers. *Biochemical Engineering Journal* **97**: 50–58.

[9] Huang, C. J., G. Wei., Y. C. Jie., J. J. Xu., S. Y. Zhao., L. C. Wang., dan S. A. Anjum. 2015. Responses of Gas Exchange, Chlorophyll Synthesis and ROS-Scavenging Systems to Salinity Stress in Two Ramie (*Boehmeria nivea* L.) *Cultivars. Photosynthetica* **53** (3): 455-463.

[10] Jamsari. 2007. Bioteknologi Pemula: Prinsip Dasar dan Aplikasi Analisis Molekuler. UNRI Press.Pekanbaru.

[11] Kipriotis, E., X. Heping., T. Vafeiadakis., M. Kiprioti., dan E. Alexopoulou. 2015. Ramie and Kenaf as Feed Crops. *Industrial Crops and Products* **68**: 126–130.

[12] Liu, F. H., Z. Li., Q. Liu., H. He., X. Liang., dan Z. Lai. 2003. Introduction to the Wild Resources of the Genus *Boehmeria* Jacq. in China. *Genetic Resources and Crop Evolution* **50** (8) : 793-797.

[13] Liu, L. J., Z. Q. Meng., B. Wang., X. X. Wang., J. Y. Yang., dan D. X. Peng. 2009.Genetic Diversity among Wild Resources of the Genus Boehmeria Jacq. FromWest China Determined Using Inter-simple Sequence Repeat and Rapid Amplification of Polymorphic DNA Markers. *Plant Production Science* **12** (1): 88-96.

[14] Liu, T., S. Tang., S. Zhu., dan Q. Tang. 2014. QTL Mapping for Fiber Yield-Related Traits by Constructingthe First Genetic Linkage Map in Ramie (*Boehmeria nivea* L. Gaud). *Molecular Breeding* **34** (3): 883-892.

[15] Liu, T., S. Zhu., Q. Tang., dan S. Tang. 2015. Identification of a *CONSTANS* Homologous Gene with Distinct Diurnal Expression Patterns in Varied Photoperiods in Ramie (*Boehmeria nivea* L. Gaud). *Gene* **560**: 63–70.

[16] Mayerni, R. 2006. Prospek dan Peluang Tanaman Rami di Indonesia.Andalas University Press. Padang.

[17] Mayerni, R, T. Maideliza. D. Rezki. 2017. Comparative Study of Length and Growth Rate of Ramie (*Boehmeria nivea* L. Gaud.) Bast Fiber of Indonesian Clones. *International Journal on Advanced Science Engineering Information Technology*.7 (6) : 2273-2278.

[18] Mayerni, R, Gustian, SROS Chan. 2018. Morphology and Molecular Identification Of Indonesia Ramie Clones (*Bohmeria nivea (L.)* Gaud.). Accepted in *International Journal on Advanced Science Engineering Information Technology*

[19] Mitra, S., M. Kumar., M. Saha., dan B. S. Mahapatra. 2014. Effect of Irrigation and Nutrient Management on Growth, Fibre Yield and Water use of Ramie (*Boehmeria nivea*). *Indian Journal of Agricultural Sciences* **84** (5): 595–601.

[20] Rohlf, F.J. (1998) NTSYS-pc: Numerical Taxonomy and Multivariate Analysis System Version 2.0. Applied Biostatistics Inc., Exeter Software, Setauket, New York.

[21] Satya, P., S. Mitra., D. P. Ray., B. S. Mahapatra., M. Karan., S. Jana., dan A. K. Sharma. 2013. Rapid And Inexpensive Naoh Based Direct PCR For Amplification Of Nuclear And Organelle DNA From Ramie (*Boehmeria nivea*), A Bast Fibre Crop Containing Complex Polysaccharides. *Industrial Crops and Products* **50**: 532–536.

[22] Subandi, M. 2011. Budidaya Tanaman Perkebunan. Gunung Djati Press. Bandung.

[23] Syukur, M. , Sujiprihati and R. Yunianti. 2012. Teknik Pemuliaan Tanaman. Penebar Swadaya. Jakarta.

[24] Tenda, E., M. Tulalo., dan Miftahorrachman. 2009. Hubungan Kekerabatan Genetik Antar Sembilan Aksesi Kelapa Asal Provinsi Sulawesi Utara. *Jurnal Littri* **15** (3): 139-144

[25] Yang, B., M. Zhou., W. S. Shu., C. Y. Lan., Z. H. Ye., R. L. Qiu., Y. C. Jie., G. X. Cui., dan M. H. Wong. 2010. Constitutional Tolerance to Heavy Metals of a Fiber Crop, Ramie (*Boehmeria nivea*), and its Potential Usage. *Environmental Pollution* **158**: 551–558.

[26] Zhu, S., S. Tang., Q. Tang, dan T. Liu. 2014. Genome-Wide Transcriptional Changes of Ramie (*Boehmeria nivea* L. Gaud) in Response to Root-Lesion Nematode Infection. *Gene* **552**: 67–74.

IOP Conf. Series: Earth and Environmental Science **327** (2019) 012015 doi:10.1088/1755-1315/327/1/012015

Optimization of Composition, Temperature and Time on Production of Particle Board Made from Pine Bark and EFB

A Saad[1], A Kasim[2], Gunawarman[3], Santosa[2]

[1] Departement of Agricultural Engineering-Mechanical Engineering Andalas University, Padang Institute of Technology, Padang
[2] Department of Agricultural Engineering, Faculty of Agricultural Technology, Andalas University, Padang
[3] Mechanical Engineering-Andalas University, Padang 25166- Indonesia

Corresponding author's e-mail address: asfarizals@gmail.com

Abstract. The study is aimed to find out the feasibility on the Empty Fruit Bunches (EFB) and the *Merkusii Pine* Bark as a raw material for particle board production. The research carried with varied composition, times and temperature, testing according to the standards of SNI-3-2105-2006 and JIS A 5908-2003. The fiber length of the Empty Fruit Bunches is 0.1-20 mm and particle size of the pine bark passes sieve 40 mesh. The experiment result shows that density, moisture content and MOR meet SNI and JIS but MOE does not meet the standard. The optimal condition is obtained on the composition 70/30 and 75/25.
Keywords - EFB, pine bark, composition, press time, press temperature, properties

1. Introduction

Be aware of adhesives from Urea Formaldehyde (UF), Phenol Formaldehyde (PF), Melamine Formaldehyde (MF), Resorcinol Formaldehyde (RF) and Cresol Formaldehyde (CF) that are not environmentally friendly and unhealthy, on the other hand the lack friendly fiber and decreased wood products has prompted the producers to look for alternative sources of fibers containing cellulose for decades [1][2]. This technology product is known as binder less particleboard, and has developed since the mid1980s until now [3]. Particle boards without synthetic and environmentally friendly adhesives have been developed, a number of particle board studies have been carried out with different materials namely corn biomass [1], biomass pineapple [2] Palm tree trunk [3], rice straws [4], Neolamarckia wood *(Neolamarckia cadamba)* [5], date palm branches [6], coir pith [7] particles of young and old oil palm trunks[8], Jatropha curcas [9] , acacia wood [10], Palm Oil Trunk and gambir *(Uncaria gambir Roxb.)* [11], durian skin fibers [12], Mahogany wood particle [13], Chitosan [14], corn biomass [15], bamboo [16].

Interested in the development of eco-friendly organic adhesive, then the pine bark is used as an adhesive in the manufacture of organic composite particle board made from EFB. The chemical composition of the EFB is 57.17% cellulose, lignin, 5.68% hemiselulose 31.89%, holoselulose 62.87%[17]. On the other hand, pine tree merkusii flourishing on the island of Sumatra (Aceh, West Sumatra, Riau, Jambi and South Sumatra). Pine wood is treated to use in buildings (interior and unique) and its skin produces waste. Pine bark that has elements of tannin-containing compound complex of polyphenols (Cl_2HgN_2. H_2O), yield point temperature 101,6°C. EFB and pine bark is organic waste and is not maximum used.

The particle board which is made of EFB and pine merkusii bark is aimed to produce eco-friendly particle board. It should meet the standards, optimizing function the empty palm bunches and pine bark waste. This will reduce the use of formaldehyde adhesives and costs. The manufacture of particle boards is also an effort to meet the needs of the board, whose production is reduced every year. In this study, the physical properties test (density, moisture contents) and mechanical properties test (modulus of rupture, modulus of elasticity and internal bonding)

2. Material and Method

2.1. Prepare of palm oil fiber

EFB is taken from PT. AMP PLANTATION in Agam Regency, West Sumatra which is 110 km from the city of Padang. The EFB itself was still mixed with soil and sand, so it needed to be cleaned. Cleaning was done by soaking in clean water containing 1-2% NaOH for 1 hour at a temperature of 105 ° C [21]. Then it was soaked in flowing clean water for one hour. The next step, drying it until the moisture content was 10-20%. The drying step was held with the sun radiating 400-770 watts. Furthermore, EFB fiber was cut with cutting machine and filtered. The size of the filter was 16 mesh, meanwhile the length of the fiber after filtering was 0.1-2.0 cm. At last, the fiber was dried until the moisture content was 5-6%.

2.2. Prepare of fine bark particle

Merkusii pine bark was obtained from the pine plantations of Tanah Datar district, Batu Sangkar, 120 km from the city of Padang. The pine bark was separated from the wood then dried until the water content was 5-6%. Pine bark was cut to a length of 5 mm and smoothed with ball milling and then filtered the particles which passed 40 mesh sieve.

2.3. Particle Board

The dimension of the particles board was l x b x t: 30x30x1.1 cm. The composition of EFB and pine bark were 90/10; 85/15; 80/20; 75/25 and 70/30. The working temperature was 150°C; 160°C; 170°C; 180°C and 190°C. The press time was 10, 15, 20, 25 and 30 minutes. The samples were created into 125 pieces. Target density was 1.0 $g.cm^{-3}$. After mixing the adhesive evenly fibers, it was put in molds measuring 30 x 30 x 8 cm. The mold then was closed and cold compressed for 5 minutes. The plate itself was made of 5 mm thick steel plate. After mixing EFB with pine bark, cold pressing was done at 28ºC for 5 minutes followed by hot pressing, then the panel was cooled in a plywood box for 7 days, with temperature was around 28ºC. Standard particle board reference, namely JIS 5908-2003 and SNI 03-2105-2006[18]

2.4. Density

The test specimen density of size lxbxt: 10x10x1.1 cm. Density particle board was determined by dividing the weight (W) and volume (V) of a specimen as shown below:

$$\rho = W/V \ g.cm^{-3} \qquad (1)$$

2.5. Moisture Contents

All of the specimens with a dimension of 10 cm × 10 cm × 1.1 cm were measured by using a weighing balance and recorded as M_1. The samples were dried in the oven at 105ºC for 24 hour and reweighed (M_2). The moisture content(φ) was calculated using formula shown below:

$$\varphi = ((M_1 - M_2)/M_2) x 100\% \qquad (2)$$

2.6. MOE and MOR

The specimen dimension for examination MOE and MOR was 25 x 4 x 1.1 cm. The three-point bending test method was applied with a computer control system namely Universal Testing Machine (UTM) with a type of UH 300 kN. The bending test referred to ASTM D1037. The data was analysis using version 10 of the design expert programs.

$$MOE = F_{pl}S^3/4wt^3y_1 \quad (3)$$

$$MOR = 3F_{pl}S/2wt^2 \quad (4)$$

2.7. Internal Bonding (IB)

IB is calculated using equation:

$$IB = \Delta F/A \quad (5)$$

Where W: Weight of the specimens (gram), V: Volume (cm^3), M_1 and M_2: weight of the initial and final weight (gram), ΔF: pressure difference, the end and the begining(N), L: distance buffers (cm), w: width of the specimens (cm), t: thickness of the spesimens, y_1: Deflection (cm), F; Pull style (N), A: cross-sectional area (cm^2)

3. Result and discussion

3.1. Density and moisture contens

The particle board of EFB and pine bark which has finished the hot press process is then cooled for 7 days at room temperature. This is to lower the temperature on the surface and in the middle part of the particle board which must be equal to room temperature. Density test and moisture content were taken from all particle board compositions, sample sizes and methods referring to SNI 03-2105-2006. The results are as shown in Figures 1 and 2, that the particle board density in the composition of EFB and pine bark 70/30 is 0.72-0.78 g.cm^{-3} at all hot press temperatures, it meets the standards of SNI 03-2105-2006 for medium density, medium density value 0.4-0.9 g.cm^{-3} [18], the same is true for compositions 75/25; 80/20; 85/15 and 90/10.

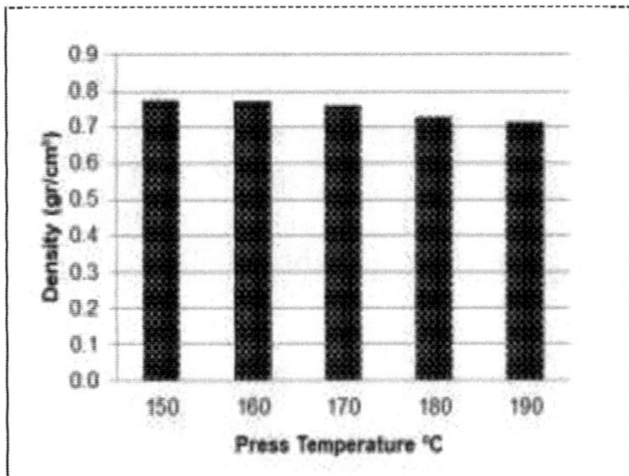

Figure 1. Density on the composition 70/30

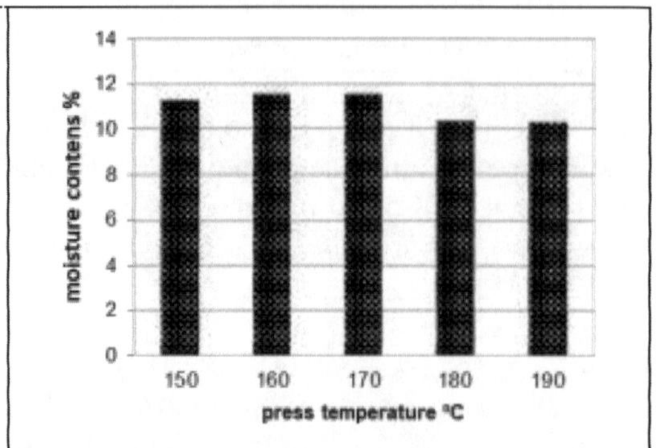

Figure 2. Moisture contents on the composition 70/30

Table 1. The particle board quality for medium density[a)]

Properties	Units	Value
Moisture contens (φ)	%	≤ 14
Density (ρ)	$g.cm^{-3}$	0,4-0,9
Modulus of Rupture (MOR)	MPa	≥ 8.04
Modulus of Elasticity (MOE)	MPa	≥ 2001
Internal Bonding (IB)	MPa	≥ 0.147

[a)] Indonesian National Standard (SNI) for Particle Boards, *BSN*, SNI 03-210-2006, p-15

The experiment of dry based moisture content on the 70/30 compositions in all temperature levels shows the value in 10.2-11.5%. This moisture contents value is smaller then the maximum value in the table 1 which is 14%. So the value meets the particle board standard. Moisture content in the composition 75/25; 80/20; 85/15 and 90/10, the value 10-12%

3.2. Modulus of Rupture (MOR)

The important mechanical properties of the particle board are Modulus of Rupture and internal bonding. Experimental data were analyzed using software "expert design version 10" The results of data analysis were displayed in three-dimensional graphs, each composition analyzed the MOR and IB values with hot press temperature variables and hot press time.

Figures 3, 4, 5, 6 and 7 each show MOR values for each particle board composition. In the composition of 70/30 the maximum value of MOR is obtained: 8.57 MPa, temperature of hot press 165-175°C and the time of hot press 20-23 minutes. the maximum value of MOR, temperature and time of each composition is shown in Table 2.

Table 2. Maximum MOR value, Temperature and time for each composition

Composition	Max.MOR (MPa)	Temperature (°C)	Time (minutes)
90/10	6.99	160-180	20-25
85/15	7.22	170-180	17-25
80/20	7.55	160-180	15-25
75/25	**8.15**	165-178	18-24
70/30	**8.57**	165-175	20-23

From Table 2, it is known that the optimum MOR that meets SNI 03-2105-2006 is 8.15 and 8.57 MPa, compositions of 75/25 and 70/30, temperature of hot press 165-175 and time of 18-23 minutes. The MOR value of the composition of 90/10, 85/15 and 80/20 has not met SNI 03-2105-2006

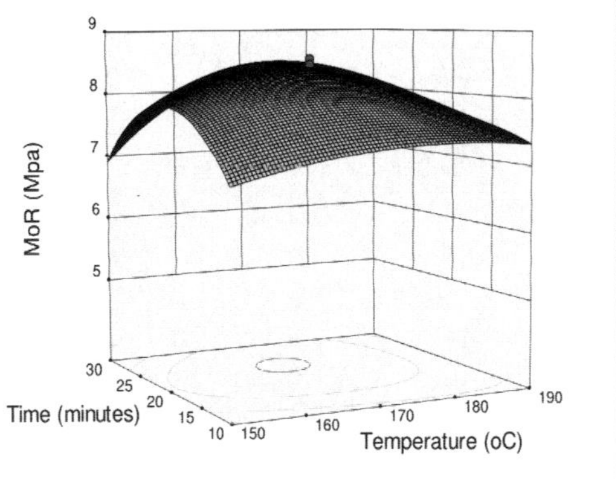

Figure 3. Modulus of Rupture on the composition 70/30

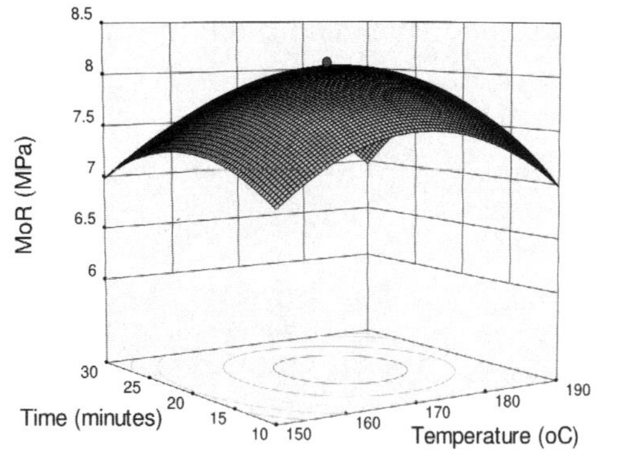

Figure 4. Modulus of Rupture on the composition 75/25

IOP Conf. Series: Earth and Environmental Science **327** (2019) 012015 doi:10.1088/1755-1315/327/1/012015

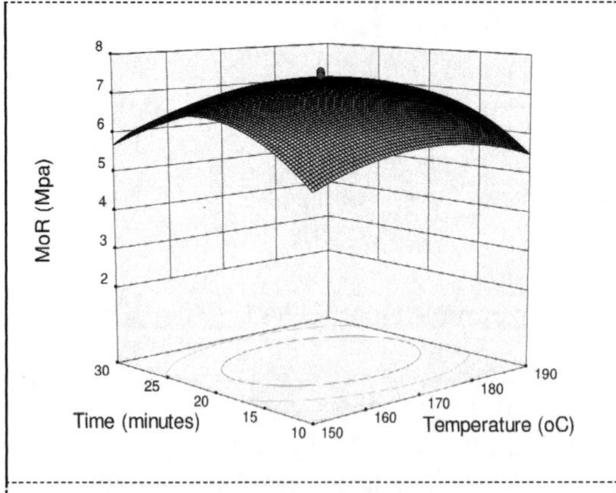

Figure 5. Modulus of Rupture on the composition 80/20

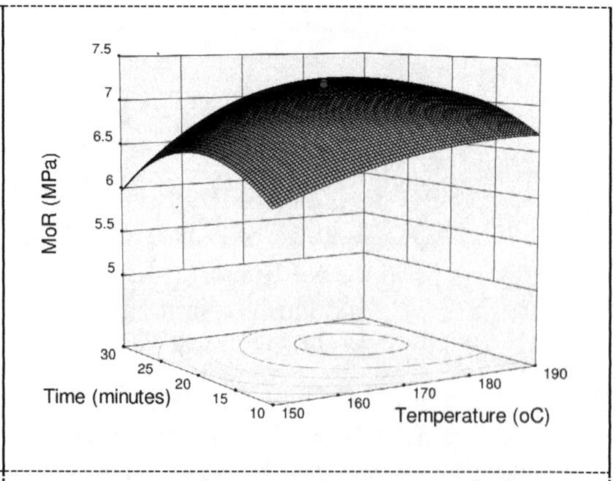

Figure 6. Modulus of Rupture on the composition 85/15

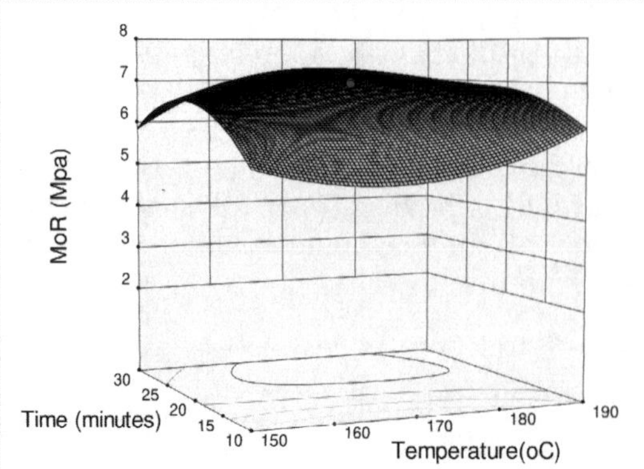

Figure 7. Modulus of Rupture on the composition 90/10

3.3 Internal Bonding (IB)

Using the same analysis method, the internal bonding obtained is shown in figure 8, 9, 10, 11 and 12. The figures illustrate the optimum value of internal bonding with two parameters i.e time and temperature, different conditions for each composition, time and temperature, shown in table 3. Internal bonding at the optimum conditions of each composition meets SNI 03-2105-2006.

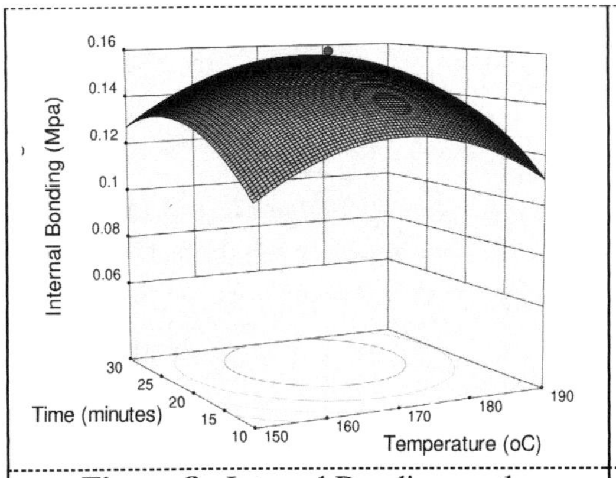

Figure 8. Internal Bonding on the composition 90/10

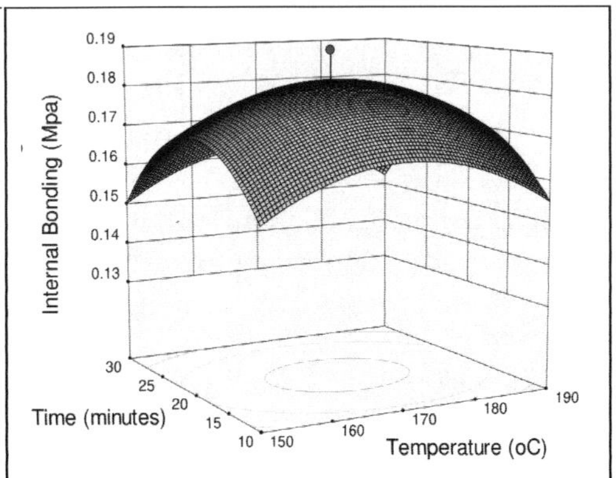

Figure 9. Internal Bonding on the composition 85/15

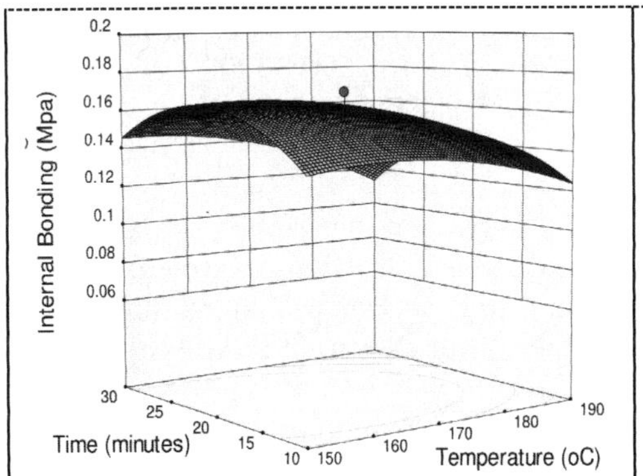

Figure 10. Internal Bonding on the composition 80/20

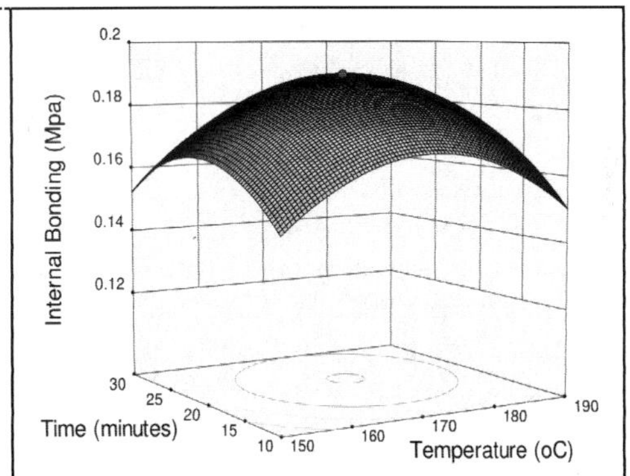

Figure 11. Internal Bonding on the composition 75/25

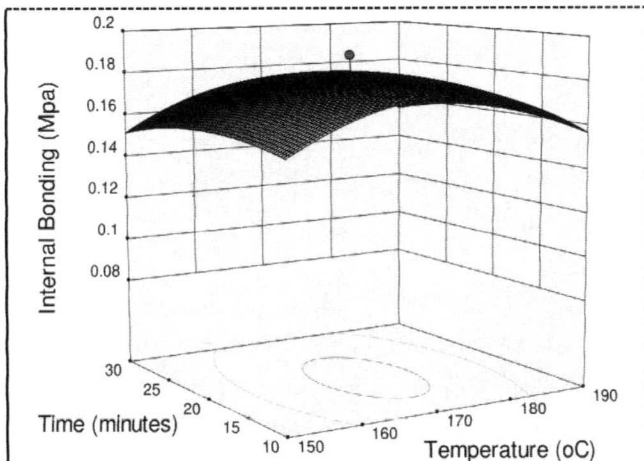

Figure 12. Internal Bonding on the composition 70/30

Table 3. Maximum IB value, temperature and time for each composition

Komposisi	Mak. IB (Mpa)	Temperature (°C)	Time (minutes)
90/10	0.156	160-180	17-25
85/15	0.17	160-180	15-25

80/20	0.16	160-175	15-25
75/25	**0.19**	170-175	19-23
70/30	**0.19**	165-177	15-24

Although the internal bond meets the standard, the highest value is obtained in the compositions of 75/25 and 70/30, the temperature of the hot press is 165-177°C and the hot press time is 15-24 minutes

4. Conclusion

From the data analysis and discussion of the experiment on particle boards made from Empty Fruits Bunch and pine bark some conclusions can be taken:

1. The pine bark waste and empty fruits bunches can be processed into particle boards without synthetic adhesives, this absolutely can reduce the use of synthetic adhesives using, reduce empty fruits bunches waste and *pine merkusii* skin and meet the needs of local boards and national.

2. The optimum value of the properties of particle board (density, moisture content, MOR and IB) without synthetic adhesives meet SNI 03-2105-2006 and JIS 5908-2003 with Empty Fruits Bunches compositions and *pine merkusii* skins is 70/30 and 75/25.

Acknowledgment

The author is a doctoral candidate in the field of natural fiber composite materials at Andalas University, Padang. Thank you for being conveyed to Prof. Anwar Kasim, PhD, Prof. Gunawarman, PhD and Prof. Dr. Santosa as promoters who had given advice, the Head of the workshop, Padang State Polytechnic received a thank you for being allowed to use the universal tensile testing machine (UTM) and all friends in the laboratory who had helped smooth this research.

References

[1] M. Baskaran *et al.*, “Properties of binderless particleboard from oil palm trunk with addition of polyhydroxyalkanoates,” *Composites Part B: Engineering*, vol. 43, no. 3, pp. 1109–1116, 2012.

[2] Y. Indrayani, D. Setyawati, T. Yoshimura, and K. Umemura, “Mechanical and Physical Properties Ofmedium Density Fiberboard Produce from Renewable Biomass of Agricultural Fiber,” vol. 3, no. 3, pp. 66–68, 2013.

[3] M. Baskaran, R. Hashim, O. Sulaiman, S. Hiziroglu, M. Sato, and T. Sugimoto, “Optimization of press temperature and time for binderless particleboard manufactured from oil palm trunk biomass at different thickness levels,” *Materials Today Communications*, vol. 3, pp. 87–95, 2015.

[4] X. Li, Z. Cai, J. E. Winandy, and A. H. Basta, “Selected properties of particleboard panels manufactured from rice straws of different geometries,” *Bioresource Technology*, vol. 101, no. 12, pp. 4662–4666, 2010.

[5] H. Lias, J. Kasim, N. Atiqah, N. Johari, I. Lyana, and M. Mokhtar, “Influence of Board Density and Particle Sizes on the Homogenous Particleboard Properties from Kelempayan(Neolamarckia cadamba),” *International Journal of Latest Research in Science and Technology ISSN*, vol. 3, no. 6, pp. 173–176, 2014.

[6] M. Ghofrani, A. Ashori, and R. Mehrabi, “Mechanical and acoustical properties of particleboards made with date palm branches and vermiculite,” *Polymer Testing*, vol. 60, pp. 153–159, 2017.

[7] E. Ahmed, A. Das, M. Hannan, and M. Shams, “Particleboard from coir pith,” *Bangladesh Journal of Scientific and Industrial Research*, vol. 51, no. 3, p. 239, 2016.

[8] J. Lamaming, R. Hashim, O. Sulaiman, T. Sugimoto, M. Sato, and S. Hiziroglu, “Measurement of some properties of binderless particleboards made from young and old oil palm trunks,” *Measurement: Journal of the International Measurement Confederation*, vol. 47, no. 1. pp.

813–819, 2014.

[9] D. Iswanto, A.H., Febrianto, F., Hadi Y.S., Ruhendi, S., & Hermawan, "Sifat fisis dan mekanis papan partikel dari kulit buah jarak (Jatropha Curcas) diperkuat partikel kayu.," *Jurnal Ilmu dan Teknologi Kayu Tropis,* vol. 10, no. 2, pp. 103–111, 2012.

[10] I. Palle, N. Hori, T. Iwata, and A. Takemura, "Optimization of polyol production via liquefaction from Acacia mangium and analysis of the polyols by traditional methods and two-dimensional correlation spectroscopy," *Holzforschung*, pp. 1–8, 2018.

[11] A. Kasim and A. Fuadi, "Influence of Temperature and Pressing Time on Particleboard Processing from Palm Oil Trunk (Elaeis guineensis Jacq.) and Gambir (Uncaria gambir Roxb.) Adhesive on Particleboard Properties," *Wood Science and Technology*, pp. 17–21, 2007.

[12] S. Charoenvai, "Durian Peels Fiber and Recycled HDPE Composites Obtained by Extrusion," *Energy Procedia*, vol. 56, pp. 539–546, 2014.

[13] R. Widyorin and F. E. Puspitasari, "The influence of treatment extraction and time press on the particle board without adhesive from mahogany sawn powder," *Biokomposit*, pp. 225–232, 2010.

[14] X. Ji, Y. Dong, R. Yu, W. Du, X. Gu, and M. Guo, "Simple production of medium density fiberboards (MDF) reinforced with chitosan," *Holzforschung*, vol. 72, no. 4, pp. 275–281, 2018.

[15] T. Wu, X. Wang, and K. Kito, "Effects of pressures on the mechanical properties of corn straw bio-board," *Engineering in Agriculture, Environment and Food*, vol. 8, no. 3, pp. 123–129, 2015.

[16] R. Liu, M. Liu, S. Hu, and A. Huang, "Comparison of six WPCs made of organo-montmorillonite-modified fibers of four trees , moso bamboo and wheat straw and poly (lactic acid) (PLA)," *Holzforschung*, pp. 1–10, 2018.

[17] F. Kasim and A. Kasim, "Hydrolysis of Oil Palm Empty Fruit Bunch Fibers to Produce Sugar Hydrolyzate as Raw Material for Bioethanol Production," *International Journal on Advanced Science, ...*, vol. 3, no. 3, pp. 24–27, 2013.

[18] Indonesian National Standard Agency, "Indonesian National Standard (SNI) for Particle Boards," *BSN*, vol. SNI 03-210, p. 15, 2006.

IOP Conf. Series: Earth and Environmental Science **327** (2019) 012016 doi:10.1088/1755-1315/327/1/012016

Prediction Carbohydrate Content of Sweet Passion Fruit Seeds (140 Days After Flowering) Using NIR Spectra with Artificial Neural Network

D P Daulay[1], Andasuryani[2], K Fahmy[2], M Viyona[3]

[1, 3] Magister Student of Biosystem and Bioprocess Engineering, Andalas University, Padang, Indonesia
[2] Department of Agricultural Engineering, Andalas University, Padang, Indonesia

Corresponding author's e-mail address: dianapebrianidaulay@gmail.com

Abstract. This study purposed to develop estimation methods using NIR spectroscopy in determining carbohydrate content of sweet passion fruit seeds at the level maturity at 140 days after flowering and developing an NIR calibration model using an ANN (Neural Network) chemo metric technique. The research method used several stages, labeling the material preparation stage, the measurement stage using NIR, the chemical analysis stage and the data processing stage with artificial neural networks. At the reflectance input the highest R-Total value which the number of neurons in the hidden layer is 7 (R-Total is 0.870) and R-SNV pretreatment as the input of the highest R-Total value is found in the number of neurons with hidden layers which is 9 (R- Total is 0.949). Based on these results the development model is suitable as a predictive value, this can be seen from the regression value approaching 1.
Keywords: Passion Fruit Seeds; Artificial Neural Networks (ANN); Carbohydrate; Near Infrared (NIR)

1. Intruduction

Passion fruit production increased in 2016-2017 from 108,145 tons to 141,195 tons [1]. That passion fruit production was 64,151.80 tons with the highest production center in Lembah Gumanti Sub-district at 63,365.80 tons and the lowest production center was in Gunung Talang Sub-district [3]. Passion fruit cultivation in Indonesia is quite large, therefore there are three types of passion fruit cultivated, known as yellow fruit passion fruit (Passiflora edulis Sims f. Flavicarpa Deg.), Acid passion fruit with purple fruit peel (Passiflora edulis f. Edulis Sims), and passion fruit konyal (sweet passion fruit) (Passiflora ligularis Juss) [3, 4]. Superior varieties of passion fruit in Sumatra Province include Markisa Gumanti and Markisa Solinda viewed by type of flower [5]. Markisa konyal variety (sweet passion fruit) has bright potential and market potential with its markets, which are in Jakarta, Bandung, Pekanbaru and Batam [6]. According to Interviews result with the farmers, known that sweet passion fruit (Passiflora ligularis Juss) has five phases of fruit development that can be seen physically based on fruit skin color but for passion fruit that worthy to harvest there are three levels of maturity one of which is maturity level 140 days after flowering (100% orange).

Passion fruit is consumed in fresh and processed forms (making juice, syrup, jelly and others). The amount of processing of passion fruit which is done continuously will cause large waste which can even damage the environment. The passion fruit processed waste reaches 60% of the weight of the fruit, which are 45% of fruit peels and 15% of seeds [7]. There is no utilization of seeds even though passion fruit seeds contain some useful chemical content, especially in the pharmaceutical field [8,9,10]. Passion fruit seeds contain several components, like carbohydrate 48. 73% [11] and 1.11 g / 100 g [12]. Carbohydrate is one of the nutrients that needed by humans to produce energy, give sweetness to food, save protein, help remove feces and regulate fat metabolism. The absence of observations and information on the sweet passionate carbohydrate content, for this reason, it is necessary to test the laboratory with chemical methods.

Observations in laboratories using chemical methods require considerable time and require compounds are expensive, cause waste and damage materials so that they are not in accordance with the quality of measuring sweet passion fruit seeds quickly and easily. One of the alternative is to measure the chemical content of passion fruit seeds non-destructively, using NIR (Near Infrared Reflectance). (Near Infrared Reflectance) has several advantages compared to observations using chemical methods that do not damage the material to detect various analysis samples to reach a depth of 2-5 millimeters and can detect various components using one spectra data [13]. NIR can predict chemical and physical ingredients without requiring much treatment of samples and does not cause waste that can pollute the environment. The method of analysis using NIR (Near Infrared Reflectance) can be done by processing various methods including PCL, PLS, Spepwise Multiple Linear Regression (SMLR) and Artificial Neural Networks (ANN) [14].

Artificial Neural Networks (ANN) have advantages for non-linear models. Artificial neural networks (ANN) can use layers by layers that greatly determine the performance of the resulting model because the more layers are used, the better the performance of the model is produced [15]. The aim was to develop an estimation method using NIR spectroscopy in determining the carbohydrate content of sweet passion fruit seeds at the maturity level at 140 days after flowering (100% orange) and developing an NIR prediction model using ANN (Neural Network) chemometric techniques.

2. Materials and methods

The study began with the preparation of sweet passion fruit seeds consisting of three processes, the harvesting process, the cleaning process (separation of seeds from mucus) and the drying process of sweet passion fruit seeds at maturity 140 days after flowering (100% orange). Sweet passion fruit from maturity level 140 days after flowering (100% orange) and sweet passion fruit seeds are shown in Figure 1.

The second stage of measurement uses 60 NIR samples of sweet passion fruit with a maturity level of 140 days after flowering (100% orange) by placing samples of sweet passion fruit seeds into the petri dish until they are full until there is no light gap at the time of measurement. Measured samples will be scanned three times in different positions. The seeds that have been placed on the petri dish will be irradiated near infrared (NIR) with a wavelength of 1000-2500 nm. Spectrum data was collected by reflectance measurements from samples using 4 cm-1 data intervals with a scale of 10000-4000 cm-1 so that data obtained from 1501 variables.

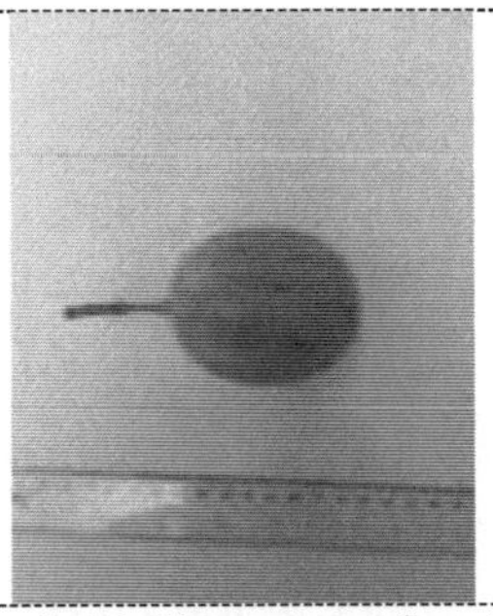

Figure 1 (a). Sweet passion fruit from maturity level 140 days after flowering (100% orange)

Figure 1 (b). Sweet passion fruit seeds

The third stage is chemical analysis to produce data on carbohydrate content of sweet passion fruit seeds, the results of the measurement data as reference data (output) to produce the NIR estimation model. Carbohydrate is one of the ingredients found in passion fruit seeds that have considerable value. the formula for finding carbohydrates is found in **equation 1**. The NIR reflectance data was analyzed using SPSS 19 Statistical software for main component analysis (PCA) to produce a PC produced as a factor of representing reflectance data. The input is 1501 spectral (R) spectra or reflectance is pretreatment. This study was built with two ANN models with pretreatment data as treatment of input data, namely reflectance spectra and standards normal variate of reflectance spectra (R-SNV).

Carbohydrate (%BK) = 100% - (Protein Content + Ash Content + Fat Content + Water Content) (1)

PC generated input data and carbohydrate data from chemical measurements in the laboratory as target data. Data processing is done using artificial neural networks (ANN). Artificial neural networks consist of several processes that are interconnected. Layer is the nerves usually organized into groups, input is connected with output through the hidden layer [16, 17]. The statistical parameters used in evaluating the model are the correlation coefficient (R) and mean square error (MSE) [18].

3. Results and Discussion

3.1 Characteristics of spectra sweet passion fruit seeds at maturity level 140 days after flowering (100% orange)

The process of transmitting NIR (Near Infrared) spectra consists of two parts, where some will be reflected (reflectance) and partially absorbed (absorbance) by the material in vibrating C-H, N-H and OH bonds. Reflectance and absorbance waves can be seen in Figure 2 and Figure 3. In the picture it can be seen that the shapes of the two waves are the same, where the shape of the wave is characteristic of NIR spectra of sweet passion fruit maturity level 140 days after mixed (100% orange), but the reflectance wave intensity and absorbance are different in each sample. Stated that the higher the reflectance value, the smaller the absorbed wave energy (absorbance) [19]. This means that the concentration of organic matter at these wave lengths is small, and vice versa. Absorbance waves that are seen several wave peaks, this states that there is a dominant chemical element. One of the ingredient that found in sweet passion fruit seeds at the maturity level of 140 fruits after flowering (100% orange) is carbohydrate. In absorbance waves there are peaks at wave lengths of 1203, 1449, 1672, 1727, 1783, 1931, 2119, 2275 and 2485. Wave lengths at the peak (absorbance) wave states that there are chemical elements that dominating include starch (1,200 nm, 1,450 nm, 2,100 nm, 2,276 nm, and 2,500 nm), cellulose (1,780 nm and 2,336 nm) and water (1,450 nm and 1,940 nm) [20]. Based on this, the value of starch in sweet passion fruit seeds was between (1203, 1449, 2119, 2275 and 2485), cellulose (1672, 1727, 1783 and 1931) and there was a water content at 1449 wave length.

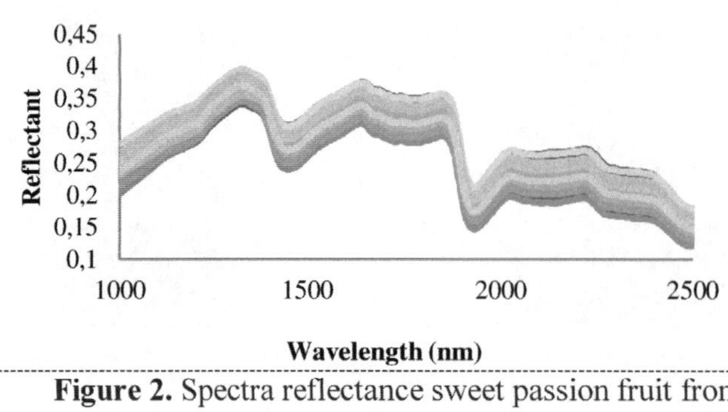

Figure 2. Spectra reflectance sweet passion fruit from maturity level 140 days after flowering (100% orange)

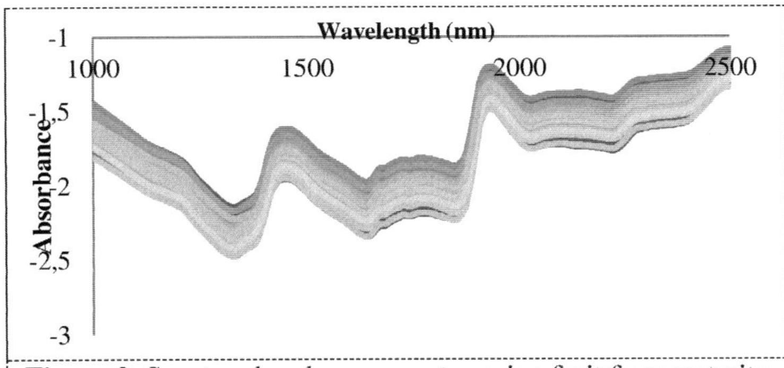

Figure 3. Spectra absorbance sweet passion fruit from maturity level 140 days after flowering (100% orange)

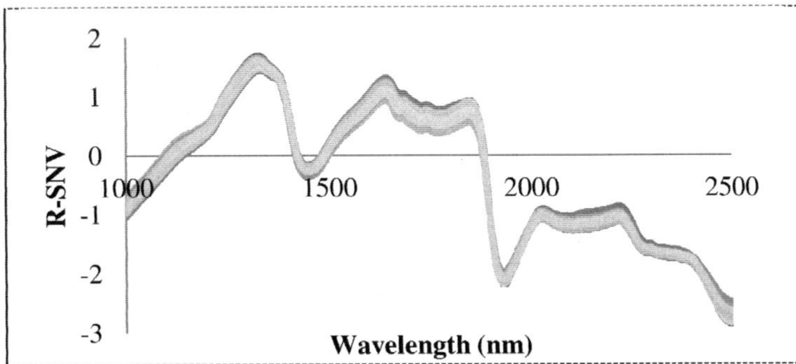

Figure 4. Spectra R-SNV sweet passion fruit from maturity level 140 days after flowering (100% orange)

In addition, reflectance data is given a normal standard variety treatment to see the difference in effect on the original data (reflectance), shown in **Figure 4**. It is seen that after the pretreatment the result of the wave is getting better. States that the SNV method is a transformation that removes scatter effects from the spectrum by concentrating and scaling the individual spectrum. Like MSC, the difference in NIR waves in each sample is one of the characteristics of differences in chemical content. Based on these differences it is thought to be a determinant of the amount of chemical elements in sweet passion fruit seeds [15, 21,22].

3.2 Set Data Set and Training Model

Development of neural network models (ANN) as a simulation model of estimating carbohydrate content in NIR spectra data on sweet passion fruit seeds at the level of orange (100% orange) 140 HSB, one of the nutrients that needed by humans. In addition the chemicals found in passion fruit seeds are particularly useful in the pharmaceutical field [8,9,10]. This study has several networks used to train and predict the carbohydrate content of sweet passion fruit seeds at the maturity level of 140 fruits after flowering (100% orange) using Matlab R2014a software to obtain artificial neural networks. Training of learning model prediction data algorithm used is Leevenberg-Marquardt. The type of network used is feed propagation. The network that was built consist 7 PCs which were produced as factors that represented reflectance data as input data and carbohydrate content produced from laboratory measurements as output data, thus obtaining a network. MATLAB Toolbox opens a data manager window for artificial neural networks that allows users to import, create, and then export artificial neural networks and generated data [18]. Based on this, the training of artificial neural networks in the network illustration window is selected as follows:

Network Input : 7 NIR Data PCs
Network Target : Carbohydrate Chemical Data (Reference Data)
Network Type : Feed-Forward Back-Propagation.
Function of Training : TRAINLM
Adjustment of Learning Functions: LEARNGDM.
Functions of Performance : MSE
Number of Hidden Layers : 1
Number of Neurons Hidden Layer: 3, 5, 7.9, 11, 13 and 15

Determination of the number of inputs and hidden layers and neurons in the hidden layer needs to be tried and assessed. Network training is using the above parameters by entering network input and output values. Network combinations using diverse input data are trained to produce maximum regression values (coefficient R). Conduct some of training with several variations of neurons in the hidden layer on each layer of the network to build up to the maximum regression value (R) using the type of feed-forward back-propagation network by applying the iterative principle calculated based on each layer. Pretreatment as a variation of input used are trained to use networks with illustrations of the same network. The activation function used is binary sigmoid (Logsig) and MSE (Mean Square Error) obtained from the reference output target with the output of neurons on the network, so the results obtained are the gradient value and the best validation value of each network, then it can be determined network for the best training.

3.3 Results, Discussion and Validation

Training on artificial neural networks is using (trainlm) Levenberg-Marquardt as a training algorithm on networks that have input and targets. Nntraintool JST window during training results of training that can be checked and allow to stop training when there is interference. Besides that there is a regression button to see linear regression between the reference target and the resulting network output reaching a high value. This study uses 2 treatments, which are using reflectance and R-SNV as input data. It was used to treat the number of hidden layer neurons (3, 5, 7.9 11, 13 and 15), aiming to see the effect of the use of pretreatment and the number of neurons in hidden against the predicted results. Table of R values, MSE best validation and gradient values on Passion Fruit Sweet maturity level 140 days after flowering (100% orange) can be seen from **Table 1**.

Table 1: R values, MSE best validation and gradient values on Passion Fruit Sweet maturity level 140 days after flowering (100% orange)

Pre-treatment	PC	Neuron (Hidden)	NILAI R				Mse Best Validation	Gradient
			Train	Validation	Test	All		
Reflectance	7	3	0.763	0.762	0.793	0.773	0.653	0.220
		5	0.811	0.832	0.724	0.795	0.515	0.122
		7	0.882	0.876	0.746	0.870	1.542	0.363
		9	0.842	0.877	0.909	0.859	1.744	0.193
		11	0.825	0.871	0.856	0.835	0.989	4.100
		13	0.820	0.863	0.883	0.831	0.803	0.319
		15	0.822	0.711	0.686	0.796	1.284	4.172
R-SNV	7	3	0.853	0.821	0.805	0.827	2.090	2.601
		5	0.840	0.856	0.882	0.849	0.531	0.278
		7	0.930	0.982	0.975	0.940	0.136	3.469
		9	0.975	0.869	0.990	0.949	2.153	0.237
		11	0.914	0.868	0.859	0.893	1.107	0.074

13	0.863	0.925	0.872	0.869	0.694	0.271
15	0.945	0.898	0.819	0.915	0.499	0.289

In the table above seen the differences that occur in each treatment both in the pretreatment of input data and the number of neurons in the hidden layer which are seen from the R correlation value, MSE best validation and gradient. Reflect input The highest R-Total value is found in the number of 7 neurons (R-Trainning value is 0.882 (y=0.79x+13), R-Validation is 0.876 (y=0.84x+9.4), R-Test is 0.746 (y=0.57x+25) and R-Total is 0.870(y=0.77x+13) with the best validation MSE value is 1.542 and the gradient is 0.363) , While the input using pretreatment (R-SNV) the highest R-Total value is found in the number of 9 neurons (R-Trainning value is 0.975 (y=0.93x+4.3), R-Validation is 0.869 (y=1.3x-20), R-Test is 0.990 (y=1.1x+4.7) and R-Total is 0.949 (y=0.96x+2.5) where the value of MSE is best validation amounting to 2,153 and a gradient of 0.237). The relationship between output and target is seen from the R value produced, if the value of R (Regression) obtained is almost close to 1, it means the relationship between output and target is appropriate [18]. The results with the R-SNV pretreatment produced the best guess. Pretreatment on NIR spectra data can minimize bias in data retrieval [19]. The use of R-SNV pretreatment is one method used to reduce deviation caused by the nature of scattering (dispersing) material [23].

The carbohydrate content uses NIR data on sweet passion fruit seeds with a maturity level of 140 days before flowering (100% orange) based on treatment carried out on reflectants as input of the best number of neurons, 7 while in the input data using the best SNV pretreatment with 9 neurons. The network the estimated R value can be increased by conducting training continuously or with different training functions to produce maximum R values such as adding hidden numbers, number of neurons in hidden or using different training functions [24]. In addition, as a comparison of the results of research with methods similar to different studies and using artificial neural networks, among others, namely in the study of Mahmoud et al. (2017) about Modeling artificial intelligence from bio sorption of cadmium (II) using rice straw (R-Total = 0.92) [24], Mahmoud et al. (2012) regarding the application of Artificial Neural Networks (ANN) for predictions of MESIR-EL-AGAMY wastewater treatment plants (R-Total = 0.9) [25], Ali et al. (2018) about the prediction of the power of a small hydroelectric power production at the Lake Himreen Dam (HLD) using artificial neural networks (R-Total = 0.96) [18] and others.

4. Conclusion

The conclusion in this study is the characteristics of sweet passion fruit seeds at the maturity level 140 days after flowering (100% Orange) has a chemical content which is carbohydrates that very useful as one of the nutrients needed by humans to produce energy, give sweetness to food , protein saver, helps remove feces and regulates fat metabolism. The reflectance input of the highest R-Total value is the number of neurons in the hidden layer is 7 (R-Trainning value is 0.842, R-validation is 0.876, R-Test is 0.746 and R-Total is 0.870 with the best validation MSE value is 1.542 and the gradient value of 0.363 an and reflectance using the standard normalize variate (R-SNV) pretreatment as the input of the highest R-value is found in the number of neurons with hidden layers is 9 (R-Trainning value of 0.975, R-Validation of 0.869, R- The test is 0.990 and R-Total is 0.949 that the best value of MSE validation is 2.153 and the gradient is 0.237. The result of development model is suitable as a predictive value, this can be seen from the regression value that approaching 1. The results obtained by pretreatment R-SNV compared with without pretreatment the best results are found in the input data using pretreatment (R-SNV).

Acknowledgments

The authors express their gratitude to to supervisors and families who have participated in making this paper

References

[1] BPS. Official Statistical data, "*Fruits and vegetables production in Indonesia (2012-2017),". Agency of Center for Statistics*, Indonesian, 2016.

[2] Solok District Agriculture Office. 2017. Annual Report of Solok District Office 2017. olok Regency Government Solok Agriculture Service.

[3] Winks, CW., Menzil, CM dan Simpson, DR, "*Passion fruit in queensland. 2. botany and cultivars*," Queensland agric. J, vol. 114, no. 4, pp. 217-25, 1988.

[4] Marpaung, AE., Karsinah dan Bina, Br Karo, "*Characterization and Evaluation of Passion Fruit Hybrid Results of Purple and Red Passion Fruit crosses*," Research Institute for Vegetable and Tropical Fruit Plants, J. Hort, vol. 26, no. 2, pp. 163-170, 2016.

[5] Buharman, B., Yanti, Mala dan Edial, Afdi, "*Passion Agribusiness Development Perspective in Solok Regency, West Sumatra,*" West Sumatra Agricultural Technology Study Center, vol. 7, no. 2, pp. 54-68, 2014.

[6] Andraini, H, "*Development of sweet passion fruit (passiflora ligularis) in the highlands in solok regency, west sumatra,*" Farming, vol.1, no. 1, pp. 22-26, 2002.

[7] Supriyatna dan Erwin, S, "*Process of Passion Fruit Skin Waste Processing as a Mixture of Ruminant Feed,* "Center for Animal Husbandry Research and Development, Indonesian, 2006.

[8] Silva, JK., Cazarin, CBB., Colomeu, TC., Batista, AC., Meletti, LMM., Paschoal, JAR., Junior, SB., Furlan, MF., Reyes, FGR., Augusto, F., Junior, MRM dan Zollner, RL," *Antioxidant activity of aqueous extract of passion fruit (passiflora edulis) leaves: in vitro and in vivo study," Food research international,* vol. 53, no. 2, pp. 90-882, 2013.

[9] Silva, RM, Placido, GR, Silva, MAP, Castro, CFS, Lima, MS & Caliari, M, "*Chemical characterization of passion fruit (Passiflora edulis* f. *lavicarpa) seeds,* " *African journal of biotechnology*, vol. 14, no. 14, pp. 3-1230, 2015.

[10] Correa, EM., Medina, L., Monteiro, JB., Valle, NO., Sales, R., Magales, A., Souza, FCA., Carvalho, TB., Lemos, JR., Lira, EF., Lima, ES., Galemo, DML., Morales, L., Ortiz, C dan Carvalho, RP, "*The intake of fiber mesocarp passion fruit (passiflora edulis) lowers levels of triglyceride and cholesterol decreasing principally insulin and lepti, " J. aging res. clin. Pract*, vol. 3, no. 1, pp. 5-31, 2014.

[11] Malacrida, CR and Jouge, N, "*Yellow passion fruit seed oil (passiflora edulis* f. *flavicarva): physical and chemical characteristics*, " vol. 55, no. 1, pp. 127-134, 2012.

[12] Chau, C.F and Huang, Y.L, "*Characterization of passion fruit seed fibre-a potential fibre source,*" Food chemistay, vol. 85, pp. 184-194, 2003.

[13] McLeod, G., Clelland, K., Tapp, H., Kemsley, EK., Wilson, RH., Poulter, G., Coombs, D., Hewitt, CJ, "*A coparison of variate pre-selection methods for use in partial least square regression: A case study on NIR spectrosopy applied to monitoring beer fermentation,* "J food engineering, vol. 90, pp. 300-3007, 2009.

[14] Büning-Pfaue, H, "*Analysis of water in food by near infrared spectroscopy,* " *J food chemistry,* vol. 82, pp. 107–115, 2003.

[15] Cen, H and He, Y, "*Theory and application of near infrared reflectance specroscopy in determination of food quality,* " *Trends food sci* technol, vol. 18, pp. 72-83, 2007.

[16] Patterson, DW, "*Artificial neural networks; theory and application*, "Singapore, Prentice Hall, 1996.

[17] Adrizal, "Estimating the Composition of Fish Flour Nutrients with Artificial Neural Networks Based on Near Infrared Absorption, " Dr. Dissertation, Bogor Agricultural Institute, Bogor, 2007.

[18] Ali, T.H; Mohd, H. B. S; and Ahmed N. A, "Prediction of small hydropower plant power production in Himreen Lake dam (HLD) using artificial neural network," Alexandria engineering journal, vol. 57, pp. 211–221, 2018.

[19] Jonni, F and Usman, A," *Detection of rice seed varieties using near infrared spectroscopy and artificial neural network,"* Food farming research, vol. 1, no. 1, pp. 29-36, 2017.

[20] Osborne, B. G., T. Fearn, and P. H. Hindle, "*Partial NIRS, with applications in food and beverage analysis*," 2nd eds. longman *scientific and technical*, United Kingdom, 1993.
[21] CAMO, *Method Reference The Unscrambler® X 10.2*, 2012.
[22] Munawar AA, "*Multivariate Analysis and Artificial Neural Network Approaches of Near Infrared Spectroscopic Data for Non-Destructive Quality Attributes Prediction of Mango,*" PhD, Dissertation, Georg-August University, Goettingen, 2014.
[23] Rinnan, A., N. Lars, V.D.B. Frans, T. Jonas, B. Rasmus, and B.E. Soren, "*Infrared spectroscopy for food quality analysis and control*," Data Pre-processing, Elsevier, pp. 29-50, 2009.
[24] Mahmoud, N; Alaa, E. D. M. Manal, F; Ahmed, R., "*Artificial intelligence modeling of cadmium(II) biosorption using rice straw*," Appl Water Sci, vol. 7, pp. 83-823, 2017.
[25] Mahmoud, S. N., Medhat, A.E., Moustafa, Hamdy, A.E., Seif, Galal, E. K., "*Application of artificial neural network (ann) for the prediction of EL-AGAMY wastewater treatment plant performance-EGYPT*," Alexandria engineering journal, vol. 51, pp. 37–43, 2012.

IOP Conf. Series: Earth and Environmental Science **327** (2019) 012017 doi:10.1088/1755-1315/327/1/012017

Utilization of Artificial Neural Network (ANN) To Predict Fat Passion Fruit Seed Content (*Passiflora ligularis*) Based On NIR-S Value

M Viyona[1], Andasuryani[2], R E Putri[2], D P Daulay[3]

[1, 3] Magister Student of Biosystem and Bioprocess Engineering, Andalas University, Padang, Indonesia

[2] Department of Agricultural Engineering, Andalas University, Padang, Indonesia

Corresponding author's e-mail address: melvieviyouna@gmail.com

Abstract. The research was conducted in July-October 2018 at the Laboratory of Food Processing Techniques for Agriculture, Center Instrumentation and AMGIS Laboratory of Agricultural Engineering, Andalas University, Padang. Using the experimental method between wavelength data and fat chemical data of passion fruit seeds. PC utilization as a grouping of data and input data on Artificial Neural Networks (ANN). ANN uses the matlab R2015a application with the TRAINLM training method, Dividerand as a grouping of data and performance based on Mean Squared Error (MSE). Random use of data for set train, set test and set validation. The data set is 70% of the total data, 15% of the other data is used for the set test model, the rest is used for testing the model through a set of validations of 15% of the total sample. The number of samples used was 180 samples. The best value from PC treatment is PC 6 in both original data and SNV. The best value of neuron treatment is on 5 PC 6 neurons in both original data and SNV. The correlation value resulting from the estimation model of passion fruit seed content is 0.85958 with a very strong level of relationship between predictive fat content and laboratory fat content. The best value for validation of MSE on PC 6 neurons 5 with SNV data is 0.95881. *Keywords:* artificial neural network (ANN); fat content; passion fruit seeds; spectrum NIR

1. Introduction

Passion fruit is a horticultural plant that lives by propagating. Sweet passion fruit producing centers in Indonesia, namely West Sumatra, especially in Solok Regency. [1] stated that sweet passion fruit is grouped into three types, namely super solinda sweet passion fruit, gumanti and local sweet passion fruit. Sweet passion fruit production in Indonesia is 77190 tons in 2017 [2] Based on the study of Solok [3] passion fruit production was 92983.7 tons, with the largest production centers in the Valley Gumanti District as many as 87449.7 tons and the smallest production in Pantai Cermin District.

Passion fruit is not only consumed fresh but is also often used as a processed product in the form of syrup, pudding or flavorings. Passion fruit passion generally causes waste. According to [4] the seed waste and passion fruit skin ranges from 65-70%. The high waste generated results in a buildup of waste, therefore it is necessary to process waste. Passion fruit skin waste treatment has been carried out to avoid the accumulation of passion fruit skin waste which is used for complete pellet feed mixtures conducted by [5] besides that passion fruit skin can be processed as alternative energy.

The chemical content of passion fruit seeds can be used as raw material for making various processed products. The chemical content of passion fruit seeds which is very possible to be processed products is one of the fat content. According to [6]) there are 30.39% fat content in passion fruit seeds, another opinion about the fat content of passion fruit seeds according to [7] where there are 23.8%. This high fat content in passion fruit seeds can be used as raw material for medicines, oil extraction and so on. [6]) state that the fat characteristics of passion fruit seeds are similar to other vegetable fats with high antioxidant activity, thus the fat of passion fruit seeds can be used as raw material for food products, detergents, cosmetics, vitamin supplements and biodiesel.

The fat content in general can be known through the chemical test phase of passion fruit seeds. Chemical tests take a long time, other chemical compounds and damage the material and cause the rest of the analysis process in the form of waste and pollution. Another way to observe the fat content of passion fruit seeds can use Near Infrarad Spectroscopy. Near Infrarad Spectrocopy Method is easy to do, does not damage materials, does not cause pollution or waste. The use of Near Infrared Spectroscopy method is supported by quantification analysis using chemometrics in the form of PLS, PCR, Multiple Linear Regression (MRL) and Artificial Neural Networks (ANN), this was stated by [4]. Artificial Neural Networks (ANN) have the advantage of analyzing non-linear data while other methods for linear data analysis. Estimating the fat content using ANN is expected to represent chemical analysis data.

Research on analyzing using ANN has been widely carried out including Estimation of Fish Flour Composition [9] Chemical Composition of Corn[10], Prediction of Quality of White Crystal Sugar [11] Estimation of Violence and Total Dissolved Solids [12] The development of ANN is not only used for estimating the content but also used as a prediction of various needs as done by [13], [14] and [15] The purpose of this study is to develop a measurement method with NIR-S to determine the fat content using Artificial Neural Networks (ANN) in passion fruit seeds.

2. Method

2.1 Time and Place

The research was conducted in July 2018- February 2019 at the Laboratory of Food Processing Techniques for Agriculture, Center Instrumentation and AMGIS Laboratory of Agricultural Engineering, Andalas University, Padang.

2.2 Materials and Tools

The material in this study is sweet passion fruit (Passiflora ligularis) in Green (120 HSB), Green Orange (130 HSB) and Orange (140 HSB). Sweet Passion Fruit (Passiflora ligularis) is harvested in the Aie Sonsang area, Nagari Aie Dingin, Lembah Gumanti District, Solok Regency, West Sumatra. Other materials in the form of Aquades, Water, Plastic Clips, Work Sheets, Lebel Paper, Filter Paper, Diethyl (petroleum ether) and Tissue solvents. The tools used in this study were bucket, sieve, handkerchief, KERN PLI 360-3 NM digital scale serial No. WL 122169, Buchi NIRflex N-500 Solids spectrophotometer, petri dish, laptop set, Vivo Y53 camera, Soxhlet, condenser, fat pumpkin.

2.3 Material Preraration Process

Passion fruit harvested based on age after flowering was observed based on fruit skin color to determine the level of fruit maturity, the fruit maturity level used was three levels of maturity namely green (120 HSB), orange green (130 HSB) and orange (140HSB). The process of separating passion fruit pulp from passion fruit seeds. The process of drying seeds to avoid fungal growth in passion fruit seeds. Passion fruit is used as seen in Figure 1.

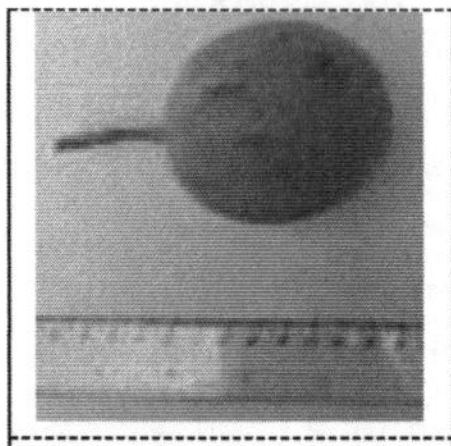

Figure 1 (a) Passion Fruit Orange (140 HSB)

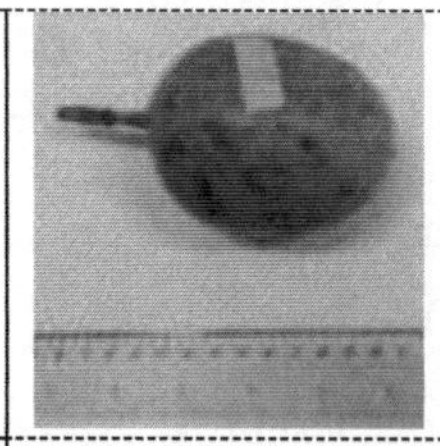

Figure 1 (b) Green Passion Fruit (130 HSB)

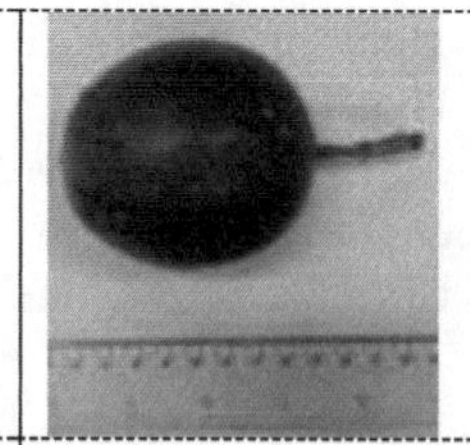

Figure 1 (c) Green Passion Fruit (120 HSB)

Note: HSB = (Day After Flowering)

2.4 Measurement of the passion fruit seeds spectrum
Collecting data spectrum by measuring sample reflectance uses intervals on 4 cm-1 data on a scale that ranges from 10000-4000 cm-1 or 1000-2500 nm which produces 1501 variables. The spectrum results obtained in each sample are averaged to build a model with calibration data and then the data is stored as reflectance (R) value. Reflectant spectra data obtained from NIR-S as much as the samples were processed original and pre-treatment of SNV treatments. The original spectra of spectra and the pre-treatments obtained were extracted in the form of PCA. PCA used in this analysis is processed using IBM SPSS 19 so that PCA data is obtained as ANN input. PCA used is PC 6 and PC 9 as PC treatment.

2.5 Chemical analysis
Development of models with Artificial Neural Network (ANN) methods must use laboratory analysis. Chemical analysis of fat is used as reference data in the form of actual fat content data. According to [16] fat content can be measured in the Laboratory using extraction methods using Soxhlet. This method is the easiest method for measuring fat content. The fat content can be obtained after the measurement so that the buffer is cured using the following equation.

$$L = {}^{BL}/_{BS} \times 100\% \qquad (1)$$

Information:
L = percentage of fat content (%)
BL = fat weight (grams)
BS = sample weight (gram)

2.6 Artificial Neural Networks (ANN)
The structure and architecture of artificial networks in general can be seen in the study of [9]. The Artificial Neural Network used in this study is the backpropogation network and trained using a network of backpropogation algorithms. Hidden layer used as many as one hidden layer with three neuron treatments. The treatment of neurons used is 3, 4, and 5 neurons in each PC treatment.

Analysis of Artificial Neural Networks uses a sample of 180 samples where 70% of the sample is set train data, 15% is a data set test and another 15% is a data set validation. Samples used 126 for train 27 tests and 27 validations. Construction of ANN model using Matlab R2015a. The matlab application selects randomly the data used for set train, test and validation using dividerand, the training used is Levenberg-Marqurादt (Trainlm), looking at the quality of the model based on Mean Squared Error (MSE). The results obtained in the form of regression values, training state and performance.

3. Resullts and Discussion

3.1 Sample Spectrum and Chemical Analysis

The sample spectrum is obtained by scanning samples using NIR-S which has a wavelength of 4000-10000 cm^{-1} or 2500-1000 nm. Data obtained from NIR-S is reflected data from passion fruit seeds in the form of reflectance data. The reflectance data produced in each sample is 1501 wave variables with wave intervals of 4 cm^{-1}. So that the wave is obtained as shown in Figure 2.

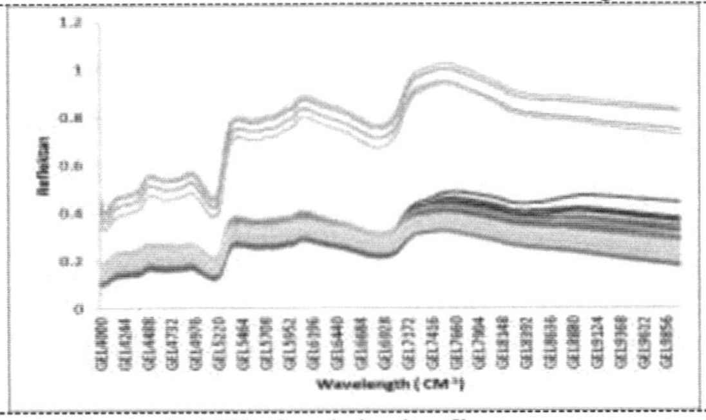

Figure 2. NIR-S original reflectance wave

The picture above explains that the wave valley has a chemical content detected. The pre treatmean treatment used aims to avoid the noise that is too high. This can be seen in Figure 3. Explaining that the SNV pre-treatment makes the wave in the same position, this reduces the impact of the nois. Figure 2 and Figure 3 show the difference between the two original wave data and after standardization (SNV) wave changes are seen. The original wave value is more diverse than the wave value in the SNV treatment. Data waves after SNV are gathered in one data stack so that data grouping is easier.

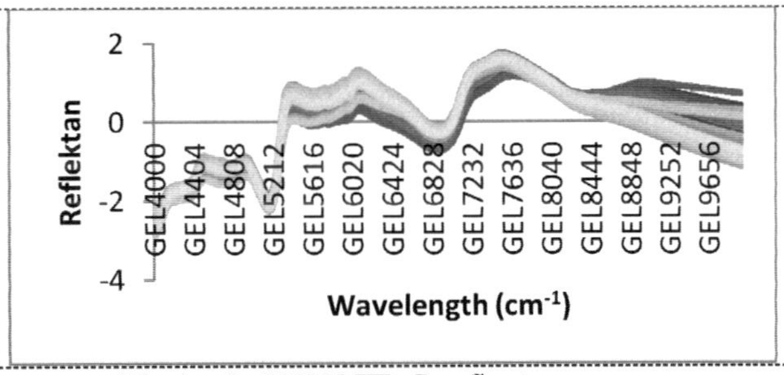

Figure 3. SNV NIR-S reflectance wave

The fat content of sweet passion fruit seeds harvested at three levels of maturity based on skin color after flower bloom ranges from 13% - 24%. The average fat content of the 180 samples used was 19.711%. The maximum value of fat content obtained from laboratory tests is 23,866%. The highest fat content is found at the orange maturity level (140 HSB). The minimum value of fat content obtained is 13,396%. The lowest fat from laboratory tests is at the green maturity level (120 HSB). The chemical content obtained is in accordance with the reference according to [7] that the passion fruit fat content is 23.8%.

3.2 Artificial Neural Networks (ANN)

Artificial neural networks are used to construct non-linear models. The artificial neural network used in this study is a back propagation network with alternating model flow. Data input used in this study is reflectance data from NIR-S waves. Reflective data is used with two treatments of original data and pre-treatment data using SNV. Both of these data are made in the form of PCA which is used as input to ANN exercises. The PC treatment in this study is PCA 6 and PCA 9. The formation of PCA uses the IBM SPSS 19 application. How to make PC A by entering the wave transpose results on the IBM SPSS 19 then selecting analyze is then copied by the factor and determining the desired PC.

The data target in ANN training is chemical data from fat. ANN training uses matlab R2015a. The matlab R2015a application uses a non-tool tool that simplifies work from ANN. The network formation on ANN uses the Feed-Forward backprop type with the training function in the form of Levenberg-Marquardt (Trainlm) and performance of Mean Squared Error (MSE). The following training parameters are used:

ShowWindow : true
Show Cammand Line : False
Show : 25
Epochs : 1000
Time : Inf
Goal : 0
Min-grad : 1e-07
Max_fail : 6
Mu : 0.001
Mu_dec : 0.1
Mu_inc : 10
Mu_max : 10000000000

Artificial neural networks have three main layers, namely the input layer, hidden layer and output layer. Hidden Lapisa in this study uses 3 treatments, namely based on the neurons in the hidden. The number of neurons used is 3, 4, and 5 neourons. So that results are obtained as found in Table 1.

Tabel 1. Data on Artificial Neural Network Processing Results

Pre Treatment	Treatment		R Trains	R Tests	R Validations	R All	Best Validation MSE	Gradient
	PC	Neuron						
Original	6	3	0.80223	0.82661	0.84968	0.8122	1.1678	0.001038
		4	0.86642	0.80427	0.83254	0.8514	1.0886	0.005803
		5	0.85988	0.8571	0.82529	0.85537	1.083	0.033646
	9	3	0.81432	0.86026	0.83828	0.8252	0.95613	0.41365
		4	0.84973	0.84077	0.85246	0.8453	1.3933	0.25489
		5	0.83997	0.85742	0.87978	0.84302	0.90951	0.046126
SNV	6	3	0.78942	0.84845	0.85953	0.80517	1.3647	0.1812
		4	0.81177	0.87034	0.91015	0.83398	0.7467	0.60467
		5	0.86488	0.84735	0.83422	0.85958	0.95881	0.078872
	9	3	0.84232	0.77045	0.75841	0.81749	1.4389	0.041599
		4	0.82341	0.86005	0.86856	0.83663	1.2797	0.33918
		5	0.86382	0.79473	0.81772	0.84897	1.5344	0.01468

The table above explains that the r All value on the original data generated on PC 6 is higher than that on PC 9 seen in Neuron 4 and 5, whereas in Neuron 3 there is an increase in the value of R on PC 9. This proves that PC 6 is better at detection. fat content compared to PC 9 for neurons 4 and 5. The highest r all the second data PC treatment with 3 neuron treatments in neouron 5 PC 6 amounted to 0.85537, with the output equation = 0.75 * target + 4.9. This r value explains the relationship between the chemical value of fat and predictive value of ANN has a very strong relationship based on [17] the correlation values as follows: 1) 0.00-0.19 with a very low level of relationship, 2) 0.20-0.39 low relationship level, 3) 0.40-0.59 medium level, 4) 0.60-0.79 strong relationship levels and 5) 0.80-1.00 has a very strong relationship. Gradient

The r value on standardized data (SNV) at Neuron 3 and 4 on PC 9 experienced an increase from PC 6, but experienced a decrease in neurons 5. The highest value of the correlation coefficient was found on PC 6 with the number of neurons 5 with a value of 0.85958 or 85.958% , with the output equation = 0.76 * Target + 4.6. The acquisition value of PC 6 has a very strong level of relationship. The highest r value in the original data and SNV is on PC 6 with the number of neurons 5 where the values are respectively 0.85537 and 0.85958. of the two values above, it is explained that the pre-treatment of SNV can improve wave performance in detecting the fat content found in passion fruit seeds. Judging from the best value of validation, the MSE obtained MSE after pre-treatment was also decreased. The MSE value of the original data with PC 6 neuron 5 is 1,083 while the SNV value is 0.95881. Research using this method has been carried out. One of them is [18] conducting research using artificial neural networks as pendektesi and electricity production planning in this study. Another study conducted by [14] uses Artificial Neural Network to design an electric power optimization scenario with the acquisition of a correlation coefficient value of 0.87607 which produces an output equation = 0.8 * target +570.

4. Conclusion

The estimation model of the passion fruit fat content using Artificial Neural Networks (ANN) using two data in the form of original data and pre-treatment data obtained different results. The best value from PC treatment is PC 6 in both original data and SNV. The best value of neuron treatment is on 5 PC 6 neurons in both original data and SNV. Increased correlation results were seen after the data used was carried out by pre SNV treatment. The correlation value resulting from the estimation model of passion fruit seed content is 0.85958 with a very strong level of relationship between predictive fat content and laboratory fat content. The best value for validation of MSE on PC 6 neurons 5 with SNV data is 0.95881.

Acknowledgements

Praise and Gratitude to Allah SWT, thanks to Dr. Andasuryani, S.TP, Msi, Dr. Renny Eka Putri, S.TP, MP and Diana Pebriani Daulay S.TP always took the time to guide in completing this journal. Thank you to both your parents and family.

Reference

[1] Karsinah, Silalahi F.H dan Mansur A 2007 Exploration and Characteristics of Passion Fruit Germplasm. J. Hort. 17 40: 297-306 .

[2] (BPS) Central Bureau of Statistics 2016 Production of Fruits and Vegetables in Indonesia in 2012-2017. https://www.bps.go.id/. (Accessed on August 5, 2018).

[3] [Diperta]. Solok District Agricultural Service 2014 Solok Agriculture Service Database. Solok Featured Commodities.

[4] Ginting S P 2004 Challenges and Opportunities for the Use of Local Feed for the Development of Goat Farming in Indonesia. Proceedings of the National Goat Cut Workshop, Center for Animal Research and Development: 62-77.

[5] Simanihuruk K 2005 Utilization of Passion Fruit Skin (Passiflora edulis Sims f. Edulis Deg) as Complete Pellet Feed Mixture for Peanut Goat. Bogor: Bogor Agricultural Institute (IPB).

[6] Malacrida C R, and Jorge N 2012 Yellow passion fruit seed oil (Passiflora edulis f. flavicarpa): physical and chemical characteristics. Brazilian Archives of Biology and Technology, 55(1), 127-134. http:// dx.doi.org/10.1590/S1516-89132012000100016.

[7] Karsinah, Hutabarat R.C dan Manshur A 2010 Exotic Fruit Rich in Benefits. Agricultural Research and Development Center, Ministry of Agriculture, Horticulture Science and Technology. Balitbu: Aripan Solok.

[8] Cen H and He Y 2007 Theory and Application of Near Infrared Reflectance Spectroscopy in Determination of Food Quality. J. Trends in Food Sci & Technol 18: 72-83.

[9] Adrizal, Purwadaria H K, and Suroso 2007 Estimating the Composition of Fish Flour Nutrients with Artificial Neural Networks Based on Near Infrarad Absorption [Dissertation]. Bogor: Bogor Agricultural Institute (IPB).

IOP Conf. Series: Earth and Environmental Science **327** (2019) 012017 doi:10.1088/1755-1315/327/1/012017

[10] Andrianyta H, Budiastra I W, and Suroso 2006 Determination of Nondestructive Corn Chemical Composition using Near Infrared Reflectance (NIR) and Artificial Neural Networks (Dissertation) Methods. Bogor: Bogor Agricultural Institute (IPB).

[11] Silvia E, Marmin, Machfud, and Nasution M Z 2007 Artificial Neural Network Design for Quality Prediction of White Crystal Sugar [Thesis]. Bogor: Bogor Agricultural Institute (IPB).

[12] Ni'matullah, Purwanto A, and Suroso 2009 Estimation of Change in Hardness and Total Dissolved Solids (TPT) of Sawo Fruit (Manilkara zapote. L) During Storage by Artificial Neural Network Approach [Thesis]. Bogor: Bogor Agricultural Institute (IPB).

[13] Zhang K, Zhang B, Chen B, Jing L, Zhu Z, and Kazemi K 2016 Modeling and optimization of Newfoundland shrimp waste hydrolysis for microbial growth using response surface methodology and artificial neural networks, Mar. Pollut. Bull. 109 (1) : 245–252.

[14] Tahir M H,Tehzeeb ul H, and Saqib M A 2016 Optimal scheduling of electrical power in energy-deficient scenarios using artificial neural network and Bootstrap aggregating, Int. J. Electr. Power Energy Syst. 83:49–57.

[15] Pelletier F, Masson C, and Tahan A 2016 Wind Turbine Power Curve Modelling Using Artifical neural Nework. Renewable Energy 89 : 207-214.

[16] Association of Official Analytical Chemists [AOAC] 1999 Official Methods of Analysis. Ed ke-16. Maryland : AOAC International.

[17] Pratomo D S 2014 Regression Analysis and Correlation Between Visitors and Buyers Against Purchase Nominal at Indomaret Kedung mundu Semarang with the Smallest Quadratic Method. Dian Nuswantoro University: Informatics Engineering.

[18] Hammid A T, Herwan M Bin S, and Abdalla A N 2017 Prediction of small hydropower plant power production in Himreen Lake dam (HLD) using artificial neural network, Alexandria Eng. J.

.

Optimisation on Frequency of Addition to the Volume of Water by Mechanical Milling Coconut Milk in Traditional Food Processing 'Galamai'

Evawati[1], I Roza[2], R A Fadri[2]

[1] Program Studies Food Technology, Departement of Agricultural Technology, Agricultural Polytechnic State of Payakumbuh, Regency of West Sumatra, Indonesia

[2] Program Studies Food Technology, Departement of Agricultural Technology, Agricultural Polytechnic State of Payakumbuh, Regency of West Sumatra, Indonesia

Corresponding author's e-mail address: evawati72@yahoo.com

Abstract. Coconut milk is the main raw material needed for processing galamai. Galamai quality is strongly influenced by the quality of the milk used. In the mechanical processing of coconut milk, coconut milk produced is influenced by the amount of additional water and frequency milling. The process of adding the coconut milk in the manufacture of galamai there are two phases to dissolve "saka" and thick coconut milk to dissolve rice flour and Glutinous rice flour. Not yet known how much additional water volume and frequency of the mechanical squeezer milling machine to produce coconut milk coconut meat that is suitable for the manufacture of galamai. The research method is design completely randomized design (CRD) with two factors based on the results of the study treatment Gives the best results of treatment A2B3 (weight ratio and weight addition of grated coconut water is 1: 1 and frequency milling 3 times).

Keywords: Coconut Milk; Galamai; Frequency Mechanical Milling; Glutinous Rice Flour

1. Introduction

Coconut milk, white-colored liquid, is an emulsion of oil in water extracted from fresh coconut meat. Maturity coconuts is a critical factor in the extraction of coconut milk. Coconuts which are not old, when extracted will produce milk in small amounts and low quality [8]. Added [1] Coconut milk is obtained by extortion coconut meat that has been shredded by the addition of water in a certain amount. According to [13] ratio of water to the grated coconut 1: 1 and 1: 2. While [6] Addition of water in the extraction of coconut done gradually (with a frequency of 2-3 times) until the milk obtained clearly no longer contain oil. The total addition of water 2 times the weight of grated coconut.

According [12] Scar and extortion coconut using this machine will be able to produce milk with the quality and quantity is better than scar and extortion with traditional tools. Added by [11] average capacity squeezer machine 29.532 Kg milk / hour.

Coconut milk is the main raw material needed to manufacture galamai, galamai quality is strongly influenced by the quality of the milk used. At extortion mechanical coconut milk, coconut milk produced in addition affected by palm varieties are also determined by the amount of additional water and frequency milling. Coconut milk is used for the processing of galamai consists of two kinds of thick coconut milk and thin coconut milk. Thick coconut milk contains a lot of fat so that the resulting galamai which has a delicious taste and form a smooth texture. Thin coconut milk is used to dilute the flour, so the dough and to dissolve the sugar.

Galamai is a typical food of the Payakumbuh city which has been widely known to the public as souvenirs and culinary tours, pertained to food semi-wet (Intermediate Moisture Food-IMF), galamai for Java known as porridge or dodol , According [1], porridge is a kind of food that have the nature of a rather wet so it can be eaten without moistened beforehand (rehydration).

Coconut milk in dodol serves to obtain a certain firmness, flavor and aroma. Coconut milk used in the manufacture of porridge is taken from an old coconut, fresh, and clean. Aging will affect the level of coconut oil produced. The use of oil is enough to produce a porridge with good quality and ease in packing. Oil shortage to the frame will cause sticky when packed, while the excess oil will also accelerate the process of rancidity [16] According [18, 13] foodstuffs that are high in fat susceptible to rancidity, especially because hydrolyzed and oxidated.

Based on the above, we need a study that aims to assess the optimization of additional water volume and frequency of coconut milk by mechanically grinding extortion especially traditional food processing such as galamai.

2. Research Method

The experimental design used in this research is to use the design of completely randomized design (CRD) with three times repetition. Two factors in this study, namely A (Comparison of grated coconut to the volume of water addition), A1 (1 ½), A2 (1: 1), A3 (1: 1½), A4 (1: 2) and B (Frequency addition of water) B1 (one time), B2 (twice), B3 (three times) with repeat three times. Combination treatment is A1B1, A1B2, A1B3, A2B1, A2B2, A2B3, A3B1, A3B2, A3B3, A4B1, A4B2, and A4B3. Results were analyzed by ANOVA and continued with DNMRT test (Duncan's New Multiple Range Test) with a significance level of 5% by the method of SPSS System). The analysis performed in this study is the analysis of fat content, water content, protein content, ash content, and carbohydrate content using AOAC method [2], as well as yield and long ripening process galamai.

The processing is done mechanically ranging from extortion coconut milk and kneading galamai. According [6] Dimensions of machine specifications extortion made coconut milk has a specification as follows:

Materials made of angle iron frame 40 mm ST 37 material with dimensions of length 100 cm and width 60 cm.

a. The electric motor is used 1.5 HP / 1450 RPM.
b. Sieve mesh size of 20 mesh.
c. 25 cm diameter screw press.
d. Worm Gear with reduction ratio 60 WPX lap of 1: 30 ROHA.
e. Inlet and exit is made with stainless steel plate 10 mm thick.
f. Material wall plate press machine using 2 mm thick stainless steel.
g. Pulley used with size 3 in diameter and 7 in.
h. Engine capacity 20 Kg / h.

While the workings of milk squeezer machine is as follows; Enter the shredded coconut into the mouthpiece of the machine (according to treatment), then grated coconut will be rotated and inserted into the screw / screw in a tube press, then threaded / screw running with the principle of the force of pressure, then the milk will separate exit through the filter, and the pulp will come out through the sewer coconut pulp.

Specifications galamai kneading machine which is the length x width x height 160x 20 x 130 cm, motorcycle engine: a 1.5 HP - 4 P - 1 PAS with a working capacity of 20 kg for a stirring (2.5 h) Further galamai making process, namely: Mixing (thick coconut milk, sugar saka, and palm sugar), heating (until sugar is dissolved), the inclusion of glutinous rice flour and rice flour (previously dissolved with thick coconut milk), cooking I (cooked until thick and glossy), cooking II (insert syrup glucose and margarine, printing and packaging). This study was conducted in Small and Medium

Enterprises (SMEs) 'Nita' located in Sicincin Payakumbuh, and in the laboratory Polytechnic State Agriculture Payakumbuh.

3. Result and Discussion

3.1 Fat Content

From Figure 1 it can be seen that the addition of water volume and frequency of milk mechanically grinding extortion especially traditional food processing such as galamai affect the fat content. The lowest fat content on A4B1 treatment which grated coconut comparison: the volume of additional water = 1: 2, with the frequency of grinding one time. Levels of fat in the treatment of A4B1 was 7.81%. While treatment has the highest fat content is in treatment A2 B3 is grated coconut comparison: the volume of additional water = 1: 1, with three times the frequency milling. According [15] The more water that is used in making coconut milk, the milk is more dilute so the water level is higher and the lower the dissolved solids such as fat content. Besides, the lower the viscosity of the viscosity also in comparison with grated coconut milk: the volume of additional water = 1: 1 was 0.055 kg / ms Added by [6] To obtain optimum extraction results in extortion milk needs to be done gradually.

Besides, in the processing of galamai of 12 treatments over turns to the treatment comparison grated coconut and additional volume of water = 1: 1/2 shortage of oil so that the processing time is rather sticky. Added by [16] on the oil shortage will cause a sticky porridge packed time, while the excess oil will also accelerate the process of rancidity.

The destruction of fat mainly caused by the rancidity due to the autoxidation of unsaturated radical fat acid in fats. Free radical of oxygen forms active peroxide that able to form hydroperoxide that is unstable, easy to break into substance with shorter carbon chain by high radiation, high heat level, metal catalyst or enzymes. Substances with shorter carbon chain are fat acids, aldehydes, volatile ketones and produces rancidity in fats [16].

The process of fat breakdown in a material can be hydrolytic and oxidative. Hydrolytic damage that is fat can increase the amount of free fatty acids. Rancidity that occurs as a result of damage to fat, caused by release of fatty acids that have evaporated, especially of fatty acids short ber C atoms, such as fatty C4, C6, C8 and C10 [5]

Unsaturated fat acid can chain the oxygen in its complete chain thus produce peroxide. The existence of peroxide in long term can cause vitamin destruction content in the oil [14]. Added by [12]; the rancidity level increase as the peroxide level, peroxide can cause the destruction of vitamin such as vitamin A, C, D, E, K and few of Vitamin B.

Overall levels of fat from galamai still meet the Indonesian National Standard, with a range of fat content was 7.81% - 10.6%. Fat content galamai / porridge according to SNI 01-2986-1991 is at least 7%. For more details can be seen in Figure 1.

3.2 Protein Content

From Figure 1 it can be seen that the addition of water volume and frequency of milk mechanically grinding squeeze on traditional food processing galamai no effect on protein content galamai. Range protein content of 12 treatments is 3.147% - 3.707%. The addition of coconut water volume in extortion and blackmail frequency milk provide protein levels decrease tendency galamai. Based on the National Standard SNI 01-2986-1992 protein content of porridge / galamai / dodol min 3%.

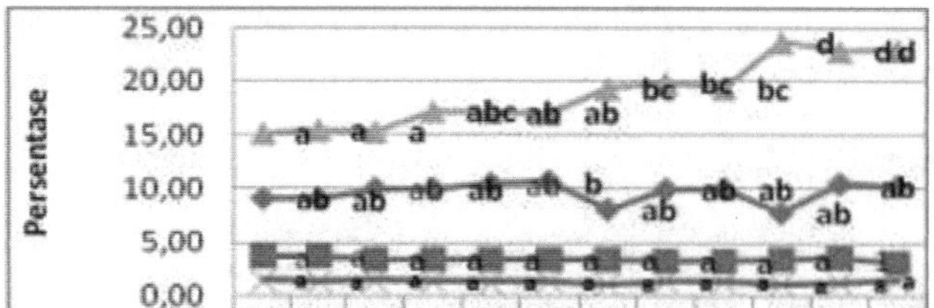

	A1B 1	A1B 2	A1B 3	A2B 1	A2B 2	A2B 3	A3B 1	A3B 2	A3B 3	A4B 1	A4B 2	A4B 3
Fat	9,13	9,04	10,0	9,96	10,4	10,6	8,14	9,96	9,96	7,81	10,4	10,2
Protein	3,65	3,71	3,42	3,42	3,45	3,41	3,44	3,29	3,23	3,49	3,51	3,15
Moisture	15,2	15,5	15,3	17,2	17,1	17,0	19,3	19,9	19,3	23,7	22,8	23,0
Ash	1,46	1,37	1,43	1,39	1,30	1,33	1,14	1,41	1,37	1,11	1,27	1,47

Figure 1. Levels of Fat, Protein Content, Moisture Content and Ash Content Galamai

3.3 Water Content

Of the various levels of volume addition of water and frequency of grinding extortion coconut mechanically on traditional food processing galamai affect the water content galamai, range water content of 12 treatments over is 15.220% - 23.733% According to [3]. The range of moisture content expressed galamai belong to the semi-wet foods (intermediate moisture food-IMF). Porridge is one type of wet semi food which has a water activity (Aw) ranged from 0.70 to 0.90. This is because the presence of water activity in the porridge contained moisture content of approximately 15-50%.

Based on the National Standard SNI 01-2986-1992 water content of porridge / galamai / dodol max 20%. Various levels of additional water volume and frequency of milk mechanically grinding squeeze on traditional food processing treatment galamai adding water volume to double the grated coconut with the addition of water to produce a wide range of frequencies galamai with water content passes ISO standard. Water can hydrolyze the oil into glycerol and free fatty acids.

This hydrolysis process is aided by the presence of acid, alkaline, moisture, heat and enzymes such as lipase lipolitic [12, 13]. Hydrolysis of oil caused a rancid taste in the oil. Triglisserida decomposition would be perfect if the available water is quite a lot. The effect of this hydrolysis process is a fatty acid that is shaped autocatalysis accelerate the reaction and the action of the enzyme on oil. For palm oil hydrolysis process have significance, because almost 90% of triglycerides having short hydrocarbon chain fatty acids are C6 to C14. The short-chain fatty acids when removed from the ester bond would cause unpleasant odors, especially the smell of caproic acid, capric acid and caprylic acid [4].

3.4 Ash Levels

Various levels of additional water volume and frequency of milk mechanically grinding squeeze on traditional food processing galamai effect on ash content, ash content ranges from 12 treatment above is 1.110% - 1.473% Based on the National Standard SNI 01-2986-1992 ash content of porridge / galamai is max 1.5%, this means galamai has a number of macro and micro mineral elements that the body needs. According to [17]. Ash content has to do with mineral materials. Mineral-shaped inorganic salts. In the combustion process to a temperature of 600°C combustible organic material, whereas inorganic substances not burn. An organic substance that does not burn is called ash, which consists of minerals such as Ca, Mg, Na, P, Fe, Mn, and Cu which determine the quality of a food ingredient.

3.5 Level of Carbohydrate

Galamai carbohydrate content of various levels of additional water volume and frequency of milk mechanically grinding squeeze on processing gives a significantly different effect, range water content of 12 treatments over is 62.317% - 70.533%. Carbohydrate analysis done by different methods, of 12 treatments over there is a tendency many additional water volume the lower the carbohydrate content of galamai. Based on the National Standard SNI 01-2986-1992 there is no absolute requirement for carbohydrate content, it's just sugar calculated as sucrose Min. 40%.

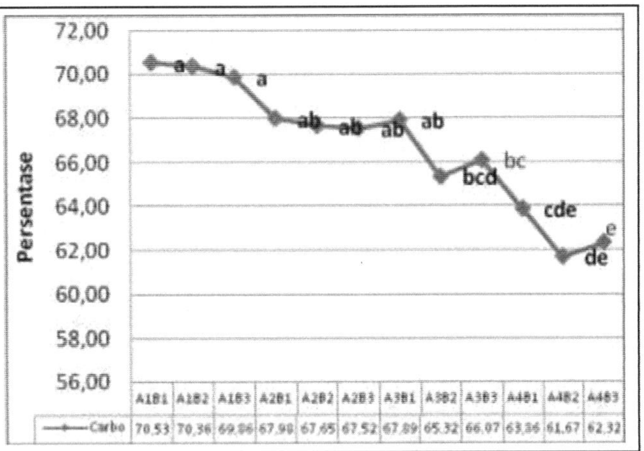

Figure 2. Levels of Carbohydrates Galamai

3.6 *Cooking Duration*

Duration of cooking of various levels of additional water volume and frequency of milk mechanically grinding squeeze on traditional food processing galamai of 12 treatments give significantly different effect. For more details can be seen in Figure 3.

From Figure 3. It can be seen that there increase the length oftime of cooking along with the increasing addition of additional volume of water. In the treatment A1 grated coconut comparison with the volume of water with the addition of various frequencies addition of water tend to be shorter time compared to the other treatment. Is the higher the water content increases long cooking time galamai, because it takes considerable time to evaporate the water content of the material in part bound either freely or bound water. According to [17] Water in food acts as a solvent of some components besides that it participated as a reagent, while the shape of water is found as free water and bound water. Free water can easily be lost in case of evaporation or drying, while the bound water is difficult freed in this way. Actually, the water can be physically bound by kapiller system and chemically bound water, such as water and water-bound crystal dispersion system.

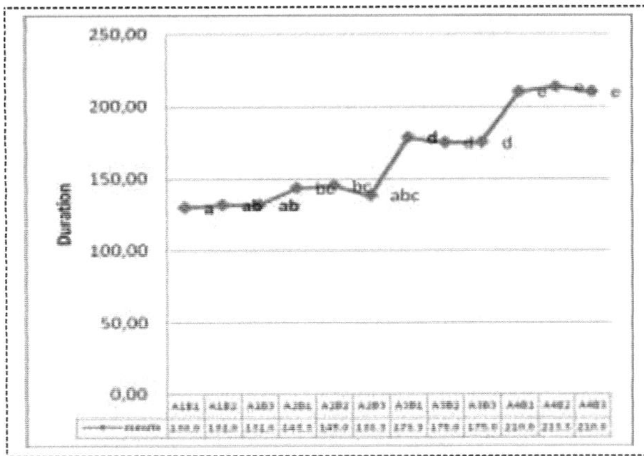

Figure 3. Cooking Duration of Galamai

3.7 Yield

Based on Figure 4 it can be seen that there is a tendency to increase the yield in line with the increasing addition of additional volume of water. In the treatment A1 grated coconut comparison with the volume of water with the addition of various frequencies addition of water tend to be shorter time compared to the other treatment. Besides, the low yield of the treatment due to the lack of a required amount of coconut milk to dissolve sticky and to finalize the dough. So some sticky dough on the container and the resulting galamai less shiny. In contrast is the higher the water content semangkin long cooking time and tends semangkin galamai high water content galamai. High water levels lead to rancidity due to hydrolysis reaction.

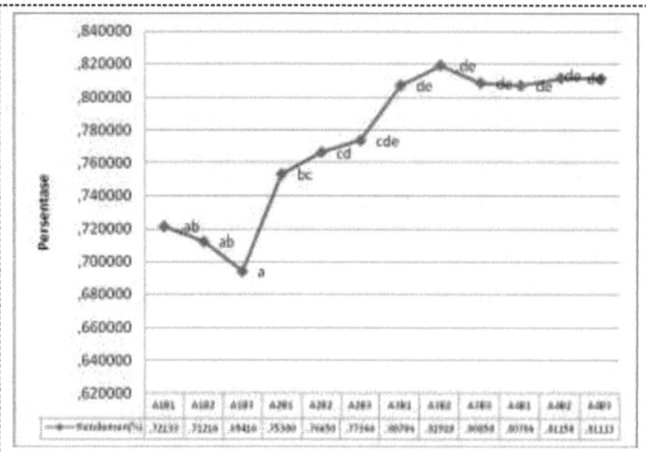

Figure 4. Rendemen of Galamai

4. Conclusion

From this study it can be concluded that the various levels of additional water volume and frequency of cocounut milk mechanically grinding squeeze on traditional food processing galamai significantly affect the fat content, water content and level of karbohydrat, but has no effect on protein content and ash content of galamai. There is a tendency of increase is the higher volume of water and the lower the frequency of extortion causing galamai the lower the levels of fat and water content is the higher. Of the 12 treatments have turned out with the addition of water up to twice the weight of grated coconut give the highest water content and pass National Standard SNI 01-2986-1992.

As for treatment with grated coconut comparison: volume addition of water (1: 1/2) and various frequency grinding the resulting yield is relatively lower than other treatments because of insufficient milk to disperse the dough so that the dough is sticky.

The treatment provides the best value is A2B3 treatment with grated coconut comparison: volume addition of water (1: 1) and the frequency of milling three meet the Indonesian National Standard SNI 01-2986-1992. Results of analysis of 10,600% fat content, protein content of 3.410%, 17.660% moisture content, ash content of 1.330%, and the carbohydrate content of 67, 5%. With long cooking 138.3 minutes and the yield of 67.5%

References

[1]. Astawan, M., dan M. W. Astawan. 1991. Teknologi Pengolahan Pangan Nabati Tepat Guna. Jakarta: CV. Akademika Presindo.
[2]. AOAC, 1984. Official Metods of Analysis of the AOAC. Inc., Arlington Virginia.
[3]. Buckle, K.A., R.A. Edwards, G.H. Fleet, and M. Wooton. 1987. Ilmu Pangan. Penerbit Universitas Indonesia. Jakarta.
[4]. Djatmiko, B., Irawadi. 1981. Pengolahan Kelapa. F T P. Institut Pertanian bogor. Bogor
[5]. Eskin, M. A. N., Henderson. M., Townsead, R. J. 1971. Biochemistry of food. Academic Press.

New York
[6]. Evawati, 2013. Teknologi Pengolahan Lemak dan Minyak. Politeknik Pertanian Negeri Payakumbuh. Payakumbuh
[7]. _______, Hasman. E, dan Jamaluddin. 2013. Peningkatan Produksi Dengan Mesin Peras Santan dan Pengaduk Mekanik Pada Usaha Makanan Tradisional di Payakumbuh. Laporan Akhir Program Hi-link tahun ke I Politeknik Pertanian Negeri Payakumbuh
[8]. Hangenmaier, R.O. 1980. *Coconut Aquaeous Processing*. University of San Carlos. Cebu City. Philipine
[9]. Husni. H. 1991. Efek Pengolahan Minyak Kelapa Secara Peragian Pada Beberapa adonan ragi terpilih terhadap rendemen dan mutu minyak kelapa. Fateta. Pusat Penelitian Universitas Andalas. Padang.
[10]. Irwan. A., I. Nazwir, I. Roza, E. Syafri. 2014. Peningkatan Produksi Dengan Mesin Peras Santan dan Alat Cetakan Mekanik Pada Usaha Galamai di Payakumbuh. HUPT. Politeknik Pertaanian Negeri Payakumbuh. Payakumbuh
[11]. Kasifalham,F., B. D. Argo, dan M. Lutfi. 2013. Uji Performansi Mesin Pemarut Kelapa dan Pemeras Santan Kelapa. Jurnal Keteknikan Pertanian Tropis dan Biosistem Vol. 1 No. 3, Oktober 2013, 204-212
[12]. Ketaren, S., 2012, Pengantar Teknologi Minyak dan lemak Pangan, UI-Press, Jakarta.
[13]. Oktaviani. N.T., 2009. Hubungan lamanya pemanasan dengan kerusakan minyak goreng curah ditinjau dari bilangan peroksida. Jurnal Biomedika. Vol. 1 No.
[14]. Patty, P.V.,2015. Pengaruh lama fermentasi terhadap ranciditas minyak kelapa yang diproduksi secara tradisional. Biopendix, Volume 1, Nomor 2, Maret 2015, hlm. 137-142
[15]. Srihari,E., F.S. Lingganingrum, R. Hervita, H. Wijaya. 2010. Pengaruh Penambahan Maltodekstrin Pada Pembuatan Santan Kelapa Bubuk. Seminar Rekayasa Kimia dan Proses. Jurusan Teknik Kimia Fakultas Teknik Universitas Diponegoro. Semarang
[16]. Widyaningrum, E. 2011. Pengendalian Mutu Dan Rencana HACCP Terhadap Produk Jenang Kudus “Rizqina”. Program Studi DIII Teknologi Hasil Pertanian Fakultas Pertanian Universitas Sebelas Maret.Surakarta
[17]. Winarno, F. G. 2004 Kimia Pangan dan Gizi. Gramedia. Jakarta.
[18]. Woodroof, J.G. 1978. Coconut; Production, Processing, Products. 2nded. The AVI Publishing Company, Inc., Westport, Connecticut.

Design of Portable Coffee Roaster for Home Industry

Radi[1], Bambang Purwantana[1], Rio Priambudi Alamsyah[2], Hafidz Dwi Prawira[2]

[1]Department of Agricultural and Biosystem Engineering, Faculty of Agricultural Technology, Universitas Gadjah Mada, Yogyakarta, 55281, INDONESIA
[2]Bachelor student at Department of Agricultural and Biosystem Engineering, Faculty of Agricultural Technology, Universitas Gadjah Mada, Yogyakarta, 55281, INDONESIA
Corresponding author:
E-mail: radi-tep@ugm.ac.id

Abstract. At present, the availability of an appropriate roaster for home coffee processing industry is very limited. This study aims to design a portable coffee roaster with a maximum capacity of 2 kg per batch. The roasting machine is designed with six main parts, namely as a roasting cylinder, an outer cylinder equipped with an inlet-outlet funnel, a main frame, a heat source with the LPG fuel, a cooling unit, and a temperature control system. The roasting machine has been tested to roast an amount of green bean at a constant temperature of 225°C with three targets of roasted degree, namely as light, medium and dark. The test result shows that the roasting machine can work properly. The temperature control system is able to regulate the roasting temperature as same as the temperature target during the process. With the stable temperature, the machine can roast the green coffee sample to reach the target roasted degree with good result. The test shows that the roaster can roast the sample of 1 kg of green bean at light roast degree with an average time of 13.2 minutes and an energy consumption of 4708.92 kJ. For the same amount of sample, the coffee roaster can work properly for the roast degree target of medium with an average time of 16.9 minutes and an energy consumption of 4708.92 kJ. Meanwhile, for dark roasted target, the roaster requires an average time of 28.59 minutes and consume an energy of 9417.85 kJ.

Keywords—design, coffee roaster, portable, home industry, performance.

1. Background

Secondary coffee processing is a series of operation to process coffee beans (green coffee bean) as the main raw material into ground coffee (ground coffee powder) and its derivative products. This processing includes a series of stages, starting from the preparation of raw materials, followed by the roasting, mixing, grinding process, and ending with the development process of coffee derivative products. Roasting is the most crucial stage on the coffee processing because the flavor formation process occurs therein. The coffee flavor is produced through a series of physical and chemical changes in the flavor candidates substance inside coffee [1] [2]. Roasting is a heat treatment process that aims to form the flavor and aroma of coffee [3] [4]. Coffee roasting is a very important process to form some organoleptic parameters (taste, aroma, and color) that underlie coffee quality and guarantee the quality of drinks [5]; [6]. The process takes place at high temperatures (> 200 °C).

In practice, coffee roasting equipment in the community is quite diverse. Traditional people generally use a pan that is burned from the bottom side. The heat will be transferred into the upper side where the coffee beans are placed. The beans will be heated directly with such conduction and confection procedure. Such roasting process is considered unhealthy, besides being wasteful of energy due to the very low absorption efficiency of heat. In its development, several roaster machines were later developed. Unfortunately, commonly roaster is originally developed to support medium and medium scale coffee processing industries. Today, some coffee roaster machines for the small scale are available even though they are generally very simple. Besides it often yields the unsatisfactory results, commercial roaster is very wasteful of energy and is considered less practical. The considerable demand for small coffee roaster machines and the limitations of the commercially available coffee roaster machines have encouraged researchers to develop these roasted machines in the last two years. This study aims to design and construct make small-scale of portable coffee roaster with high efficiency of energy for home coffee industry.

2. Material and Method

2.1. Tools and material

This research used material for roaster construction and material for testing. The construction material included plate of stainless steel, perforated plate of stainless steel, L-profile of steel for frame, shaft of stainless steel, bearing, motor and corresponding driver, controller components (thermocouple, thermo-control, display, power supply, solenoid valve, servo motor), and a heating element; while material for testing included coffee bean sample and LPG gases. Besides, it also applied some equipment for construction such as welding machines, lathes, roller machines, milling machines, drilling machines, sewing machines, grinding machines, bending machines, etc. In addition to construction equipment, several test apparatuses were also used such as multimeters, oscilloscopes, meters, scales, stop watch, etc.

2.2. Research sites

The research was conducted in the Lab. of Energy & Agricultural Machinery, Department of Agricultural and Biosystems Engineering, Faculty of Agricultural Technology, Universitas Gadjah Mada. This work was done in nine months, starting on March until November 2018.

2.3. Research procedure

This research was conducted in three stages, namely designing, manufacturing, and testing. The design phase aims to realize the design in the form of technical drawings that are ready to be realized or realized at the manufacturing stage. The manufacturing phase is a process to construct a roasting machine in accordance with the technical drawings that have been made in the previous stage. The manufacturing phase begins with determining the construction material requirements, purchasing these materials, and making roasted machine parts. After all the parts are constructed, the parts are combined with one another so that it becomes a roasting machine as a predetermined design. After the construction phase, the research continues with the testing phase. Tests are carried out in two stages, namely functional tests that aim to ensure that all engine components can work as their functions, and performance tests that aim to evaluate the work ability of the engine based on the function that has been set.

2.4. Roaster design

This research aimed primarily to design and construct a prototype of a small-scale portable coffee roaster with high energy-efficiency for the home coffee industry. Developing of the roasted machine has actually begun through a series of previous works. Our initial work was concentrated on designing a hygienic roasting machine. In this work, we focused on improving tool of pan in traditional roasting process. For hygiene purpose, the primary roaster was constructed with stainless steel material. The initial design had a capacity of 0.5 kg / batch, which was consisted of two main parts, namely as a roasting cylinder and outer cylinder (cover). This first-generation roaster machine was operated manually with a gas stove as heating source. The manual roaster had been tested. The result showed that it could work well with good performance. One of the roaster disadvantages was the unsatisfactory roasted coffee, especially when it is viewed from the color uniformity of the roasted coffee. The manual stirring process was suspected to be the cause of the roast maturity level. Therefore, at the next development work, a roasting machine with the same capacity was designed by adding an electric motor to rotate the roasting cylinder continuously. The test results showed that the roaster prototype worked better than previously design. It was able to roast the sample of green coffee with uniform roasting level. The second prototype was kept to develop such roaster for the next experiment. On the third roasting generation, we focused on improving the second design, especially by adding the capacity up to 2 kg per batch and improving the heating source using special burner for LPG fuel. Design of this roaster is presented in Figure 1.

Figure 1. Design of Portable Coffee Roaster for Home Industry

3. Results and Discussion

3.1. Prototype of portable roaster

The roaster (as shown in Fig. 1) has six main parts, namely roasting cylinder, an outer cylinder equipped with an inlet-outlet funnel, a main frame, a heat source with the LPG fuel, a cooling unit, and a temperature control system. Besides, it is also equipped with supporting parts such as dirty-storage, supporting wheel, and small window for monitoring purpose. The roasting cylinder is designed by perforated plates with the hole diameter of 3 mm. Besides to accelerates the roasting process, the perforated tube is selected to improve the quality of roasted coffee, especially that relates to the cleaning aspect. Hopefully, the dirty component of roasting, such as pericarp, silver skin usually peals during roasting will release through to the perforate. The roaster cylinder is driven by a DC motor. The rotation of cylinder functions as a stirrer which will guarantee the quality of the roasted beans. While, the cover tube (cylinder) serves to maintain heat energy inside the roaster. With the cover, the speed of heat transfer from roaster to environment can be minimized. The cover is also equipped with an input funnel that functions to hold the coffee beans (raw materials) temporarily before being put into the roasting cylinder. The cover is also supported with an output funnel which can be opened and closed as necessary. The unit of heating serves to supply the heat source for roasting. In this design, a special stove model is constructed to optimize the roaster. To guarantee that the roasting process is stable, the heating unit is equipped with a control unit. The control functions to regulate the supply of fuel injected to the heating element. The control unit consists of a thermocouple, solenoid valves and a thermo-control unit. While the cooling unit consists of a tube with a device made of perforated stainless plate. This part is also equipped with a blower unit which serves to suck heat when the cooling process takes place. The main frame part is designed from elbow-profile iron, functions to install all components of the roaster.

After the final design of the roasting prototype is completed, the next step is calculating the construction materials. Based on the design, necessary construction materials can be estimated. After determined, the next step is buying the materials. In this procurement, materials are purchased in stages based on the level of urgency. A number of prioritized materials include stainless steel plate, perforated stainless steel plate, and elbow iron. After all materials have been received, the next activity is manufacturing. The manufacturing activity begins with preparing of construction equipment that will be used to construct the roaster. The production machines used for manufacturing activities include cutting machines, grinders, lathes, milling machines, bending machines, roller machines, and welding machines. Constructing the roaster starts with part by part. Starting with six main parts construction, the activities is then continued to create other supporting parts. After all parts are manufactured, then they are assembled. The results of the portable roaster is presented in Figure 2.

Figure 2. Portable coffee roaster machine for home industry scale

3.2. *Test result*

After the roaster machine is assembled, the next stage is functional and performance testing. Functional test is carried out to ensure that the roaster can work properly. This work is carried out to test all of the roaster parts and ensures that all components can work as expected. After the roaster works properly, performance testing is then conducted. In this test, the roaster is used to roast the prepared coffee beans. There are three targets of roast level, i.e. light, medium and dark. A number of parameters are measured during the process. The parameters include roasting time, energy consumption, water content of beans (before and after roasting), roasted coffee bean colour, uniformity of roasted bean, percentage of grain cracking, and density of beans. The performance test results is presented on Table 1. Based on the result, it can be inferred that the roaster can work properly, especially when it is viewed from the parameter of uniformity of roasted bean.

Table 1. Result of performance test of the portable coffee roaster

No	Parameter	Light	Medium	Dark
1.	Roasting time (min)	13.20	16.90	28.59
2.	Energy consumption (gr of LPG)	100	100	200
3.	Moisture content (% wb)	7,15	4.07	1.09
4.	Roasted bean color L*ab			
	L	14.994	9.664	6.204
	a	6.893	4.789	3.372
	b	2.966	-4.044	-7.987
5.	Uniformity of roasted bean (%)	92.22	94.44	98.56

No	Parameter	Light	Medium	Dark
7.	Percentage of grain cracking (%)	17	39	88
8.	Density (kg/liter)	0.54	0.42	0.31

4. Conclusion
Research on design and construction have been carried out, resulted a prototype of small-scale portable coffee roaster for home industry of coffee. The prototype was operated with LPG gas, using perforated roaster cylinder, and had a capacity of 2.5 kg per batch. The roaster consisted of six main parts, i.e. namely as a roasting cylinder, an outer cylinder equipped with an inlet-outlet funnel, a main frame, a heat source with the LPG fuel, a cooling unit, and a temperature control system. This roaster could be operated easily. The material of green coffee was inserted through the input funnel and was channelled to the roasting cylinder after the roast temperature was reached. The roasting temperature could be adjusted easily, making it possible to roast the green coffee according to the desired temperature condition. The performance test showed that the portable roaster machines worked well. With a green coffee sample of 1 kg per batch, the machine was able to roast the sample till to the target of light, medium, and dark with a roasting time of 13.2; 16.9; and 28.59 minutes respectively.

5. Acknowledgements
This work was supported by many contributors; thus, we would like to thank to Faculty of Agricultural Technology, Universitas Gadjah Mada, and Indonesia Toray Science Foundation (ITSF). We also thank to CV. Faathira Alam for anything support for this design and construction.

References
[1] Sri Mulato, Edi Suharyanto, and Kaswanto, *Products Development of Coffee*. Indonesian Coffee and Cocoa Research Institute, Jl. PB. Sudirman No. 90, Jember 68118 - Indonesia, 2010.
[2] H.-S. Chung, D.-H. Kim, K.-S. Youn, J.-B. Lee, and K.-D. Moon, "Optimization of roasting conditions according to antioxidant activity and sensory quality of coffee brews," *Food Sci. Biotechnol.*, vol. 22, no. 1, pp. 23–29, Feb. 2013.
[3] O. Gonzalez-Rios *et al.*, "Impact of 'ecological' post-harvest processing on coffee aroma: II. Roasted coffee," *J. Food Compos. Anal.*, vol. 20, no. 3, pp. 297–307, May 2007.
[4] H.-D. Belitz, W. Grosch, and P. Schieberle, *Food Chemistry*. Springer Science & Business Media, 2009.
[5] J. A. Hernández, B. Heyd, C. Irles, B. Valdovinos, and G. Trystram, "Analysis of the heat and mass transfer during coffee batch roasting," *J. Food Eng.*, vol. 78, no. 4, pp. 1141–1148, Feb. 2007.
[6] J. A. Hernández, B. Heyd, and G. Trystram, "Prediction of brightness and surface area kinetics during coffee roasting," *J. Food Eng.*, vol. 89, no. 2, pp. 156–163, Nov. 2008.

IOP Conf. Series: Earth and Environmental Science **327** (2019) 012020 doi:10.1088/1755-1315/327/1/012020

Study on Physical Feature Extraction of Fermented Cocoa Bean with Digital Image Processing

Radi[1,*], Anggoro C. Sukartiko[2], M. Prasetya Kurniawan[2], Agus Pamudji R.[3], Gabriel C. Saragi[1], Bayu P. Bangun[1], and Dena P. Putri[2]

[1]Department of Agricultural and Biosystems Engineering, Faculty of Agricultural Technology, Universitas Gadjah Mada, Yogyakarta, Indonesia
[2]Department of Agro-Industrial Technology, Faculty of Agricultural Technology, Universitas Gadjah Mada, Yogyakarta, Indonesia
[3]Department of Food and Agricultural Products Technology, Faculty of Agricultural Technology, Universitas Gadjah Mada, Yogyakarta, Indonesia
Corresponding author:
E-mail: radi-tep@ugm.ac.id

Abstract. The study aimed to develop software for estimating the size of the fermented cocoa bean. The size of cocoa bean is one of the quality attributes needed in secondary cocoa processing, especially those related to the activities of sorting and preparing raw materials. This system was developed by adopting digital image processing techniques, starting from sample preparation and followed by capturing the image of cocoa bean samples, converting the color image into monochrome, segmentation, improving image results with a combination of dilation and erosion processes, to the feature extraction as the final stage. The extracted features of the beans' size are length, width, circumference, and area of the samples, all measured in pixel unit. The features, called as estimated size, were then regressed to the real samples' size, called as physical features, measured physically in the laboratory. The obtained regression formula showed a strong relationship between the estimated and physical features. The strong correlation between the two features enables the development of a rapid estimation of the physical features of cocoa beans based on digital image processing

Keywords—fermented cocoa bean, physical parameter, estimation, digital image processing

1. Background

Different from coffee or tea, which are consumed in the form of extracts, cocoa is consumed in the form of a solution (suspense). Beside stimulating alkaloids, especially theobromine, cocoa products also contain several nutrients, such as fats, carbohydrates, and proteins. Unlike coffee and tea, cocoa must be consumed in large quantities in order to experience the stimulating effects [1]. Because of its distinctive flavor, the cocoa powder and its derivatives are very popular.

Ready-to-serve cocoa products are obtained from the processing of dried cocoa beans through a series of stages, including beans sorting, roasting, milling, and packaging. The sorting stage is the initial stage in the processing of secondary cocoa, which aims to select good beans to produce high quality of cocoa products. At the farmer level, this sorting stage is generally conducted manually, as depicted in Figure 1, illustrating the workers were sorting the fermented beans into several classes according to the specified quality criteria. Generally, the fermented cocoa beans are divided into four categories, large, medium, small, and bad classes. The sorting is based on the visual appearance of the beans, which is undoubtedly difficult to be quantified, the sorted can vary due to elements of subjectivity, and can take long processing time.

Figure 1. Manual Sorting of Fermented Cocoa Beans

In contrast to the sorting system at the farm level, the cocoa processing industry generally uses standard sorting. In this case, a number of producers implemented a quality classification system normally used in its trading. In Indonesia, the classification and quality standards of cocoa beans are regulated by Indonesian National Standard SNI 2323-2008, which includes definitions, classifications, quality requirements, methods of sampling, and testing methods. Based on these standards, cocoa grains are grouped based on two approaches. The first approach uses bean size, where the beans are grouped into five classes, i.e., AA, A, B, C, S, ranged from showing the biggest/ superior to the smallest classes [2]. The second approach uses the basis of seed defect rates, in which cocoa beans are classified into three classes. The standard has set quality attributes as a basis for sorting standard to reduce the risk of non-uniformity in the previous process. Although farmers and industries use different standards, both apply physical parameters for classification.

Because the manual sorting process has several weaknesses, this study aims to estimate the physical parameters of cocoa beans that are useful in developing a method of sorting cocoa beans quickly. The digital image processing method approach was used to develop the sorting system. This study focused on the initial study to see the potential for digital image analysis in estimating the physical quality parameters of cocoa beans, especially those related to size quality parameters.

Various efforts to quickly detect the quality of cocoa beans have been carried out. Some quality parameters used as a basis include physical parameters of seeds and fermentation index. One study to classify the quality of cocoa beans has been done by [3]. The study used the basis of image processing to identify the quality of cocoa beans, consisting of two parts, namely the image processing module and classification module. The image processing module extracts the image of cocoa beans to get the shape and color characteristics such as area, circumference, length, segment width, roundness, R, G, B, color values, hue, saturation, and intensity, while the classification module includes two artificial neural network structures to classify cocoa beans into damage categories and fermentation status. The test results showed that the classification level for damage criteria reached accuracy of 79.25%, which consisted of 84.38% accuracy for intact seed, 51.72% accuracy for broken seeds, and 98.29% for seed fraction, and 20% for damaged seeds. While the classification of fermentation status has an accuracy of 88.54%, which consists of 95.62% for fermented seeds, and 71.43% for unfermented cocoa beans.

Meanwhile, [4] investigated a quantitative procedure based on simple image analysis. The analysis applied color measurements and artificial neural networks (ANN) to predict the fermentation index of cocoa beans. The ANN model based on color measurement was tested to predict fermentation index (FI) of fermented cocoa beans. The RGB values were measured from the surface and middle area of fermented seeds on images obtained with a camera and desktop scanner. FI is defined as the ratio of total free amino acids in fermented and non-fermented samples. The ANN model process inputs of the RGB value of cocoa surface and R/G ratio of wet cocoa beans to predict the FI. The model was able to predict the FI without statistical differences compared to the experimental value. The performance of the ANN model was evaluated by the coefficient of determination, Bland-Altman analysis and Passing-Bablok regression analysis.

Another study was carried out by [5], which examined nondestructive detection using FT-NIR spectroscopy. In this study, cocoa beans were prepared with fermentation variations for 0, 7, 14, and 21 days. The cocoa seeds are fermented by the heap method. The fermentation index of all treatments ranged from 0.535 and 1.242, the pH value between 4.26 to 6.13, and the total polyphenol levels varied between 6.48 and 15.58 mg/g. Samples were scanned with the NIR apparatus for spectral analysis. The results obtained from the conventionally analysis of cocoa quality was correlated with the wave spectrum for each treatment. The partial squared regression technique was then used to develop a calibration model. The scanned spectrum was then processed by vector normalization, multiplication correction scatter, and the first derivative technique. The calibration model turned out to be able to predict the total fermentation index and total polyphenols (R2> 0:80), while for pH obtained the value R2 <0:80. With this good result, this method can be used as an alternative of FI determination of fermented cocoa grains with advantages of detection time (<1 minute), more quickly than conventional methods (28 hours).

2. Material and Method

2.1. Sample Preparation and Image Capture

The sample was taken from the local farmers association at Gunungkidul, Yogyakarta. The measurements were carried out at the Energy and Agricultural Machinery Laboratory, Universitas Gadjah Mada, from April to November 2018. The sample used for this study was fermented dry cocoa beans. The beans consist of three categories, according to the sorting system at the farmer level, i.e., large, medium and small classes. The image was captured in a closed container, equipped with constant lighting, and a camera mounted on a sample board with a distance of 10 cm. The cocoa bean scanning box was made similar to [6]. To simplify the image processing stage, cocoa beans were placed on a sample board with a blue background. The image scanner wall was coated with a dark (black) cloth to minimize the amount of reflected light to the camera, as described in [7]. Other equipments used were vernier calipers and digital scales.

2.2. Research Procedures

In this study, several samples were taken randomly (each category as many as 60 beans) from the farmers. The cocoa beans were then taken to the laboratory for further analysis. Each cocoa bean was then taken into its image. Every bean was scanned once to produce an image with a bitmap format. The collection of images was then analyzed using a program that has been designed. Other equipment used was vernier caliper and digital scales which were used to measure the physical quality parameters of cocoa beans, i.e., length, width, circumference, area, and weight. Detail description is shown below, and the illustration is depicted in Figure 2.

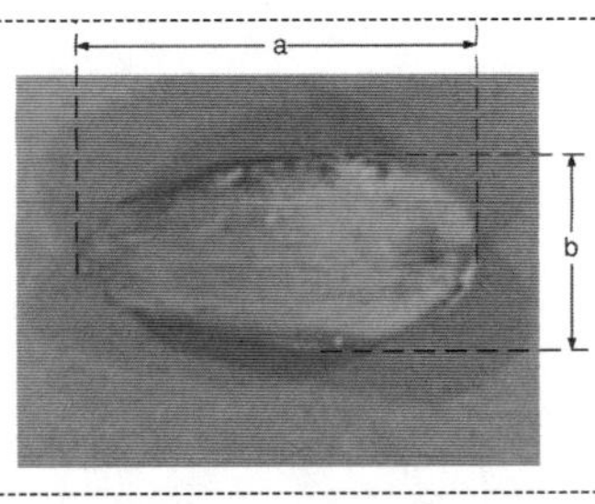

Figure 2. Physical Parameter of Cocoa Bean

$$Grain\ length = a \qquad [1]$$
$$Grain\ width = b \qquad [2]$$
$$Grain\ circumference = \pi . a . b \qquad [3]$$
$$Grain\ area = \frac{1}{4}\pi . a . b \qquad [4]$$

Image analysis of cocoa beans aims to determine these physical parameters automatically. To get the features in question, the image of cocoa beans was analyzed by a series of stages, including the conversion of color images into gray images (monochroma), image segmentation, image improvement, feature extraction. The conversion of color images into gray images was intended to simplify the segmentation process. In this case, the gray image is formed from the blue channel (B). While the segmentation process is carried out by the method of thresholding. The results of thresholding generally still contain a number of unwanted noise. For this reason, the closing process with a combination of dilation-erosion processes is applied to improve the segmented image. Image features are then extracted from the repaired image. The stages of image processing are presented in Figure 3. The image features are then analyzed for their relationship to the measurement parameters.

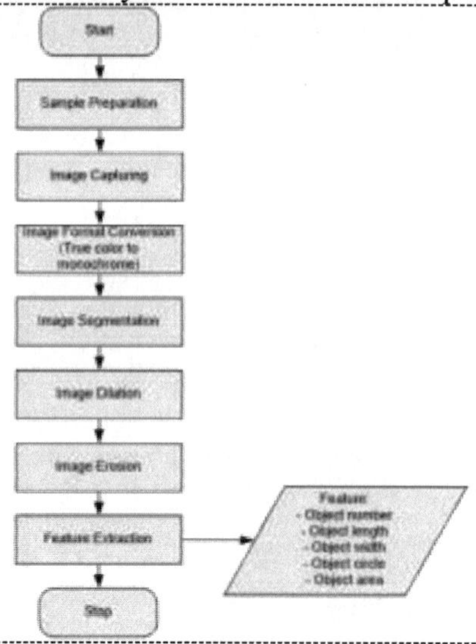

Figure 3. Stages of Digital Image Processing and Extraction of Physical Features of Cocoa Beans

3. Results and Discussion

3.1. The physical feature extraction program of cocoa beans

The program for estimating the physical parameters of cocoa beans is presented in Figure 4. The program was developed to analyse the image of cocoa beans offline. The program interface consists of three part, namely the main panel which contains buttons for uploading images, setting parameter for analysis and saving the result; the second part is image panel which includes three spaces (original image, monochrome image, and segmented image); and a display panel to show extracted features.

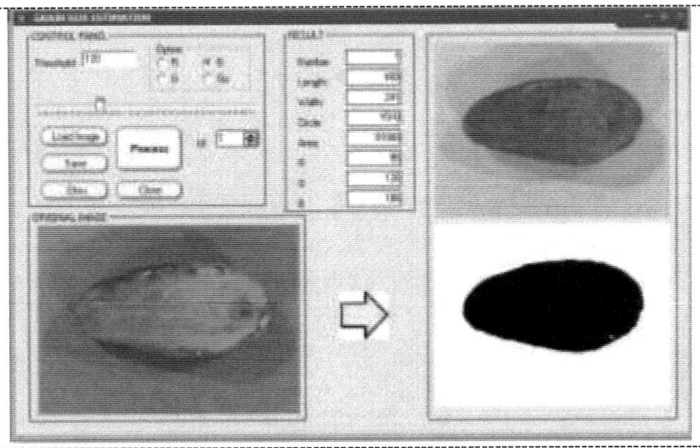

Figure 4. The Interface of The Program for Extracting Physical Features of Cocoa Beans

3.2. Features of length and width of cocoa beans

The main physical parameters of cocoa beans are length and width. Based on the algorithm, those parameters measured in pixels were then compared to the real measurement feature. Both data of those features are plotted in a two-dimensional graph. The plotting result of the length parameter is shown in Figure 5, while data of cocoa width are scattered in Figure 6. The result shows that linear regression analysis obtains relation of length feature with equation of $y = 0.0047x + 0.5178$, while relation of length feature with equation of $y = 0.0033x + 0.56$. The result also shows that the regression coefficient of length feature ($R^2 = 0.5992$) is higher than the width feature ($R^2 = 0.2669$). It means that the proposed methods have the ability to extract length feature better than width feature. From the image data, noise of object shadow appears on the side of cocoa bean. Its may be caused due to the position of lighting of the chamber. From the Figure 5, we can see that a number of points deviates from other data. It may cause the low value of the regression coefficient.

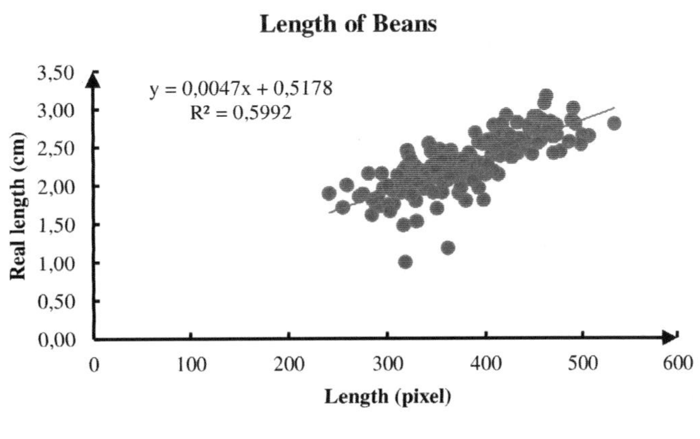

Figure 5. Plotting Data of The Length Parameter of Cocoa Beans

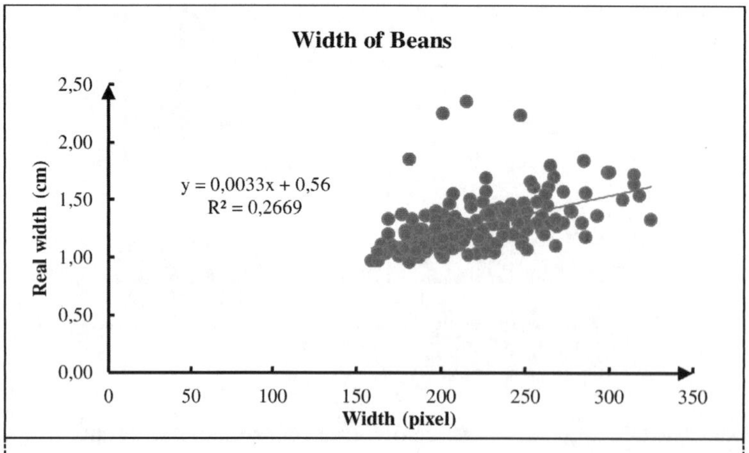

Figure 6. Plotting Data of The Width Parameter of Cocoa Beans

3.3. Features of circumference and area of cocoa beans

Circumference and area of cocoa bean are other useful physical features. Those features are directly related to the size of cocoa beans, which is one parameter for the classification. In this case, circumference and area features are calculated by assuming that the grain is an ellipse (Figure 2). The relationship between the circumference data from measurement and the circumference data obtained from segmentation process is given in Figure 7. While relationship between area data from real measurement and from object segmentation is provided in Figure 8. Based on these results, it can be inferred that the circumference parameters provide a weak relationship indicated from the small regression coefficient. The low regression coefficient may be caused by different measurement method. In the y-axis component (Figure 7), data was obtained through circumference calculations based on eq. 3, while measurement of the feature on image processing approaches the number of outermost points of segmented object. For this reason, it is necessary to investigated more the both methods. Compare to the circumference feature, area parameter provides better distribution scatter. It has strong relation indicated from the value of the regression coefficient (R^2 = 0.6148). It means that the number of pixels counted on the object strongly correlates with the area of cocoa beans.

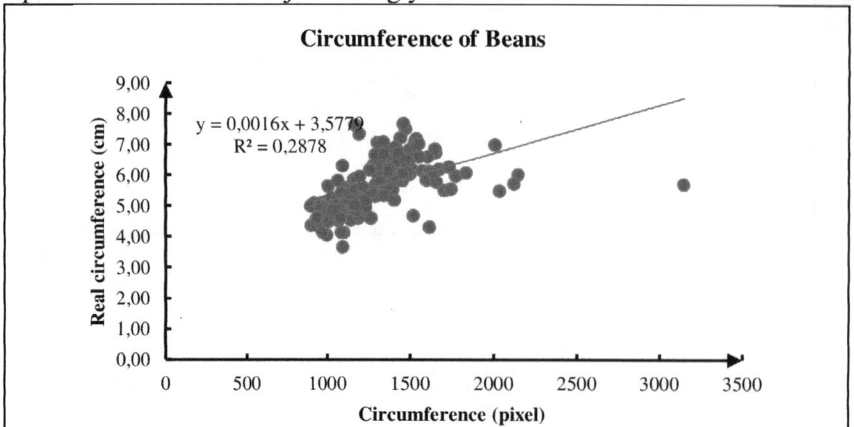

Figure 7. Plotting Data of The Circumference Parameter of Cocoa Beans

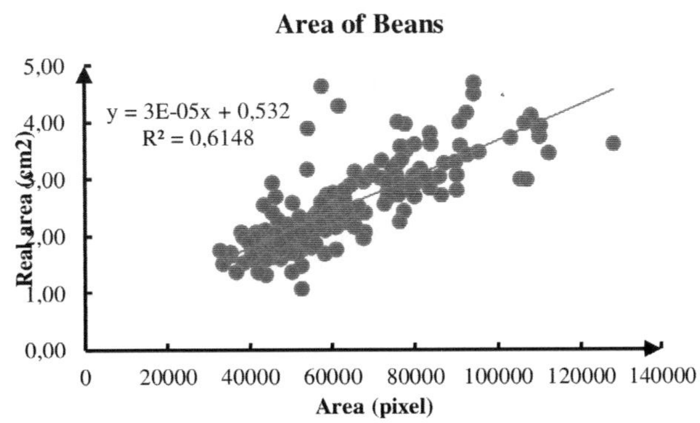

Figure 8. Plotting Data of The Area Parameter of Cocoa Beans

4. Conclusion

The cocoa bean physical feature extraction program has been developed on the basis of digital image processing. These physical features include seed length, seed width, seed circumference, and seed area. The regression results between features extracted with feature measurements show a fairly varied distribution which causes low regression coefficients obtained. Based on the regression results, the feature area provides the best results with a regression coefficient of 0.615. the distribution of area data that is still varied is very likely caused by imperfect segmentation results. For this reason, further studies related to adaptive segmentation techniques will be developed.

5. Acknowledgements

This work was supported by the Ministry of Research, Technology and Higher Education of the Republic of Indonesia [Schema of PTUPT with no. of agreement: 1882/UN1/DITLIT/DIT-LIT/LT/2018]. We also would like to grateful to Universitas Gadjah Mada and Farmer Association of Ngudi Raharjo II, Plosokerep, Bunder, Patuk, Gunungkidul, Yogyakarta for their support.

References

[1] H.-D. Belitz, W. Grosch, and P. Schieberle, *Food Chemistry*. Springer Science & Business Media, 2009.

[2] Anonim, "Biji Kakao (SNI 2323:2008)." Badan Standarisasi Nasional, 2008.

[3] I. W. Astika, M. Solahudin, A. Kurniawan, and Y. Wulandari, "Determination of Cocoa Bean Quality with Image Processing and Artificial Neural Network," *Computer Based Data Acquisition and Control in Agriculture*, 2010.

[4] N. León-Roque, M. Abderrahim, L. Nuñez-Alejos, S. M. Arribas, and L. Condezo-Hoyos, "Prediction of fermentation index of cocoa beans (Theobroma cacao L.) based on color measurement and artificial neural networks," *Talanta*, vol. 161, pp. 31–39, Dec. 2016.

[5] S. Sunoj, C. Igathinathane, and R. Visvanathan, "Nondestructive determination of cocoa bean quality using FT-NIR spectroscopy," *Comput. Electron. Agric.*, vol. 124, pp. 234–242, Jun. 2016.

[6] Radi, Muhammad Rivai, and Mauridhi Hery Purnomo, "Combination of First and Second Order Statistical Features of Bulk Grain Image for Quality Grade Estimation of Green Coffee Bean," *ResearchGate*, vol. 10, no. 18, pp. 8165–8174, Oct. 2015.

[7] F. Faridah, G. O. F. Parikesit, and F. Ferdiansjah, "Coffee Bean Grade Determination Based on Image Parameter," *TELKOMNIKA Telecommun. Comput. Electron. Control*, vol. 9, no. 3, pp. 547–554, Dec. 2011.

IOP Conf. Series: Earth and Environmental Science **327** (2019) 012021 doi:10.1088/1755-1315/327/1/012021

Development of "Wheel Hoe" Appropriate Tool for Supporting Organic Farming

O Hildayani[1], R E Putri[2], Andasuryani[2]
[1] Student of Agricultural Engineering, Agricultural Technology Faculty, Andalas University, Padang, West Sumatra, Indonesia
[2] Lecture of Agricultural Engineering Program Agricultural, Technology Faculty, Andalas University, Padang, West Sumatra, Indonesia

Corresponding author's e-mail address: okprimahildayani7@gmail.com

Abstract. Development of "Wheel Hoe" is aimed to produce the multifunction tools according to the agriculture field condition in West Sumatra. Wheel Hoe has three implements which used to trim weeds (blade assembly), to cultivate a soil (tine cultivator), to make the drainage (furrower) on the dried soils. This research did at The Laboratory of Production and Agricultural Machinery Management, Agricultural Engineering Program, Agricultural Technology Faculty, Andalas University. The research method is designing a prototype wheel hoe and testing in West Sumatra through observations involved: the speed, wheel slip, work capacity, spacious efficient, and time lost of wheel hoe. Wheel hoe used an engine 5 HP by Dymos. In this research have been produced a wheel hoe that has a dimension length: 180cm, width: 60cm and height: 100cm and weight: 145.35 kg. Wheel hoe test result used tine cultivator implement with an average speed of 0,01 m/s. The width of the plowed land with tine cultivator implement is 50cm. the spacious efficient of wheel hoe using the implement of tine cultivator is high on a red-yellow podsolic soil as much as 97,67% and tine cultivator on andosol soil and litosol soil as much as 96,12 % and 96,17%. Base on the spacious efficient has been got, so we got a conclusion that wheel hoe using tine cultivator implement can be used on all of the type soil in West Sumatra.
Keyword: wheel hoe, implement and the type of soil

1. Introduction

The organic food system or often mentioned the organic agriculture is one of the Indonesian Government Program to decrease usage of chemical fertilizers and pesticides which is causing reduced soil fertility and an environmental damage. An organic agriculture is defined as the holistic and integrated system of agricultural productions, through optimizing the health and productivity of agro-economy so, it will be producing the good quality of food. There are principles of organic agriculture that is the fertility and soil biological activity must be maintained, the soil must behave right humus and loose, trimming weed is using mechanical or semi-mechanical.

Organic agriculture development in West Sumatra has been a government program since 2006. It seems in the West Sumatra strategic plan. An area of organic agriculture reaching 1360 ha located on the several regencies in 2014 in West Sumatra [8]. West Sumatra farmer problems are lack of agricultural mechanization to support the productivity of agricultural product. The West Sumatra farmers are still using the traditional tools in their agricultural activities.

Tillage is one of the important things in the agricultural activities by both organic agriculture and the inorganic agriculture. It's a process preparing the land to plant and taking care of the plant in order to be apart from weeds for the plant growth. The main purpose of tillage divided into 3 phase that is to open a planting path, to eradicate weeds, and to repair the land conditions. [6] had done tiller rotary modification as implement on a tractor, [2] had made a weed cutters design on SRI and having a power capacity 3 HP/6000 rpm, and [6] had modified the agriculture engines and tools of weeds cutters. Base on the last researches, the agriculture engines, and tools are used to tillage, to open drainage path, to cut the weeds and in its tool operating we must have the skill to master it. So, wheel hoe is a problem-solving. It can be used on the small farm area, and it can be used to cultivate land, to cut the weeds, and to open the drainage path.

Wheel hoe is an appropriate tool for supporting farmers in their activities. The wheel hoe benefit is nearly the same as a hoe and a tractor. It has several implements according to user requirement.

In developed countries, wheel hoe had so many been used in the farming activities and the household scale, but it is a semi-mechanical tool. West Sumatra farmers have not used the wheel hoe, therefore, it is required developing semi-mechanical tools to support organic farming in West Sumatra.

The hypothesis of research are: 1) the small wheel hoe is more efficiently used in a small area 2) wheel hoe will be able to decrease the cost productions. The general aim of research is to develop the multifunction tools (wheel hoe) according to the tillage in West Sumatra while the particular aims in this research are 1) to design a prototype wheel hoe 2) to do technical testing on wheel hoe for each of its implementation.

2. Research Method

2.1 Tools and Materials

This research divided into 3 phase are the design phase, making a prototype phase and performance test phase. In the design phase, tools and materials used are computers including SolidWorks. To make a prototype, tools used are a drill, grinder, electric welding, lathe, etc while materials are the iron plate, iron shaft, rubber wheel, mor, bolt, paint. In a performance test phase, tools and materials used are the prototypes, wheel hoe, camera, stationary and meter.

2.2 Establishment Wheel Hoe

To make wheel hoe we required a few of implement that is to trim weeds (blade assembly), to cultivate a soil (tine cultivator), to make the drainage path (furrower) while the wheel hoe component that is following:

2.2.1 *Drive Motor.* To determine the power of the wheel hoe, the formula can be used that is

$$Pl = \frac{Ds\ x\ (d\ x\ L)\ x\ V}{75} \qquad (1)$$

Where:

P1	: The power for a tillage, tine cultivator (HP)
Ds	: Draft spesific land (kgf.m/cm²)
d	: Depth of tillage, tine cultivator (cm)
l	: Widht of tine cultivator (cm)
V	: Speed (m/s)
75	: unit convertion, 1 HP = 75 kgf.m/second

To determine the power to drive the wheel's Wheel Hoe:

$$P2 = \frac{Crr\ x\ w\ x\ V}{75} \qquad (2)$$

Where:

P2	: The power to drive the wheel (HP)
Crr	: coefficient of rolling resistance (without unit)

W : Weight tractor (kg)
v : Speed (m/second)
76 : Unit convertion, 1 HP = 75 kgf.m/second

Design of the drive motor power (Engine) Wheel Hoe

$$PE = \frac{P1}{\eta 1} + \frac{P2}{\eta 2} \quad (3)$$

Where:

PE : The power of motor *wheel hoe* (HP)
P1 : The power of the tillage using *tine cultivator* (HP)
P2 : The power of the drive motor *wheel hoe* with a implement *tine cultivator* (HP)
η1 : the efficiency of the power from motor to *tine cultivator* (desimal)
η2 : the efficiency of the power from motor to the whell's *wheel hoe* with an implement *tine cultivator* (desimal)

To determine the minimum weight of Wheel Hoe

$$B\ min = \frac{Ds\ x\ d\ x\ l}{TR} \quad (4)$$

Where:

B *min* : the minimum weight of wheel hoe (kg)
DS : *Draft spesific* land (kg/cm²)
l : widht of the weeds cutter (cm)
d : depht of the weeds cutter (cm)
TR : *Traction ratio* (without unit)

To determine the maximum weight of Wheel Hoe

$$B\ maks = (Gp\ x\ 0.78\ x\ l\ x\ 2\ x\ \sqrt{R^2} - (R\text{-}\ S)^2\ x\ 2) \quad (5)$$

Where:

B*maks* : the maximum weight of *wheel hoe* (kg)
Gp : *Bearing Capacity* (kg/cm²)
l : the wheel site widht(cm)
R : wheel spokes (cm)
S : the depth of wheel into the land (*zinkage)* (cm)
0,78 : Unit convertion, π / 4
2 : the amount of the wheel's *wheel hoe*

2.2.2 The Implement. Tine cultivator has a function to become loose the soil. In a design, tine cultivator has 4 blades with a slope of 15%. The opening tool of drainage path has a function to make a drainage path on soil. The weeds cutter (blade assembly) has a function to trim the weeds. In a design, its shape and size is made of thin steel with an angle of 15% to make it easy a back and forth motions.

2.2.3 The wheel hoe frame. Frame has a function as the place of the engine sustains and the implements of wheel hoe, besides it, frame is used to hold and control the wheel hoe. Wheel hoe size will be made of anthropometry test using 70 samples (35 men and 35 women). Anthropometry test is the numeric data having a related with the characteristic of human body that is a shape, size and power that used to the tools design or the workspace [12]. The wheel hoe frame is moved by two wheels, the space between the wheels is 120 cm, it's based on the width. To determine the wheel shaft diameter is used equation:

$$d \geq [\left(\frac{5.1}{Ta}\right) \sqrt{(km\ x\ M)^2 + (kt\ x\ T)^2}]^{1/3} \quad (6)$$

Where:

d: The wheel shaft diameter (cm)
Ta : shear stress (kg/mm^2)
Km : key factor of Flexural moment
M : Flexural moment (kg.mm)
T : Twisting moment (kg.mm)
Kt : Correction factor of Twisting moment

The rotation on the design wheel shaft is calculated by equation:

$$RPM\ roda = \frac{60\ x\ Vteo}{2\ \pi\ R} \tag{7}$$

$$Vteo = \frac{Vakt}{1-S} \tag{8}$$

Where:

Vteo : Theoretical speed (m/second)
Vakt : Actual speed (m/ second)
RPM : the rotation of wheel shaft (rotation/minute)
R : wheel spokes (m)
S : wheel slip (%)

The power transmission system has a function as a power distributor to run wheel hoe. In its design, the transmission system is used like sprocket-chains. Sprocket has been a function as a power conductor from the shaft speed reducer to the first shaft. The purpose is to avoid the high slip on the shafts because it needs more power.

The color of the wheel hoe framing is orange and for the engine is yellow while the implements of wheel hoe is black.

2.3 Technical test of the wheel hoe

The performance test on the wheel hoe is done in an organic farming area with five repetitions. Testing is using all the component of the wheel hoe implements. To prototype testing of wheel hoe for the first time is using the tine cultivator implement to cultivate soils on an area as wide as $15m^{2,}$ then the furrower is used to open the beds and irrigation path on the areas as long as 120 cm and 30 cm. after the plant is planted on the areas, we will cut the weeds with the blade assembly implement. Test pattern of the prototype path of the wheel hoe is zigzag according to a clockwise. There are the observations done on the wheel hoe technical testing that is:

2.3.1 A noise. One of the important things from the technical test is to quantify a noise rate because to know the quantified a noise rate, when the user is operating wheel hoe, they feel comfort.

2.3.2 Work speed. The working speed can be calculated by using a formula:

$$Vakt = \frac{S}{t} \tag{9}$$

Where:

Vakt : Actual speed (m/second)
S : mileage (m)
t : time (second)

2.3.3 Theoretical Work Capacity. Before calculating the theoretical work capacity, we would calculate a wheel slip, with a formula:

$$S = \frac{(\pi\ x\ D\ x\ (N-L))}{(\pi\ x\ D\ x\ N)}\ x\ 100\ \% \tag{10}$$

Where:

S : wheel slip (%)
D : Diameter (m)
N : the number of wheel turns that is 10 times
L : mileage of the wheel hoe when rotated N times (m)

$$Vteo = \frac{Vakt}{(1-S)} \qquad (11)$$

Where:

S : wheel Slip

2.3.4 The Theoretical Work Capacity. The theoretical work capacity can be calculated using the formula:

$$Kt = Vteo \; x \; W \; x \; 0{,}36 \qquad (12)$$

Where:

Kt : theoretical capacity (ha/hours)
Vteo : Work speed (m/second)
W : work width of singkal (m)
0,36 : conversion rate from m^2/second to Ha/hours.

2.3.5 *Effective Working Capacity.* Effective working capacity can be calculated using the formula:

$$Ke = \frac{TL}{TW} \qquad (13)$$

Where:

Ke : Effective Working Capacity (ha/hours)
Tl : Total of area plowed (Ha)
TW: Total of time (jam)

2.3.6 *Field Efficiency.* Field efficiency can be calculated using the formula:

$$E = \frac{Ke}{Kt} \; x \; 100\% \qquad (14)$$

Where:

E : Field efficiency (%)
Ke : Effective working capacity (ha/hours)
Kt : Theoretical working capacity (ha/hours)

2.3.7 *Losses Time When Turned.* Losses time when turned can be calculated using the formula:

$$Lo = \frac{Wb}{(Wl + Wb)} \; x \; 100\;\% \qquad (15)$$

Where:

Lo : *Losses* time when turned (%)
Wb : time for turned (second)
Wl : time to the plow fields (second)

2.3.8 Operator Power. Operator power could be measured through an operator's heart rates. It's measured before and after doing to the plow fields. The classification of people work rates aged 20 to 50 years can be seen in Table 1.

Table 1. The classification of people work rates aged 20 to 50 years

Work level	Power Requirements (kW)	Heart Rate (minutes)
Very light	Less 0,17	Less 75
Light	0,17-0,33	75-100
Average	0,33-0,55	100-125
Weight	0,55-0,67	125-150
Very weight	0,67-0,84	150-175
Beyond the limits	Up 0,84	More 175

3. Result

3.1 Design Result

This research has produced an appropriate tool "wheel hoe" with a size length 180 cm, width 60 cm, and height 100cm. It could be operated by an operator. The number of a weight of wheel hoe (included all of the implements and engines) is 145,35 kg, it's according to the theoretical calculating on the minimum weight of Wheel Hoe as much as 119.07 kg and the maximum weight of Wheel Hoe as much as 154,74 kg. wheel hoe image can be seen in Figure1.

Figure 1. Wheel Hoe

The wheel hoe is the tillage engine consists of two wheels having a function to become loose the soils, to make drainage path on the beds, and to cut the weeds on the dried soils. To do all of the activities used several implements are tine cultivator, furrower, and blade assembly. The development aimed of wheel hoe is to increase the work capacities of farmers in various farming activities through an appropriate tool in various soil types.

The engine construction of wheel hoe refers to the design consists of several components:

- The main frame
- Wheel
- Machine
- Implements

The final result of the development wheel hoe and the machine testing could be described through that following table:

Table 2. Specification of *wheel hoe*

	Specification of *wheel hoe*	
1	Price	Rp. 7.000.000,-

1	Dimension (L x W x H)	180 cm ×60 cm × 100 cm
2	Machine	Gasoline Motor
3	The number of Operator	1 person
4	Work capacity of *wheel hoe*	0,01 ha/hours
5	Efficiency	88,15 %
6	Frame construction	Strip Iron, Elbow Iron, Pipe Iron
7	The number of *Implements*	3
8	Weight	145,35 kg
9	Wheel Diameter	51,5 cm
10	Construction of Tine Cultivator	Round Iron
11	Construction of Furrower	Plat Iron
12	Construction of Blade Assembly	Strip Iron

*3.2 The Main Frame of Wheel Ho*e

The main frame of wheel hoe is made of the pipe and strip iron with length 180 cm and width 60 cm. the machine place is made of the elbow iron with length 80 cm. Base on the anthropometry test, we got an average of the height people in regency West Sumatra is 155-160 cm so, the height of wheel hoe is as much as 100 cm in order to the operator could be easy operating the wheel hoe.

Wheel on the wheel hoe is made of the round iron with diameter 51.5 cm, that's why that wheel could handle the weight of wheel hoe.

3.3 Machine

Machine used is a gasoline motor of 5HP by Dymos. It used according to the calculating of power requirement that is 3.5809 HP. When we had been testing on the wheel hoe, we used gasoline motor of 5HP in order to wheel hoe machine can be optimally worked. Chain and sprocket used is as much as four sprockets have a measurement of 15 T and 40 T and two chains have a measurement of 104 L.

Beside it, the transmission system of wheel hoe is using a pulleys and belt. Pulleys used is on the machine have a measurement of 1 inch. On the transmission system of the chain-sprocket used pulleys as big as 7 inch and using a belt to relate the pulleys in order to the wheel hoe could be run according to the machine speed.

3.4 Testing of The Implement

In this journal, the technical test is using the implement of tine cultivator. Tine cultivator has a function to become loose soils. On the implement of tine cultivator is made of the round iron sharpened with a slope 15% , the length is 100 cm, width is 30 cm. The gear of the tine cultivators are as much as 10 tine cultivators. The image of tine cultivator could be seen in Figure 2.

Figure 2. Implement of Tine Cultivator

The technical test of wheel hoe using tine cultivator has been done on three soil types that is the litosol soil in South Solok, the andosol soil in Alahan Panjang, the red-yellow podosolik soil in Limau Manis, Padang, West Sumatra. On each area, testing had been done for five times, that's because decreasing the error at data collections. The image of the wheel hoe technical testing is using the tine cultivator could be seen in Figure 3.

Figure 3. wheel hoe technical testing is using the tine cultivator

3.4.1 The Work Speed. The work speed is the required time by an operator to finish a job. It's a comparison between the mileage and times of wheel hoe. The large of the fields used to do the technical test is 15 m^2 while the length of fields is 5 m and the width of fields is 3 m. The work speed is using the implement of tine cultivator could be seen in Table 3.

Table 3. The calculating of wheel hoe's work speed with the implement of tine cultivator

Soil types	Litosol	Andosol	Red-Yellow Podsolik
the length of fields (m)	5	5	5
Times (s)	472	359	485,2
Actual speed (m/s)	0,011	0,014	0,010

Base on Table 3, we could see that an average working speed of wheel hoe on that three soil types is different that is 0.01 m/s that means in one second, wheel hoe could run 0.01 m with using the implement of tine cultivator. It's fast more than the conservative tillage or using a hoe. Time taken by wheel hoe on the same length of fields is different, that things is happened because the fields is different as well, on the fields having the rocks and wavy soils can hamper the wheel hoe working. According to [5], on the wavy and rocks soils could cause the wheel turned that is not running so need a thrust by an operator to move a tractor and make a time taken longer.

Base on the last research, its obtained actual speed on the dried fields is as big as 0.72 m/s [5], [6] described that an average of rotary hand tractor as big as 0.61 m/s. Then, [3] said that the cultivator speed that designed depend on its face soils. The conclusion is if the speed of wheel hoe is high rather the other cultivator so will have a high slip as well.

Beside it, on [13] described that the actual speed of tine cultivator with an animal as a drive as much as 1.80 km/h. that thing is lower more than the actual speed on wheel hoe. It's because the farmers is hard when they are turned so, they need more times to operate tine cultivator.

3.4.2 Theoretical Work Capacity. Before calculating of the wheel hoe Theoretical work capacity, first we're calculating the wheel slip of wheel hoe. Slip testing had been done at a distance 10 times wheel turning. Wheel's diameter is 51.5 cm. Testing had been done as much as five times on each implement on three soil types. The result could be seen in following Table 4.

Table 4. wheel slip Testing on the tine cultivator.

	Litosol	Andosol	Red-yellow Podsolik
Wheel Diameter (m)	0.515	0.515	0.515
N	10	10	10
Mileage N time (m)	4.92	6.28	4.64
Wheel Slip (%)	69.56	61.17	71.31

Base on Table 4, we got an average of slip wheel using a tine cultivator implement that is more than 50%. It's happened because the speeds of wheel hoe when the operator is operating a wheel hoe fast. The slip value is related to the work speed, the higher the work speed the bigger the frequency of wheel rotation [9].

Base on Table 4, we got a theoretical work capacity of wheel hoe. The result could be seen in following table.

Table 5. Theoretical work capacity of wheel hoe on tine cultivator

	Litosol	Andosol	Red-yellow Podsolik
Actual speed (m/s)	0,011	0,014	0,010
Wheel Slip (%)	69,56	61,17	71,31
Theoretical work speed. (m/s)	0,035	0,036	0,036
Work Width (m)	1	1	1
Theoretical capacity (ha/hours)	0,0125	0,0129	0,0129

Base on Table 5 we know that andosol soil and red-yellow pedsolik soil are higher more than lithosol soil. the theoretical work capacity of lithosol soil is as much as 0.0125 ha/hours while on the andosol and red-yellow pedasolik soil are as much as 0.0129 ha/ hours. Theoretical work capacity is affected by the width of tillage and the actual speed of a tractor [5] He stated in his research (2005) that theoretical work capacity is the machine ability to finish a job on the fields and if it runs maximally.

In his research (2005), he got an average of theoretical work capacity is as much as 0.048 ha/hours that means it's bigger than the theoretical work capacity of wheel hoe. It's because the actual speed of the wheel hoe on the implement of tine cultivator is faster so it got an effect on the wheel slip on wheel hoe higher.

3.4.3 Effective Work Capacity. Effective work capacity is a works speed to be obtained by a tillage machine based on the field area per total times (ha/hours) and it's an effective average ability. The wheel hoe calculating of the effective work capacity using the tine cultivator implement had been done as much as five times on three soil types. There is the following table that is the calculating of the effective work capacity on the implement of tine cultivator.

Table 6. The calculating of the effective work capacity on the implement of tine cultivator.

	Litosol	Andosol	Red-yellow Podsolik
Total of tillage (ha)	0,0015	0,0015	0,0015
Total times (hours)	0,1244	0,1208	0,1194
Effective work capacity (ha/hours)	0,0121	0,0124	0,0126

Base on Table 6, we know that the wheel hoe average of speed work using the tine cultivator on three soil type is nearly the same that is 0.012 ha /hours. The number of effective speed is determined by the number of the wheel slip on wheel hoe, area conditions, operator skills. Beside it, the effective work capacity is affected by that water and soil contents. So, the smaller time taken, the higher the effective work capacity [5].

The effective work capacity is usually calculated from the area tillage per time taken. Generally, the effective time much then the time theoretical. It's because 1) the loss of time when turning 2) the loss of time cleaning the machine 3) the loss of time to take a rest [5]. While [1] stated that the effective work capacity is affected by the actual work speed, width of work, time to be turned.

3.4.4 Field Efficiency. Field Efficiency of wheel is a comparison between the effective work capacity and the theoretical work capacity stated in percent (%). There is a comparison between the effective work capacity and the theoretical work capacity of wheel hoe on the implement of tine cultivator could be seen in the Figure 4.

Base on Figure 4 we could see that the effective work capacity is smaller than the theoretical work capacity cause to get the Field efficiency of wheel hoe, the effective work capacity must be smaller than the theoretical work capacity. Field Efficiency of wheel hoe using the tine cultivator implement could see in the following Table 7.

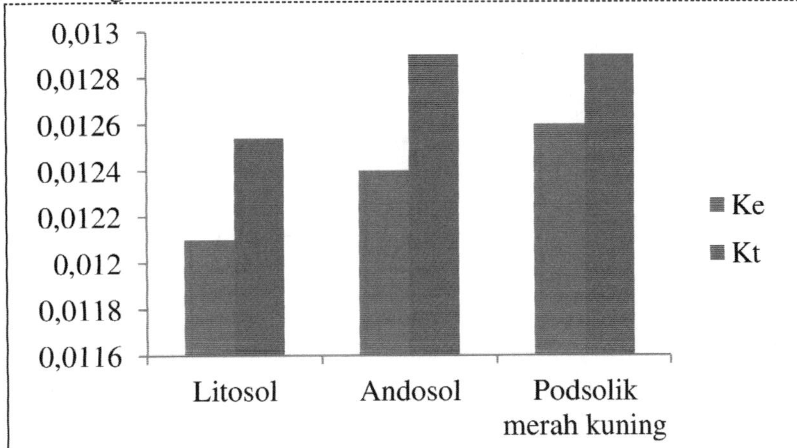

Figure 4. Effective work capacity and the theoretical work capacity of wheel hoe on the implement of tine cultivator

Table 7. The Field Efficiency of wheel hoe calculating using the tine cultivator implement.

	Litosol	Andosol	Red-yellow Podsolik
effective work capacity (ha/hours)	0,0121	0,0124	0,0126
the theoretical work capacity (ha/hours)	0,01254	0,0129	0,0129
Field Efficiency (%)	96,49 %	96,12 %	97,67 %

According to Table 7, we could see that field efficiency of wheel hoe using the tine cultivator implement is around 90-97%. It's affected by the work speed of wheel hoe to finish the tillage as wide as 15 m^2. Beside it, the field efficiency is affected by the work times that is 1) the loss of time because a wheel slip, area conditions and operator skill. Base on the analysis, the number of tillage, the field efficiency is around 75-95% [10]. There is the field efficiency using the tine cultivator implement could see in Figure 5.

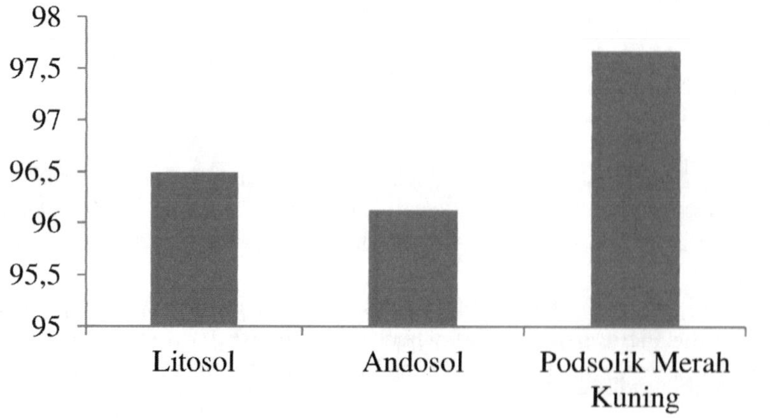

Figure 5. the field efficiency using the tine cultivator implement

3.4.5 The Percentage of The Loss of Times when Turning. Turning at the corner or the end of the field will be impacted a loss of time, particularly at the small fields. The number of turning times per width unit for a tool will be inversely proportional to the length fields. Effective times used during working, that number is determined by the ineffective time [10]. The percentage of the loss of times when turning table using the tine cultivator implement could see in the following Table 8.

Table 8. The percentage of the loss of times when turning on wheel hoe using the tine cultivator implement

	Litosol	Andosol	Red-yellow Podsolik
Times for turning (s)	10	11	11,8
Straight Walk Time (s)	20	20,2	21,6
The loss of times (%)	33,33	35,26	35,33

Base on Table 8 we could see that the loss of times of wheel hoe using the tine cultivator implement is pretty high around 33-35%. It's because the wheel slip, generally involved when wheel hoe is turning and an operator need power more to hold wheel hoe in order to the wheel hoe could walk straight so it needs time more.

If it's compared with the last research about the performance of hand tractor by [5], explained that the actual work speed of the hand tractor obtains around 0.4 -0.7 m/second. That thing is fast more than the actual speed of wheel hoe using the tine cultivator in the tillage that is 0.01 m/second while the wheel hoe is around 90-97%. It's because the work speed of wheel hoe is high so, the theoretical work capacity is high as well.

3.4.6 Noise. Noise is the sound that not wanted and disturbed the environmental comfort stated in decibel (dB). On the wheel hoe, a noise rate is measured to determine how big a sound producing when the operator is operating the wheel hoe. The noise rate could be measured using Sound Level Meter tool with a type: Extech 407736, and the noise rate of wheel hoe is 98,1 dB.

The sound in that range is very noisy. Base on the health minister regulation RI no 1405/MENKES/SK/XI/2002 about Health Requirements of Office and Industrial Work Environment, so to avoid deaf, operator is allowed operating wheel hoe without an earmuff for eight hours/day. The Noise categories and noise verges can be seen in Appendix 8.

3.4.7 The Power of Operator. The light weight of the operator's workload can be known by the operator's pulse or heart per minute. This heart rate can be measured using a stopwatch because using a stopwatch the operator's heart rate per minute can be recorded. After getting the heart rate at rest and work then it is then matched with the workload level table by Christensen (Santosa, et al 2005) which is based on the heart rate per minute of humans aged 20-50 years. Table of occupational level classification in people aged 20 to 50 years can be seen in Table 1.

Based on tests conducted on operators in Appendix 10. It can be concluded that this wheel hoe is included in the lightweight category with power requirements of 0.17 - 0.33 kW, the power value makes the wheel hoe can be operated by anyone whether male or female.

4. Conclusion

Base on the development of wheel hoe using machine 5 HP, the total weight of wheel hoe is 145.35 kg used to cultivate the dried fields, a dimension of wheel hoe is length 180 cm, width 60 cm, and height 100 cm. the average of speed on every implement is 0.01 m/s and the average wheel slip is 70%. The average of the work capacity on every implement is 0.01 ha/hours and the average Field Efficiency of wheel hoe on three soil types is 76%. Base on the resulting test, the implement of tine cultivator is more suitable for use to andosol soil

References

[1] Darmawan and Soebandi. 1999. Mechanical Equipment for Cotton Cultivation.

[2] Handoyo. 1990. Rice Weed Weeding Tools from Modification of Motorized Lawn Mowers. Faculty of Agriculture. Gadjah Mada University. Yogyakarta.

[3] Hettiaratchi, D. R. P. 1993. The Development Of A Powered Low Draugt Tine Cultivator. Soil and Tillage Research. Volume 28. Issue 2. Pages 159 - 177.

[4] International Federation of Organic Agriculture Movements (IFOAM). 2005. www.ifoam.org

[5] Santosa, Andasuryani, and V. Veronica. 2005. Performance of Hand Tractors for Soil Processing. Academic Journal. Volume 9 No. 2: October 2005, p. 1-7.

[6] Santosa, Andasuryani, Rinaldi Saputra, and Dede Pranata. 2007. Modification of Rotary Tillers as Implement in Hand Tractors. Agricultural Engineering Journal. Vol. V. Number 2: 65 - 74.

[7] Santosa. 2005a. Visual Basic 6.0 and Visual Studio applications. Net 2003 in the Field of Engineering and Agriculture. Yogyakarta. Andi.

[8] The Agriculture Census, 2015 in http://www.sumbarprov.go.id accessed on October 20, 2016 at 11.20 WIB.
[9] Suprodjo. 1980. Ways to Determine the Main Size of Tractors for Soil Processing. Agricultural Mechanization Section, Faculty of Agricultural Technology, Gadjah Mada University: Yogyakarta.
[10] www.teknoperta.wordpress.com (accessed on October 12, 2017 at 8:04 p.m. WIB).
[11] www.valleyoaktool.com (accessed on June 23, 2016 at 1048 WIB).
[12] Yokatta, 2003 in https://ukhtymj.wordpress.com/2003/10/03/antropometry accessed on October 20, 2016 at 11.56 WIB.
[13] Zainal-Arifin. 1993. Development of Tine Cultivator for Weeding Tools in Intercropping Cotton, Green Mung and Corn Cropping Patterns in Banyuwangi Regency. Essay. Faculty of Agriculture, Brawijaya University, Malang.

IOP Conf. Series: Earth and Environmental Science **327** (2019) 012022 doi:10.1088/1755-1315/327/1/012022

Analysis of POME Discharge Quality from Different Mill in Perak, Malaysia: A case study

S Sapie[2], S S Jumali[2], S Mustaffha[1], D E Pebrian[2]

[1, 2] Universiti Teknologi Mara, Melaka Branch Jasin Campus, 77300 Merlimau, Melaka, Malaysia

Corresponding Author E-mail Address: samihah096@uitm.edu.my

Abstract. Palm oil mill effluent (POME) is a major concern of oil palm mill because it contributes to numerous pollution such as soil, water and air pollution. In order to maintain a clean environment for a sustainable agricultural system in oil palm plantation, the quality of POME and its level must be monitored to ensure that the requirement set by the country is met. Hence, this paper is written based on the data of POME final discharge into waterways taken from Malaysian Department of Environment in four regions of Perak, Malaysia. The data is compared between regions and relatively different levels of discharge have been produced. Some of the mills produce final POME wastewater above the limit that has been set. The remaining mills discharge the wastewater below the maximum allowed level. This is an indicator that each mill has its own Standard of Procedures in treating POME before being discharged into waterways. It is suggested that the mills take a proactive approach to treat POME beyond the ponding system. POME can be physically processed to produce biofertilizer and biocompost for organic farming. The mills apply phytoremediation techniques in treating POME to reduce the unwanted variables that can affect the sustainability of our environment.

1. Introduction

Agricultural sustainability is a concept that refers to natural process integration, minimal use of non-renewable inputs that affected the environment and wellbeing of human through the utilization of skills and knowledge of the farmers who jointly solve agricultural problems [1]. In palm oil plantation especially in Malaysia, sustainable palm oil can be achieved by following the standard of Roundtable Sustainable Palm Oil (RSPO) and Malaysia Sustainable Palm Oil (MSPO.

There are two types of waste that produced by palm oil mill which in the form of liquid and solid. Solid waste mostly obtained from fruit bunches after being stripped and leaving the empty fruit bunches (EFB) while the liquid part is come from the whole extraction process at the mill [2]. The liquid discharge is known as palm oil mill effluent (POME), is one of the greatest issues in palm oil plantation due to its potential contribution to environmental pollution where untreated POME gives higher impact to environment compared to the treated POME. Every one ton of processing fresh fruit bunch (FFB) will produce 0.65 ton of POME [3].

The main three processes involved in producing POME are sterilization that produces 36% of POME, nut cracker separation using hydro cyclone (4%) and clarification (60%) [4]. The sterilization is taking place to "cook" the fresh fruit bunch under high pressure to loosen the fruitlets using a lot of water. This water will then be discharged into sludge pond. The nut cracker includes the usage of hydro cyclone or clay bath method to separate and sort particle like oil and kernel. The clarification process is the final process to remove all impurities from crude oil in settling tanks.

POME also has high nutrients such as nitrogen, phosphorous, sodium, magnesium and calcium that may be utilized for crops and medium for microalgae growth even though it has been treated by using facultative anaerobic [5]. As stated by [6], POME is made up of 0.6-7.0% palm oil, 95-96% water, 4-5% total solid, and 2-4% suspended solid. As stated by [2], the main components that produce the POME which high in chemical oxygen demand (COD) and biological oxygen demand (BOD) values are the high amount of wastewater and oil grease from the CPO process. Because of the extremely high content of organic pollutants, POME requires numerous treatment stages. This is to prevent problems in the biological method stages and follow with water-discharge requirements.

Table 1: The characteristics of POME sample [7]

No	Parameter	Concentration (mg/L)	Average (mg/L)
1	pH	4.15-4.45	4.25
2	COD	1350-2120	1600
3	Soluble COD	20500-24500	22000
4	BOD	300-400	330
5	Total volatile solid	27300-30150	28100
6	Total suspended solid	15660-23560	18900
7	Total phosphorus	200-600	350
8	Total nitrogen	500-800	500

*All parameters are in unit of mg/L except pH

In Malaysia, Department of Environment has imposed a regulation regarding the effluent discharge from the oil palm industry. Since Malaysia is moving forward to make sustainability for mill is compulsory under MSPO, hence the treatment of POME before discharging into waterways is crucial. Hence, this paper is presents the quality of treated POME discharge and the treatments that have been applied in the oil palm industry recently.

2. Materials and methods

The study was conducted in Perak, Malaysia since this state has the largest area of oil palm plantation area in the northern region of Malaysia. The data was collected by referring to the Department of the Environment (DOE), Perak. Generally, palm oil mills will submit their report on POME quality to the Department of the Environment quarterly.

In this case study, the data used was secondary data on POME qualities of four mills obtained from the Malaysian Department of Environment of (DOE) Perak. The raw data was intended to serve as comparison for compliance study. Secondary data is basically the data or information that was either gathered by someone else for some other purposes [8].

The parameters of this case study are biochemical oxygen demand (BOD), oil and grease, suspended solids, ammonical nitrogen, total nitrogen, temperature and pH. All of the parameters are then used to compare to the management practices of POME by the palm oil mill. The management practices include the number of pond for palm oil mill effluent, the applications of treated palm oil mill effluent and the availability of bacteria for solid degradation purposes.

2.1 POME qualities

There are many important POME qualities of palm oil mill effluent, as listed in the parameters set by DOE. The parameters are set so that the POME that will be discharged into the watercourse will not cause any pollution. There are seven most important parameters for POME qualities and the parameters are biochemical oxygen demand (BOD), oil and grease, suspended solids, ammonical nitrogen, total nitrogen, temperature and pH. The parameter limits are listed in Table 2.

Table 2: Parameter limits for POME before discharged into waterways [9]

Parameter	Units	Parameter limits for POME discharge
Biochemical oxygen demand (BOD, 3-day, 30 °C)	mg/litre	100
Oil and grease	mg/litre	50
Suspended solids	mg/litre	400
Ammonical nitrogen	mg/litre	150
Total nitrogen	mg/litre	200
Temperature	°C	45
pH	-	5 – 9

Under Principle 5, a mill must have a waste administration plan to avoid or lessen the pollution by observing the source of pollution and converting them into value added products. Subsequently, the mill must have Standard of Procedure to guarantee legitimate and safe handling technique and consent to Environmental Quality Act 1974.

3. Result and discussion

There are seven parameters of POME that should be recorded which are the reading of BOD, oil and grease, suspended solid, ammonical nitrogen, total nitrogen, temperature and pH. Unfortunately, among the four mills, two of them did not submit the reading for all parameters. This is due to the negligence of palm oil mill managers since it does not give any significant for them to know. Table 3 below shows the average reading for the seven parameters that have been listed by DOE.

Table 3: The Average Reading of Seven Listed Parameters Taken from Four Palm Oil Mills Water Discharge

Palm oil mill	Average BOD (mg/l)	Average Oil and Grease (mg/l)	Average Suspend ed Solid (mg/l)	Average Ammoni cal Nitrogen (mg/l)	Average Total Nitrogen (mg/l)	Average Tempera ture (°C)	Average pH
Kerian Region I	62	na	126	63	89	30	8.3
Perak Tengah Region I	75	2	92	44	143	30	8.4
Batang Padang Region I	48	7	233	21	38	27	8.6
Hilir Perak Region I	na	1	56	1	7	30	9.3

Only Perak Tengah Region I and Batang Padang Region I have submitted the full readings. Kerian Region 1 has not submitted oil and grease reading while Hilir Perak Region 1 has not submitted its BOD reading.

By analyzing the data, it can be seen almost all palm oil mills comply to each parameter except for Hilir Perak Region I, where pH reading of the mill exceeded the acceptable range.

For BOD, there are three different parameters required by DOE. The parameters were divided into two main categories, which are for water that will be discharged into the river and for land application, which is for fertilizer purposes. For the ones that will be discharged into the river, the lower the BOD, the better the quality. In this paper, only BOD for water discharge is considered. The BOD reading was taken from below water catchment.

Meanwhile, for oil and grease, the parameter set by DOE is 50mg/l and the parameter is set only for effluent that will be discharged into the watercourse. From the data collected, only Kerian Region 1 did not submit the reading while the others had reading below the set parameter. This parameter is important because high amount of oil and grease will block the sunlight penetration to the bottom of the river. This will lead to difficulty in degrading the effluent.

As for ammonical nitrogen, the parameter set is 150mg/l. The result shows that all palm oil mills did not exceed the limit of ammonical nitrogen. Excessive amount of ammonical nitrogen will give direct impact to aquatic lives since it will increase the respiratory level and oxygen uptake by aquatic lives. The parameter set for total nitrogen is 200mg/l. Generally the amount for total nitrogen is higher than the parameters set for ammonical nitrogen as the total nitrogen contains more elements other than ammonical nitrogen.

The temperature required by DOE for effluent that will be discharged is 45°C. All of the palm oil mills that submitted their reading for temperature show that the temperature of the treated effluent discharged follows the parameter set by DOE. High temperature reduces the oxygen holding capacity thus supplying insufficient amount of oxygen to the aquatic life. The acceptable range of pH set by DOE is 5 – 9. Only Hilir Perak Region I of palm oil mills exceeds the maximum allowable pH but with a slightly high amount.

3.1 POME Treatment Method

All the palm oil mill in this study have utilized the ponding system because it is low in cost and easy to maintain since it located near to the mills. This ponding system demands long reservation period and also requires large areas for implementation. Generally, the ponds for effluent will start with cooling pond, mixing pond, anaerobic pond, aerobic pond and facultative or stabilization pond. The number of ponds may vary as the palm oil mills may have different amount of cooling ponds, anaerobic ponds and aerobic ponds. There are some palm oil mills that have only one cooling pond while other palm oil mills have two cooling ponds. The cooling pond is the pond that contains raw POME from mills, which are produced from the process of CPO productions.

POME in the pond will undergo biological degradation where this method is effective and has high efficiency in getting rid of the suspended solids, organic matter and oil and grease. This is because it has high organic loading capability and requires low energy. POME microorganisms have discovered application in POME bioconversion into functional and demanding products. However, the trend varies between palm oil mills. The variation between the data may be due to the inefficiency of bacteria for degradation process.

Other than discharging the treated POME into watercourse, POME can undergo physical treatment to produce biofertilizer and biocompost. This is because after treating raw POME, it will produce suspended solids or sludge and disintegrates solid. Both of them have high nutrient value compared to slurry and produce fetid scent. The sludge is then can be dried up and utilized as biofertilizer, biocompost and manure after undergoing specific treatment. DOE has set an allowed BOD level for treated wastewater discharge into watercourse at only 100mg/l while the BOD level for fertilizer applications to as high as 5000mg/l. By referring to the main objective of the application, palm oil mill managers shall know the best practices that they should apply in order to comply with the required parameter.

Also, the usage of aquatic or semi-aquatic plants such as *Chlamydomonas incerta* which is also known as the microalgae and apu-apu or *Pistia stratiotes*, a small plant with wide leaves is still reassuring as the wide leaves cover larger absorption area of waste in the wastewater treatment by removing unwanted nutrient and also chemical oxygen demand (COD) and BOD. They also readily remediate toxic pollutants in freshwater and seawater. This method is called phytoremediation. It is a remediation method that use green plants to reduce the contaminants without the need to excavate the contaminant elsewhere and can be done *in-situ*. This method is very effective and inexpensive.

However, it will take a long time to fully remediate pollutants and only selected plants are suitable. It also only can be applied at the final discharge of treated POME effluent.

4. Conclusion

As can be seen, POME quality is differed from one mill to another and this is due to the management practice. Hence, in order to conform to DOE's parameter of POME quality, the best management practice should be applied by palm oil mill. The managements have to ensure the availability and the efficiency of microorganisms in the effluent pond by measuring the rate of degradation (mg/L). This is because the rate of degradation can be hastened through the addition of microorganisms in the pond, as suitable microorganism addition is efficient in degrading the compost.

As for the physical method, composting is still the most effective and economical in POME treatment. This is because composting can control the nutrients and acts as bulking agents that can be considered as an alternative option in improving the oil sludge bioremediation. The palm oil mill management should also consider treating POME and recycling the sludge to come out with value-added products, which can be reused in plantation. This will help the management to overcome the problems brought by while reducing a lot of cost.

It is also relevant to implement this biological method such as phytoremediation as it is cost-effective and do not need extensive equipment and labour. As a whole, best management practices should be applied by all palm oil mills in Malaysia. This includes the constant frequency of POME quality determination; obtaining the best qualities relevant to the application of treated POME. As a conclusion, the oil palm mill management team must be fully responsible in ensuring the conservation of environment from being damaged by the oil palm waste.

Acknowledgments

Authors wishing to acknowledge assistance and encouragement from colleagues, Department of Environmental of Perak State and financial support from Universiti Teknologi MARA.

References

[1] Pretty J, Butler C, Birch N, Andow D, Osborn D, Dickie I, Smith J, Fenske R, Shiel A, and Thompson PB 2017 *Sustainability in Agriculture.* Royal Society of Chemistry

[2] Bakar M A A, Yahya W J, and Mohamad Z 2016 Pollution Reduction Technology in Palm Oil Mill through Effluent Treatment and Management. *Akademia Baru* **1**(1) 8–15

[3] Stichnothe H and Schuchardt F 2010 Comparison of Different Treatment Options for Palm Oil Production Waste on A Life Cycle Basis *The International Journal of Life Cycle Assessment* **15**(9) 907-915

[4] Zinatizadeh A A L, Mohamed A R, Abdullah A Z, Mashitah M D, Isa Mh and Najafpour G D 2006 Process Modeling and Analysis of Palm Oil Mill Effluent Treatment in An Up-Flow Anaerobic Sludge Fixed Film Bioreactor Using Response Surface Methodology (RSM) *Water Research* **40**(17) 3193-3208

[5] Yacob S, Hassan M A, Shirai Y, Wakisaka M, and Subash S 2006 Baseline Study of Methane Emission from Anaerobic Ponds of Palm Oil Mill Effluent Treatment *Science of the Total Environment*, **366**(1), 187-196

[6] Irvan 2018 Processing of Palm Oil Mill Wastes Based on Zero Waste Technology *IOP Conference Series: Materials Science and Engineering* **309**(1)

[7] Kamyab H, Din M F M, Keyvanfar A, Majid M Z A, Talaiekhozani A, Shafaghat A, and Ismail H H 2015 Efficiency of microalgae Chlamydomonas on the removal of pollutants from palm oil mill effluent (POME) *Energy Procedia* **75** 2400-2408

[8] Sloboda B 2016 *Understanding Secondary Data in Research | University of Phoenix Research Hub. Research.phoenix.edu.* Retrieved August 11, 2017

[9] *Environmental Requirements: A Guide For Investors,* 2010. Retrieved September 24, 2018

The Effect of Fish Type and Variety on Growth and Results Through the Application of Minapadi-SRI

N Rozen[1]**, A Anwar**[2]**, and N Kristina**[2]

[1, 2] Agrotechnology Study Program, Faculty of Agriculture Andalas University

Corresponding author's e-mail address: rozennalwida@agr.unand.ac.id

Abstract. National food security must be maintained because the food need of the Indonesian population in the next few years will still continue to increase. Since there is no food diversification for their staple food, rice role will still dominate. More over, the conversion of irrigated rice fields functions is also increasing. This reality must be addressed with policies and the application of efficient lowland rice production technology so that it is still able to meet national food needs while the income of wetland rice farmers increase. This study, combined an experimental production technology and descriptive methods on the farmer income aspect institutional strengthening. The aim of the research is to increase the added value of irrigated rice fields, which area is believed to be increasingly shrinking, while on the other side still produce rice as the main national food and also produce the addition of animal food in the form of fish coulde. Data obtained in the previous study showed that combination of water dept of a 10 cm with 50 cm width is the best treatment for the development of rice and fish togethere. This treatment is used as well as basis for testing the MINAPADI-SRI pattern with more varieties and types of fish, which is expected to obtaine more practical and profitable combinations. The research was conducted in the Split plot design where fish species (Tawes, Mas, and Nila) as the main plot and the rice varieties (PB42, Batang Piaman, and Beras Merah) as sub plot. The results showed that the PB42 gave higher yield compared to two the other varieties where tilapia and goldfish were which maintained in the technically irrigated rice fields.
Keywords: minapadi-SRI, lowland rice, rice variety.

1. Introduction

Rice is still the staple food of the Indonesian people and most of the population in the world. National rice needs continue to increase every year as population increases, whereas the area of rice fields continues to decrease due to various interests. Various efforts to increase production have been carried out both through extensification and intensification. One of the last decade's efforts to give hope is the application of SRI (the System of Rice Intensification) in rice cultivation. Various research reports and crop yields of farmers show a one to threefold increase in yield.

The most suitable paddy field for the development of SRI is irrigated rice fields with more guaranteed water availability. Water on SRI does not need to inundate the entire surface of the paddy field, it is enough to guarantee that the soil remains in a wet condition (around field capacity). The results of the research hope for further modifying the SRI pattern [8]. It was reported that the best results on SRI that were tried were the water in the pot about -10 cm above the ground. Thus rice can be planted in parts that are not inundated (plots), enough water is provided in the trenches made between the plots of rice planting. Planting conditions like this provide an opportunity to allow water to remain in the ditch between paddy fields from planting to near harvest. If the age of rice planted is around 100 days, then there will be about 80 days of standing water. Enough time to keep fish.

Farmers in several regions in Indonesia are accustomed to utilizing standing water in paddy fields to maintain fish, either directly between stagnant rice clumps or making special plots in the middle of rice fields, on the edge around paddy fields or alternating with rice planting. This cultivation method is known as Minapadi. West Sumatra is one of the areas where farmers are used to doing it. I Lima Puluh Kota District around 150 ha of rice fields were used for minapadi [9].

Combining SRI with minapadi is very possible. The merger is directly a form of product diversification from irrigated rice fields. However, some issues still need to get answers so that the application at the farmer level is more practical, scientifically accountable and profitable not only for farmers' income, but also nationally beneficial because it can play a role in strengthening food security. Minapadi-SRI in paddy fields with tilapia species and the results can increase farmers' income, but it has not been studied how much water is beneficial for the life of the fish in the ditch [12].

Some of these problems can be formulated as follows: if SRI rice cultivation is combined with fish maintenance it is certainly necessary to ascertain how high the water in the ditch is beneficial for both, what is the most appropriate width of the ditch, what is the most beneficial type of fish, how good the management institution starting from rice and fish cultivation, processing the results and marketing of its products. These questions need to get answers that can be justified. For this reason, this research plan was prepared.

In general, this study aims to increase the added value of irrigated rice fields, whose area is believed to be shrinking further, while still producing rice as the main national food in addition to the addition of animal food products in the form of fish. Although some of the rice fields are used as fish maintenance, the total production will not interfere with the availability of rice because the SRI results are 2-3 times the national average, in fact the additional income from fish will be more profitable for farmers. Especially if the fish they produce make processed foods such as babyfish by adding certain flavors, it might just be a *babyfish balado, babyfish flavored with rendang* and so on.

Another advantage of SRI is the reduced use of water. As stated earlier, it shows that the application of SRI to rice cultivation in irrigated rice fields can save around 40% of water needs [15]. This happened because during its growth, rice with the SRI method did not need to be inundated, enough wet conditions even to the ground a little they were still tolerated by rice [3, 10], especially if the fertilizer used came organic matter [4, 11].

In one of the counseling materials from the [2], it was explained that minapadi is a method used by farmers by combining rice cultivation and fish-raising techniques, which are carried out simultaneously in the rice fields. For this business, the construction of rice fields is not necessary, but it is only necessary to make a *caren (kemalir)*, which is a kind of ditch around in a plot of rice fields with a diagonal or crossing a plot of rice fields. This slime serves as a fish shelter and to facilitate harvesting of fish. The wide width of the slime generally ranges from 40 - 60 cm with a water depth of 40 cm. The types of fish that are usually maintained in this way include: Mas, Carp, Tawes, Nilem, Mujair and Nila (tilapia). Carp and tilapia are the types that are best kept in rice fields because they grow well with shallow water and heat resistance.

Meanwhile, the factors that influence the choice of fish are the volume of water, availability of seeds, feed, markets, and habits of farmers [16]. In Minapadi, the inundation water level of rice plants is limited to 10-15 cm, and in the caren part the water level is 20-30 cm. Compared to Tawes, Nilem, Kancra and Carp, gold fish (ikan Mas) is the best type of fish in the Minapadi system [14]. This is reflected in the high yield of rice which reached 5.7 t / ha in the minapadi plot that used goldfish as a treatment, with the highest net profit of Rp 5.15 million / ha / season. Methane gas emissions from minapadi plots that use carp are 51.2 kg CH4 / ha / season. The strains of tilapia which were maintained in a positive way had an effect on the results obtained, the best strain was larasati tilapia [12].

Other factors that need to be considered are fish populations and feeding. Solid stocking fish that is commonly done on minapadi according to [1] is for 2-3 cm sized fish as much as 2-3 fish / m2 and for fish measuring 3-5 cm as much as 1-2 fish / m2. While the feed dosage and feeding frequency also needs to be considered [6]. In the Minapadi system, fish feed is enough only with fine bran with a dose of 4-5% of the fish's body weight.

Some of the explanations above confirm that the SRI method of rice cultivation has the opportunity to be developed by combining it with the Minapadi system into MINAPADI-SRI. The results of this development are expected to still be able to increase rice yields and be added with fish products which are processed into typical foods which will later increase farmers' income and expand employment. Food security can be maintained and farmers become more prosperous.

2. Materials and Methods

This study used an experimental method with field experiments using Split Plot Design in Randomized Block Design with two factors and three groups. The main plot is the type of fish (Tawes, Mas, and Nila) while the subplots are varieties (PB42, Batang Piaman, and Red Rice). The plot area for one experimental unit is 16 m^2.

The data obtained were then analyzed using ANOVA with the F Test at a real level of 5%. Comparison of the median value of treatment using the Honestly Significance Differences (HSD) at the 5% real level.

This field experiment has been carried out on the farmer's partner member Farmers Group Banda Langik, on the Sungai Bangek, Koto Tangah Subdistrict, Padang City in February-September 2018. The land used is technical irrigated rice fields whose farmers are accustomed to the System of Rice Intensification. Data collected for rice plants are: yield and yield components of rice which includes total and productive tillers, weight of 100 grain, grain yield per clump and per harvest plot. As for fish, the data collected is livelihood, average weight, and yield.

3. Results and Discussion

3.1 Height of rice plants

Observation data on rice plant height are shown in Table 1. The results of the analysis show that there is no interaction between varieties with fish species on the height of rice plants. The main plot or type of fish is also not significantly different, but subplots or varieties have a significance difference.

Table 1. Plant height of three rice varieties in Minapadi-SRI with three types of fish

Treatment	Variety of rice			
Type of fish	PB42	Batang Piaman	Beras Merah	Average
		---------------------- cm ----------------------		
Tawes	80,28	93,73	112,68	95,56 a
Mas	80,42	92,02	106,03	92,82 a
Nila	73,85	85,47	102,87	87,40 a
Average	78,18B	90,41A	107,19A	

IOP Conf. Series: Earth and Environmental Science **327** (2019) 012023 doi:10.1088/1755-1315/327/1/012023

Note: The numbers in the lane followed by the same lowercase letters and the numbers in the row followed by the same uppercase letters differ not significantly according to the HSD test 5%.

From Table 1 above, it can be seen that the Red Rice varieties are higher in plants than the Batang Piaman and PB42 varieties. The average height of Red Rice plants is more than 100cm while the average plant height of PB42 and Batang Piaman is below 100 cm. Plant height is more influenced by genetic factors than the plant itself. Red rice includes local rice so that the height of the Red Rice plant is higher. Unlike the case with superior varieties that have been assembled into rice plants whose plant height is lower with early maturity such as PB42 and Batang Piaman varieties. The following shows the morphology of the three rice varieties (Figure 1).

Figure 1. Varieties of PB42, Batang Piaman and Beras Merah (Red Rice) varieties

From the three rice plants in Figure 4 above, it can be seen that the Red Rice is higher in plants than the Batang Piaman and PB42 varieties which are observed at 105 days after planting. Plant height is more influenced by genetic factors, so it appears that there is a difference between superior varieties and local rice.

3.2 Total Number of Tllers

Data on the total number of tillers of three rice varieties with different types of fish in Minapadi-SRI cultivation are shown in Table 2. The results of the variance analysis on the total number of tillers showed significant interactions, as well as the main plot and subplot gave a significant effect on the total.

Table 2. The total number of tillers of three rice varieties with three species of fish in Minapadi-SRI

Treatment	Variety of rice			
Type of fish	PB42	Batang Piaman	Beras Merah	Average
	-------------------- batang ----------------------			
Tawes	34,57Aa	31,77Aa	23,25Bb	29,86
Mas	34,57Aa	32,70Aa	28,43Bb	31,90
Nila	30,05Aa	28,63Bb	21,70Bb	26,79
Average	33,06	31,03	24,46	

Note: numbers in the lane followed by the same lowercase letters and the numbers in the row followed by the same uppercase letters differ not significantly according to the HSD test 5%.

From the data in Table 2 above it can be seen that there are interactions between varieties with fish species, as well as the main plot (fish species) and subplot (variety) showing the number of tillers that are significantly different. In Tawes fish the total number of saplings of PB42 and Batang Piaman varieties is higher than Red Rice. Likewise in the type of carp, but for tilapia, PB42 varieties are better planted.

Different types of fish give different numbers of tillers, as well as varieties that show more different effects due to different genetic factors. National superior varieties will have more tillers so that the grain will be more numerous with longer panicles, but local rice is usually less formed tillers. The data obtained shows that the number of tillers formed is more than the conventional method which is an average of more than 20 sticks, this is due to the SRI method, the formed tillers multiply. As stated by Sato and Uphoff (2007) shows an average yield increase of 78% and a reduction in water use of 40%, a reduction in the application of fertilizer to 50% and a reduction in production costs of 20%.

3.3 Number of Productive Tillers

Furthermore, the data on the number of productive tillers are displayed after being analyzed by variance, which shows that there is no significant interaction between varieties with fish species, the main plot is also not significantly different, but subplots have a significant effect. The number of productive tillers is shown in Table 3 below.

Table 3. Number of productive tillers of three varieties of rice plants with different types of fish

Treatment	Variety of rice			
Type of fish	PB42	Batang Piaman	Beras Merah	Average
		---------------------- cm----------------------		
Tawes	24,28	34,56	17,80	25,55a
Mas	22,56	31,60	16,76	23,64a
Nila	27,96	26,44	16,24	23,55a
Average	24,93A	30,87A	16,94B	

Note: The numbers in the lane followed by the same lowercase letters and the numbers in the row followed by the same uppercase letters differ not significantly according to the HSD test 5%.

Based on the table above, it can be seen that the number of productive tillers is more influenced by varieties, this is due to productive tillers according to plant genetics. National superior varieties are more productive than local rice, as are the Batang Piaman and PB42 varieties, while Red Rice has fewer productive tillers.

3.4 Long panicles

The following table shows panicle length of three rice varieties with different types of fish in Minapadi-SRI cultivation. From the results of the variance analysis, it can be seen that there is a real interaction between varieties with fish species on panicle length, as well as varieties, but the type of fish is not significant. The panicle length is shown in Table 4 below.

Table 4. Panicle length of three rice varieties with different types of fish

Treatment	Variety of rice			
Type of fish	PB42	Batang Piaman	Beras Merah	Average
		---------------------- cm----------------------		
Tawes	26,74 Ba	31,47Aa	31,51Aa	29,91
Mas	27,90 Ba	27,37 Bb	31,86Aa	29,04

Nila	27,32 Ba	25,96 Bb	30,76Aa	28,01
Average	27,32	28,27	31,38	

Note: The numbers in the lane followed by the same lowercase letters and the numbers in the row followed by the same uppercase letters differ not significantly according to the HSD test 5%.

In Table 4 it can be seen that there is an interaction between varieties with fish species to panicle length. In the PB42 variety, three types of fish can be maintained, as well as Melah Rice, but in the Batang Piaman variety, Tawes fish are better maintained than the other two types of fish. This is because the panicle length is influenced by genetic and environmental interactions. In Tawes fish species are more suitable to be planted with Batang Piaman and Red Rice varieties, but in the type of Goldfish more suitable for Red Rice, as well as the types of Tilapia. The Batang Piaman variety is longer in the treatment of Tawes fish species (31.47 cm), but the Red Rice and PB42 are relatively the same in the three types of fish.

3.5 Amount of grain per panicle

The number of grain per panicle of the three rice varieties with different types of fish in Minapadi-SRI is presented in Table 5. The results of data analysis show that there is no interaction between varieties with fish species, as well as main plots or types of fish, but subplots or varieties have a real influence.

Table 5. Amount of grain per panicle three rice varieties with different types of fish

Treatment	Variety of rice			
Type of fish	PB42	Batang Piaman	Beras Merah	Average
	--------------------- cm-----------------------			
Tawes	183,68	140,84	250,20	191.57a
Mas	166,68	171,84	273,48	204,00a
Nila	201,00	152,52	303,04	218,85a
Average	183,79B	155,07C	275,57A	

Note: The numbers in the lane followed by the same lowercase letters and the numbers in the row followed by the same uppercase letters differ not significantly according to the HSD test 5%.

In Table 5, it can be seen that the type of fish affects the number of grains per panicle, as well as the varieties grown. The amount of grain per panicle is more influenced by varieties due to genetic factors of the variety itself. In addition, the panicle length is closely related to the amount of grain. Red rice has a longer panicle length compared to PB42 and Batang Piaman varieties so that it has more grain amount.

3.6 Amount of rice grain per panicle

The number of pithy rice per panicle after analysis of variance showed that there was no real interaction between varieties with fish species, but the main plot or type of fish and subplots or varieties showed significantly different results. The number of pithy rice per panicle is presented in Table 6.

Table 6. Number of pithy rice per panicle three rice varieties with different types of fish

Treatment	Variety of rice			
Type of fish	PB42	Batang Piaman	Beras Merah	Average
	--------------------- cm-----------------------			
Tawes	149,60	96,44	189,08	145,04 b
Mas	174,36	129,84	212,00	172,07 a
Nila	164,80	116,04	247,68	176,17 a
Average	162,92 B	114,11C	216,25 A	

Note: The numbers in the lane followed by the same lowercase letters and the numbers in the row followed by the same uppercase letters differ not significantly according to the HSD test 5%.

Table 6 shows that the variety treatment has a significant effect on the number of rice grain per panicle, as well as the type of fish. Red rice forms more rice grain than the other two varieties. This is because the amount of rice grain is closely related to panicle length and the number of grain per panicle. Red rice gives more number of pithy rice per panicle because the length of the pan is longer so there is also a lot of grain formed. This will also cause a lot of rice grain. In addition, the number of rice grain is also influenced by environmental factors, where the type of fish will affect the rice grain formed as Tilapia is better maintained in the rice fields because it is more resistant and well developed.

3.7 100 Grain Weight of Grain

The results of analysis of variance for the weight of 100 grains of rice showed that there was no real interaction between varieties with fish species in Minapadi-SRI, but the main plot or species of fish and subplots or varieties showed significant influence. The weight of 100 grains of grain is shown in Table 7 below.

Table 7. Weight of 100 grains of grain

Treatment	Variety of rice			
Type of fish	PB42	Batang Piaman	Beras Merah	Average
	-------------------- gram----------------------			
Tawes	2,18	2,90	2,32	2,47 b
Mas	2,40	3,08	2,44	2,64 a
Nila	2,18	2,78	2,24	2,40 b
Average	2,25 B	2,92 A	2,33 B	

Note: The numbers in the lane followed by the same lowercase letters and the numbers in the row followed by the same uppercase letters differ not significantly according to the HSD test 5%.

In Table 7 above, it can be seen that the types of fish and varieties affect the weight of 100 grains of grain. The Batang Piaman variety has a weight of 100 grain grain which is heavier than the PB42 and Red Rice varieties. This is because during the vegetative phase, Batang Piaman varieties look healthy with no pest and disease disturbances, but in PB42 varieties there are attacks by stem borer pests so that after the PB42 variety can be produced it can also produce. Most likely, the 100 grains of Red Rice are low due to many being empty, because they last longer so that the nutrients absorbed must also be more, in the case that the fertilizer given is the same, so not all grain is fully filled.

3.8 Grain Weight per Clump

The results of analysis of variance on the weight of grain per clump after being analyzed by variance showed a real interaction. The weight of grain per hill is presented in Table 8.

Table 8. Grain weight per clump of three rice varieties with different types of fish

Treatment	Variety of rice			
Type of fish	PB42	Batang Piaman	Beras Merah	Average
	------------------- cm----------------------			
Tawes	43,44Bb	36,75Cc	63,85Aa	48,01
Mas	43,88Bb	41,12Bb	48,71Bb	44,57
Nila	49,83Bb	40,80Bb	58,70Aa	49,78
Average	45,72	39,56	57,09	

Note: The numbers in the lane followed by the same lowercase letters and the numbers in the row followed by the same uppercase letters differ not significantly according to the HSD test 5%.

In Table 8 it is stated that the types of fish and varieties provide a real interaction with the weight of grain per clump. In the cultivation of Minapadi-SRI for Red Rice, it is more suitable to maintain Tawes and Tilapia fish. For Batang Piaman variety, it is more suitable for carp and tilapia, but for PB42 varieties it is more suitable for tilapia that is maintained. There is an interaction between species of fish and varieties on the weight of grain per clump due to the presence of genetic interactions with the environment. Goldfish and tilapia are more suitable to be maintained in paddy fields because these types of fish are more resistant to rice fields, sometimes the water is hot due to sunny weather. As stated by Sasa and Syahromi, (2006) states that compared to Tawes, Nilem, Kancra and Carp, carp is the best type of fish in the Minapadi system. This is reflected in the high yield of rice which reached 5.7 t / ha in the minapadi plot that used goldfish as a treatment, with the highest net profit of Rp 5.15 million / ha / season. Methane gas emissions from minapadi plots that use carp are 51.2 kg CH4 / ha / season. The strains of tilapia which were kept in the minapadi effect on the results obtained, the best strain was Larasati Tilapia [13].

3.9 Results of Grain per Plot

The results of the variance analysis on the yield of grain per plot provide unrealistic interactions, as well as the main plot, but the subplots provide tangible results. Data from grain per plot results are shown in Table 9 below.

Table 9. Grain yield per plot of three rice varieties with different types of fish

Treatment	Variety of rice			
Type of fish	PB42	Batang Piaman	Beras Merah	Average
		-------------------- gram----------------------		
Tawes	1.040,00	550,00	750,00	780,00a
Mas	1.220,00	550,00	760,00	843,33a
Nila	970,00	562,00	790,00	774,00a
Average	1.076,67A	554, 00B	766,67B	

Note: The numbers in the lane followed by the same lowercase letters and the numbers in the row followed by the same uppercase letters differ not significantly according to the HSD test 5%.

From the table above it can be seen that only the treatment of varieties gave a real effect on the yield of grain per plot. Whereas the interaction is not real as is the type of fish. This is because grain is more closely related to varieties because different varieties will cause differences in results that are influenced by the genetics of the variety itself. Although the yield or grain is influenced by the interaction between genetics and the environment, in this case environmental factors have not affected much. The PB42 variety gave the most severe grain yield (1,076.67 grams) compared to Batang Piaman varieties (554 grams) and Red Rice 766.67 grams). This is related to the number of total tillers where PB42 varieties have more total tillers than other varieties. The more saplings formed, the more grain formed.

4. Conclusions

From the results and discussion that has been described, it can be concluded that: the number of tillers formed up to 12th phyllochron on PB42 varieties is 33 tillers, Batang Piaman 31tillers and Red Rice is 24 tillers with the highest yields per plot is varieties PB42. The type of fish that is suitable to be maintained in paddy fields is the type of Tilapia and Mas. Those fish are more resistant to life in the rice field environment.

IOP Conf. Series: Earth and Environmental Science **327** (2019) 012023 doi:10.1088/1755-1315/327/1/012023

It is recommended to apply the Minapadi-SRI pattern to the cultivation of lowland rice so that the formation of the tillers increases, so the results will increase as well. The application of SRI can be combined with fish (Minapadi-SRI) so that the results obtained will multiply.

REFERENCE

[1] [BBAT] Freshwater Cultivation Center 1995 *Fish Maintenance with the Mina Padi System* [Brochure] (Sukabumi: Ministry of Agriculture, Directorate General of Fisheries, Freshwater Cultivation Center)

[2] [KKP] Ministry of Maritime Affairs and Fisheries 2011 *Minapadi Cultivation Technique* (KP Human Resources Development Agency Center for Marine and Fisheries Extension)

[3] Anwar A, N Rozen, Rusnam, Agustian, and Helmi *2007 Water-saving Testing on Lowland Rice Cultivation Using the SRI Method* (Padang: Research Report on the Cooperation of Unand-PLA of Bekasi Irrigation Center)

[4] Anwar A, A Syarif, N Rozen, Agustian, Yaherwandi, M Busniah, Azrifirwan, M Makky, Armansyah, and M Kasim 2009 *Academic Review Factors Supporting Successful Application of Organic SRI* (Padang: Unand-Medco Foundation Collaborative Research Report)

[5] Berkelaar D 2001 *Rice Intensification System (The System of Rice Intensification-SRI): Few can Give More* (USA: ECHO Bulletin Development Note, January 2001. ECHO Inc. 17391 Durrance Rd. North FtMyers FI.33917)

[6] Darini M T H 2010 Effect of Dosage and Frequency of Supplementary Feeding on the Growth of Rice and Gourami Seed Results in the Minapadi System *AGRIVET Journal* **14** *(2)*

[7] Effendie M I 1979 *Fisheries Biology Methods* (Bogor: Dewi Sri Foundation Bogor)

[8] Kasli and Effendi A R 2011 Effect of Inundation Height on the Growth of Rice Paddy Plants (*Oryza sativa* L.) in *Pot Straw Journal* **4** *(3)*

[9] Ramli M 2010 The Development of Aquaculture and Its Contribution to The Agricultural Sector in The Economy of the District of Fifty Cities in West Sumatra *Journal of Aquatic Sciences* **8** *(1)*

[10] Rozen N 2008 *The Mechanism of Tolerance of Paddy Rice to Weeds in The SRI Method (The System of Rice Intensification)* (Padang: Dissertation of Postgraduate Program of Andalas University)

[11] Rozen N, A Anwar, and Armansyah 2010 Weed Control in Organic SRI *Jerami Journal* **3** *(1)*

[12] Rozen N, Afrizal, and Sabrina 2011 *Increasing The Potential of Farmers Through Transfer SRI Agricultural Technology in The City of Padang* (Padang: Final Service Report IbW Community Program. DP2M DIKTI)

[13] Salsabila A, F Basuki, and S Hastuti 2013 Performance of Growth of Different Tilapia (Oreochromis niloticus) Strains in the Minapadi Cultivation System *Journal of Aquaculture Management and Technology* **2** *(4)*

[14] Sasa J J and O Syahromi 2006 Mina Rice System in Land Productivity, Revenue and Environment Perspectives *Journal of Food Crop Agriculture* **25** *(5)*

[15] Sato S and N Uphoff 2007 A Review of On-Farm Evaluations of the System of Rice Intensification Methods in Eastern Indonesia *CAB Reviews: Perspectives in Agriculture, Veterinary Science, Nutrition and Natural Resources 2007* **2** *(54)*

[16] Suriapermana S, I Syamsiah, P Wardana, and A M Fagi 1989 *Practical Clues to The System of Fish Rice and Duck Rice Farming in Paddy Fields* (West Java: Balittan Sukamandi)

Nutrients Dynamics in Peat Soil: Influence of Fluctuating Water Table

N Q B Kassim[1] and A Yaacob[2]

[1,2] Faculty of Plantation and Agrotechnology, UiTM Cawangan Melaka Kampus Jasin, 77300, Merlimau, Melaka, Malaysia

Corresponding author's e-mail address: qursyna@uitm.edu.my

Abstract. This study assesses the dynamics of nutrients in peat soil under fluctuating water table using designated soil columns. Results showed that the concentration of total N and mineral-N (NO3--N and NH4+-N) were maximized when the water table was maintained at 40 cm. When the water table was fluctuated between 0-40 cm depth (representing the water table during wet season), soil available P reached its highest concentration in the soil. The fluctuation of water table between 40 - 80 cm depth (representing the water table during dry season) allows the gradual release of exchangeable K, Ca, Mg and Na. The study indicated that the fluctuating water table affects the different nutrients differently.
Keywords: Nutrients, Dynamics, Peat, Fluctuating, Water Table

1. Introduction

Continuous shifting of aerobic and anaerobic condition due to fluctuating water table could lead to biogeochemical changes of the soil that could affect the cycling of nutrients in the peat soil system. Understanding such changes on the nutrient dynamics will help in the management and the agronomic practices of crops planted on peat soil. The objective of this study is to evaluate the dynamics of nutrients in peat soils under fluctuating water table. It is hypothesized that the dynamics of nutrients were varied depending on the aerobic and anaerobic environment created by fluctuating water table.

2. Research Methodology

2.1 Field Sampling

A deep peat soil profile was taken from undisturbed natural peat soils in Klang, Selangor (2°57'42.99" N, 101°30'24.20"E). The soil samples were repacked into 9 designated acrylic soil columns measuring 20 cm by diameter and 100 cm by tall. The design of the column is as illustrated in Appendix 1.

2.2 Treatments

Each of the columns was applied with 14g of Baja Campuran Nanas (BCN) consisting of ammonium sulphate (AS), Christmas Island Rock Phosphate (CIRP) and Muriate of Potash (MOP) at the ratio of 72:1:27g/100g. Three treatments of water table were established in this experiment; (i) water table fixed at 40 cm, (ii) water table fluctuated at 0-40 cm and (iii) water table fluctuated at 40-80 cm.

2.3 Sampling

Soil samples and leachate were taken 3 weeks after fertilizer application to ensure uniform distribution and dissolution, followed by week 5 and week 8. The soil samples collected were oven dried, pulverized, sieved and analysed.

2.4 Sample and Data Analysis

The analysis was carried out at the Soil Science laboratory at the Forest Research Institute Malaysia (FRIM), Kuala Lumpur. The effects of treatments on soil parameter were tested using ANOVA. The means were compared using Tukey post-hoc test at 5% level. Then, the correlation was tested using Pearson correlation test.

3. Results and Discussions

Table 1 shows the descriptive data and ANOVA of soil chemical properties for each treatment whereas Table 2 shows the concentration of nutrients in leachate samples. Table 3 demonstrates the correlation between variables.

Shifting of aerobic and anaerobic condition through continuous saturation, drying and rewetting, following water table fluctuation in peat profiles had influenced the distribution and concentration of N, P and K along the vertical peat profile. In this study, soil samples of treated soil columns were acidic, with a pH of less than pH 4.1. Highly acidic features of peat is well expected because the soil is abundance with H+ ion and other organic acids such as humic acids and fulvic acids, resulted from decomposition process [1]. It is anticipated that low concentration of major nutrients, especially N and exchangeable bases (K, Ca, Mg and Na) was attributed to acidic environment in peat profile (pH < 4) and was common in tropical lowland forests [2]. Improvement of pH in deeper layers is due to deposition of observable concentration of exchangeable bases, especially Ca and Mg.

More than 50% of total carbon was present in the form of organic C, believed to be stored in organic material. Organic C in this study ranged between 15% – 54%, which was in the range of 12% - 60% for oligotrophic peat [3]. In the column with fluctuating water table, the C content tend to fluctuate as well. This was in line with findings by many which verify that anaerobic condition allows for carbon sinks whereas shifting into aerobic condition enhance the carbon to be oxidized and become sources for CO2 [1][4][5].

For all treated soil columns, the amount of N identified was ranged between 0.1% - 1.5%. This value is in the bottom range for oligotrophic peat (1% - 4 %) [3]. With ammonium sulphate application, the amount of N identified in this study appears to be minimum. Quantification of N indicated that almost 50% of N was in unknown form, suggested to be organic N since it was prominent in peat soils [1]. Smaller amount of N was found on the upper layer compare to deeper peat, indicating continuous decomposition process as well as mineralization of fertilizers in former layer. This is supported by significant amount of mineralized-N (in the form of NH4+-N and NO3--N) detected on the surface peat. There was a consistent release of NH4+-N especially by columns that exhibit permanent aerobic condition at 0-40 cm depth (T0 and T2). Under this condition, the key players of N-cycle; the microbes; are at their optimum condition to mineralize the organic material and fertilizers into NH4+-N and the nitrifiers subsequently oxidize NH4+-N into NO3--N. The process continued with reducing of nitrate into nitrogenous gaseous compounds by denitrifiers [6][7]. In this study, small amount of NO3--Nwas traceable at 0-20 cm peat layers for T0 and T2, demonstrating the sensitivity of nitrate with oxidizing condition. Apart from that, there was a possibility of dissimilatory nitrate reduction to ammonium (DNRA), demonstrate by substantial amount of NH4+-N coupled with traceable NO3--N in anaerobic peat, high C/ NO3--N ratio, followed by significant correlation between both parameters. Even though these factors may indicate DNRA process [8], further study is needed to confirm the process of DNRA in peat soil. A study has reported that after fertilization, approximately 50% of N was assimilated by crops whereas the remainder was leached primarily as NO3- [9]. In this study, high concentration of N was leached in T1, where the water table was fluctuated between 0-40 cm compared to the better drained surface peat, as in T0 and T2. This could be due to either soluble N release from fertilizers or the organic materials [10] since assimilation by crop was absent. Hence, the treatment which represented the wet season suggested that continuous fluctuating water table on peat surface coupled with high amount of precipitation could possibly enhance leaching of N from peat system.

Concentration of available P was considerably high (approximately 1.5% - 8%) for all treated column, compared to those in natural peat which have a range of 0.04% - 0.5% [3]. The difference of about 6% is possibly contributed by CIRP fertilizer application in this study. The dynamics of P in treated soil column contrarily differs from N. Under fixed water table (T0), the amount of available P steadily decreased whereas in a column where water table was fluctuated between 40 – 80 cm depth (T2), the amount of P slowly increased throughout the study period. Highest concentration of available P was achieved when the water table fluctuated within 0 - 40 cm (T1) even though the pattern was inconsistent and concurrent with fluctuating water table. This is suggested due to variation of anaerobic soil conditions and reduction of Fe3+ compounds which advance P-mobilization in peat [11], and demonstrated by significant leached of Fe in the treated column. Observable amount of P detected on surface peat (0-20 cm) was possibly from the mineralization of CIRP fertilizer, even though other factors such as reduction of Fe3+ and desorption of phosphates [12] may occur.

Table 1. Descriptive Data and ANOVA of Soil Chemical Properties

Soil Properties	Fraction	Treatment	Mean	Std. Deviation	Min.	Max.	F	Sig.
Nitrogen	Total N	T0	0.92	0.32	0.17	1.48	0.393	0.677
		T1	0.81	0.42	0.14	1.16		
		T2	0.82	0.42	0.16	1.43		
	NH_4^+	T0	1396.83	2541.558	87.75	6767.27	0.75	0.478
		T1	822.375	489.9849	57.94	1414.88		
		T2	704.097	1238.777	36.03	3608.96		
	NO_3^-	T0	4.88	14.32064	0	55.31	0.416	0.662
		T1	1.65	4.34669	0	12.36		
		T2	3.29	7.57755	0	25.51		
Available P		T0	39.37[ab]	15.31	18.18	68.68	3.34	0.045
		T1	47.42[a]	21.70	15.22	79.53		
		T2	31.23[b]	13.35	14.90	54.34		
pH		T0	3.75	0.20	3.37	4.05	0.237	0.790
		T1	3.70	0.13	3.58	4.00		
		T2	3.74	0.21	3.37	4.07		
Carbon	Total Carbon	T0	47.68	22.00	5.82	66.26	0.035	0.966
		T1	48.49	23.27	6.10	71.56		
		T2	46.32	23.29	7.28	72.02		
	Organic Carbon	T0	35.85	16.54	4.37	49.82	0.035	0.966
		T1	36.46	17.50	4.59	53.81		
		T2	34.82	17.51	5.47	54.15		
Exchangeable Bases	Soil Exch. K	T0	0.85[a]	0.09	0.67	0.96	22.365	0.000
		T1	1.12[b]	0.17	0.76	1.26		
		T2	0.83[a]	0.13	0.62	1.10		
	Soil Exch. Ca	T0	2.97	1.28	1.46	5.48	0.357	0.702
		T1	3.15	1.23	1.46	5.49		
		T2	3.34	1.09	1.97	5.27		
	Soil Exch. Mg	T0	7.55	5.43	3.63	19.9	0.369	0.694
		T1	8.27	5.91	3.32	20.9		
		T2	9.35	5.96	3.15	21.7		

	Soil Exch. Na	T0	0.67	0.41	0.26	1.67	0.502	0.609
		T1	0.56	0.28	0.30	1.19		
		T2	0.61	0.22	0.37	0.99		
CEC		T0	52.87	15.72	24	76	0.783	0.463
		T1	46.93	13.28	22	68		
		T2	47.53	13.76	21	63		

Notes:
i) T0 = Water table fixed at 40 cm; T1 = Water table fluctuated between 0-40 cm; T2 = Water table fluctuated between 40-80 cm
ii) Means followed by the same letter are not significantly different at $p > 0.05$

Table 2. Concentration of Nutrients in Leachate Sample

	T0 (Water table fixed at 40 cm)										
Nutrients (mg/L)	W3					W5					
	D1	D2	D3	D4	D5	D1	D2	D3	D4	D5	D1
N	-	-	7.85	23.96	6.31	-	-	8.51	6.67	2.44	-
P	-	-	3.09	0.16	0.59	-	-	0.87	0.17	n.d	-
K	-	-	14.95	22.78	15.43	-	-	16.07	21.74	11.87	-
Ca	-	-	7.46	12.36	9.84	-	-	7.67	27.74	37.91	-
Mg	-	-	10.73	17.30	14.56	-	-	12.00	63.11	22.25	-
	T1 (Water table fluctuated between 0-40 cm)										
Nutrients (mg/L)	W3					W5					
	D1	D2	D3	D4	D5	D1	D2	D3	D4	D5	D1
N	165.07	155.19	117.36	190.14	7.83	-	130.66	126.82	73.79	8.64	142.9
P	7.46	6.95	1.66	3.26	0.00	-	0.69	0.17	0.61	0.04	1.98
K	160.90	155.60	100.20	142.60	24.53	-	115.30	110.80	75.41	21.90	116.2
Ca	56.43	56.30	51.28	57.01	27.61	-	30.47	29.97	29.59	23.25	30.85
Mg	52.69	53.24	57.92	57.66	88.38	-	43.43	43.28	62.37	70.99	41.49
	T2 (Water table fluctuated between 40-80 cm)										
Nutrients (mg/L)	W3					W5					
	D1	D2	D3	D4	D5	D1	D2	D3	D4	D5	D1
N	-	-	4.56	n.d	n.d	-	-	-	6.47	1.35	-
P	-	-	0.02	n.d	n.d	-	-	-	0.39	0.00	-
K	-	-	15.97	n.d	n.d	-	-	-	14.77	11.75	-
Ca	-	-	11.98	n.d	n.d	-	-	-	8.04	36.78	-
Mg	-	-	22.49	n.d	n.d	-	-	-	13.00	22.00	-

Notes:

i) T0 = Water table fixed at 40 cm; T1 = Water table fluctuated between 0-40 cm; T2 = Water table fluctuated between 40-80 cm; ii) D1= 0-20 cm depth; D2= 20-40 cm depth, D3= 40-60 cm depth, D4= 60-80 cm depth, D5= 80-100 cm depth; iii) n.d.: not detected

Table 3. Correlation between Variables

Variables	pH	Organic C	NH_4^+	Avail. P	Exch. Ca	Exch. Mg	Exch. Na
NH_4^+	-.682**	.348*	-	-	-	-	-
NO_3^-	-.507**	-	.711**	-	-	-	-
Avail. P	-	.564**	.490**	-	-	-	-
Exch. K	-	.307*	-	.446**	-	-	-
Exch. Ca	-	-.717**	-	-	-	-	-
Exch. Mg	-	-.946**	-.355*	-.536**	.814**	-	-
Exch. Na	-	-.509**	-	-.344*	.481**	.528**	-
Soil CEC	-	.823**	-	.534**	-.652**	-.854**	-.366*

Notes: i) **Correlation is significant at the 0.01 level (2-tailed); ii) *Correlation is significant at the 0.05 level (2-tailed)

Decrement of P in deeper peat layer as well as mineral soil underneath peat shows a possibility of P forming complexes, primarily with trivalent or bivalent cations such as Ca, Al and Fe which was prevalent in acidic condition [13]. The former was confirmed in this study as demonstrated by increment of Ca in deeper layers whereas the later was not determined in this study. The insignificant amount of P in the mineral soil layers could be due to P-fixation, which is expected in acidic environment and the fact that P which exist in the anion forms (PO42-) are repelled by negatively charge clay [14].

Low concentration of K in this study is suspected to be due to domination of acidic cation (H+ and Al3+), demonstrated by low soil pH, as well as K being readily leached in soil solution, due to its high mobility. The former confirmed by the significant relationship between pH and K whereas the later proven by the high amount of K leached from the column. Similar with P, highest exchangeable K was achieved when the water table fluctuated within 0-40 cm. However, this high K concentration leads to more K being leached from the system. Nevertheless, some amount of K was found to be deposited at different peat depth, showing sign of its high mobility in a peat profile. Even though it was previously reported that two cations; H and Ca occupied most exchange sites in peat soils [3], the amount of Mg in this study was comparable with Ca, for all treatments. Since different cations were associated with different degrees of organic compounds, therefore the status of exchangeable cations in organic materials seems to be complicated to observed [3]. Hence, the dominance of Ca in peat agreed to be varied, depending on nature of the peat. Most Ca and Mg were observed when the Water table fluctuated between 40-80 cm depth (T2) which represented dry season, followed by T1 and T0. The movement of water through fluctuating water table is an important pathway for the elements to move up and down the peat profile. Even though the mobility and concentration was high in T2, the amount of leached Ca and Mg was high in T1 (water table fluctuated between 0–40 cm), which represented the wet season. Consequently, deposition of these bases in mineral soils underneath the peat can be utilized through fluctuating water table as most of them were highly mobile in soil solution. Hence, it is suggested that the high amount of exchangeable bases with minimum leaching was achieved when the water table fluctuated between 40 – 80 cm, near to the mineral soil layer.

Conclusions

Results showed that the fluctuating water table in peat soils exhibit different dynamics of N, P and K as well as chemical properties of the soils. By maintaining the water table at 40 cm level, the concentration of total N and mineral-N (NH4+ and NO3-) was maximized. When the water table was fluctuated at 0-40 cm depth (which is representing the water table during wet season), soil available P reached its highest concentration in the soil. The fluctuation of water table at 40-80 cm depth allows gradual release of exchangeable bases (K, Ca, Mg and Na) accompanied by less leaching. In view of the management practice, this study suggested that the planting, maintaining and harvesting of crop could follow the fluctuating of water table in peat. The fluctuation of this water table may be predicted through the monitoring of the precipitation pattern and may be divided into wet and dry season. Based on the results of this study, the planting could be done before the wet season where the mineral N was highly available. The months following wet season will gradually release soil-P which will further boost the crop growth. The dry season may help in the fruiting stage of the crop through gradual release of exchangeable K as well as other bases which is important for fruit development. Even though the activity in the field can be predicted through this seasonal variation, however, it must be accompanied by good agronomic practices such as the fertilization and liming activities.

Acknowledgements

The authors wish to thank the Faculty of Plantation and Agrotechnology and Forest Research Institute and Management (FRIM) for providing facilities for this research.

References

[1] Andriesse J P 1988 Nature and Management of Tropical Peats *FAO Soils Bulletin* FAO Rome 165

[2] Zani N F, Suratman M N, and Yaacob A 2012 The Influence of Soil Properties on The Above-Ground Tree Biomass in Taman Negara Pahang, Malaysia *IEEE Symposium on Business, Engineering and Industrial Applications* 235–245.

[3] Lucas R E 1982 *Organic Soils (Histosols): Formation, Distribution, Physical, and Chemical Properties and Management for Crop Production* Michigan State University Agricultural Experiment Station

[4] Anton E S, Seca G, Ahmed O H, and Nik Muhammad A M 2009 Effect of Precipitation Fluctuation on Soil Carbon Storage of A Tropical Peat Swamp Forest *Journal Applied Science* **6**(8) 1484–1488

[5] Asante W and Nicholas J 2012 *Carbon Stocks and Soil Nutrient Dynamics in The Peat Swamp Forests of The Amanzule Wetlands and Ankobra River Basin*

[6] Rittmann B E and McCarty P I 2001 *Environmental Biotechnology: Principles and Applications* Mc Graw-Hill, Boston

[7] Prosser J I 2007 Microorganisms cycling soil nutrients and their diversity *Modern Soil Mirobiology, CRC Press, Boca Raton* 237–261

[8] Rutting T, Boeckx P, Muller C, and Klemedtsson L 2011 Assessment of The Importance of Dissimilatory Nitrate Reduction to Ammonium for The Terrestrial Nitrogen Cycle," *Biogeosciences* **8** 1779–1791

[9] Galloway J N and Cowling E B Reactive Nitrogen and The World: 200 Years of Change *AMBIO Hum. Environ* **31** 64–71

[10] Mohd Y S, Manas M A, Sidik N J, Ahmad R, and Yaacob 2015 An Effects of Organic Substrates on Growth and Yield of Ginger Cultivated Using Soilless Culture *Malaysian Appl. Biol*. **44**(3) 63–68

[11] Tiemeyer v, Lennartz B, Schlichting A, and Vegelin K 2005 Risk Assessment of The Phosphorus Export from A Re-Wetted Peatland *Phys. Chem. Earth* **30** 550–560

[12] Welp G, Herms U, and Brummer G 1983 Influence of Soil Reaction, Redox Condition and Organic Matter on The Phosphate Content of Soil Solution *Z. Pfl. Bodenkd* **146** 38–52

[13] N C and R R W Brady 2008 *The Nature and Properties of Soils 14th Edition.*

[14] Ahmed O H, Husni M H A, and Hanafi M M 2006 Phosphorus Loss of Phosphorus Fertilizer Applied to Tropical Peat Soils in Pineapple Cultivation *Int. J. Soil Sci.* **1** 85–90

IOP Conf. Series: Earth and Environmental Science **327** (2019) 012025 doi:10.1088/1755-1315/327/1/012025

Birding Backyard: Birdwatching in Andalas University

M N Janra[1]

[1]Biology Department, Faculty of Mathematics and Natural Sciences, Andalas University. Jalan Kampus Unand Limau Manis Pauh Padang, West Sumatra 25136, Indonesia

Corresponding author's e-mail address: mnjanra@sci.unand.ac.id

Abstract. Andalas University campus comples in Limau Manis area is renowned for its beautiful scenery and close-to-nature environment due to its attachment to the forest of Bukit Barisan mountainous chain in Sumatera. Here in this paper, we descriptively discuss the ornithological aspects of the Andalas University campus complex in its relation for being a birdwatching destination. It combined the literature study on previous ornithological publications made from this area with recent observation exhaustively conducted in some strategic birding sites in the campus complex. In overall, we recorded 160 bird species from surrounding of Andalas University with possibility for more increasing new species sighted in given time; they belong to 44 families and 14 orders. Included in species inventory 1 species with Appendix I and 15 species Appendix II of CITES; 31 species protected by Indonesian Law; 1 species with Critically Endangered status, 2 species with Vulnerable status, 20 species with Near Threatened status of IUCN Red List; 16 migrant species from North and 6 migrant species from south hemisphere. There are five major bird habitats in this area; ponds, rivers or tributaries, green spaces, secondary and primary forests. We also identified direct and indirect threats for bird diversity in this area, such as illegal logging, illegal hunting and poaching, introduced predators, as well as animal abusing. Further attention needs to be paid for protecting bird species and their habitats in Andalas University campus complex through the enforcement of campus conservation policy.

1. Introduction

Andalas University (hereinafter Unand) is the oldest university outside of Java, which officially operated on 23 December 1955. After decades of up and down following regional situation in Sumatra, Unand inaugurated its 500 ha campus complex located on Bukit Karamunting, Kenagarian Limau Manis on 4 December 1995 [1]. This comprehensive complex accommodates and units 15 faculties of the university, which previously scattered in all over places in West Sumatra Province. The campus area attached to the protected forest of Bukit Barisan mountainous chain, where more than 150 ha of forested area is kept intact for various purposes, including as water catchment area, become the extension of the protection forest of Bukit Barisan mountainous chain, as well as to be the practice area for many faculties in the university [2].

After almost three decades since its establishment on early 1990s, the Andalas University's Limau Manis campus complex (hereinafter LMCC) has become the major scientific centre in Sumatra where various researches and academic activities run. Many researchers also did their works on the field of ornithology in this area which result in a considerably concise inventory of birds in LMCC [3]. On the other hand, ecotourism based on birdwatching activity become recent trending in tourism sector with certain potential to be source of income for either the management of tourist attraction, local community and other parties [4]. Recall that ecotourism has degree of flexibility to be assimilated with various existing elements in an area, it is interesting to see the prospect of avitourism to be implemented in LMCC.

This paper outlined the ornithological aspects of the LMCC related to avitourism conduct, including the threats and opportunity for avifauna diversity in the area. The feasibility of avitourism therein was also discussed

2. Materials and Methods

2.1. Study Site

Primary data collection has been conducting in an ongoing observational survey started since 2016. Survey is carried out in and around the Limau Manis Campus Complex (0°54’46”S, 100°27’39”E). Survey and data collection were based on the standardized survey method using transect line [5] and any sighting on bird was recorded using MacKinnon species list [6]. Bird survey was arranged in accordance to the birds’ active time, which were in between 6.30-10.00 am and 03.00-05.00 pm, by walking through available hiking paths in campus and in the surrounding area. Bird sighting was enhanced using Bushnell 8x42 binocular, while Nikon P900 was in use to document any sighting through digiscoping technique. Species identification was guided with suitable references [7, 8], while species conservation and distribution status referred to proper guideline [9].

Secondary data used in this study were extracted from many previous study conducted in this area. They included published articles, bachelor thesis, project reports conducted in the forested area in adjacent to the campus complex [10-14] and those performed within the campus complex itself [15-17]. The data from these sources were sorted and reconfirmed using aforementioned proper guidebooks, in order to ensure its correctness, as of the publications from before 2000 relied more on visual encounters without proper sufficient documentation which useful for further confirmation on species identification.

2.2. Data Analysis

All sorted and confirmed bird records in LMCC were input into table along with its conservation and distribution status. The table also listed year of observation, type of habitat where the observation conducted as well as other important information. The number of species, family and order recorded in each observation were then projected to see the trend of species increment per observation period. We also sought the habitats where birds frequently observed. Other ornithological-related aspects, especially about threats, conservation efforts and others were recorded directly during the ongoing observation between 2016-2018. The result then outlined descriptively to show the current ornithological conditions in LMCC, especially in its relation to the birdwatching activity which is potentially developed further.

3. Result and Discussion

3.1. Bird Inventory

Combining the data sorted from published articles, thesis, reports, along with the sighting records made during 2016-2018 direct survey, we recorded 160 bird species from Limau Manis campus complex (see Appendix). These species belong to 44 families and 14 orders, presumably the most update and complete list of avifauna for this area. Since the first ornithological publication in Limau Manis area [10] until the ongoing research conducted by the author, there was significant increase of species recorded. Three significant leaps on species increment were shown after the three initial observations [10-12], as well as on the current study (between 2016-2018). Each of three initial studies improved species record into 47, 98 and then 129 species respectively (see Figure 1 below).

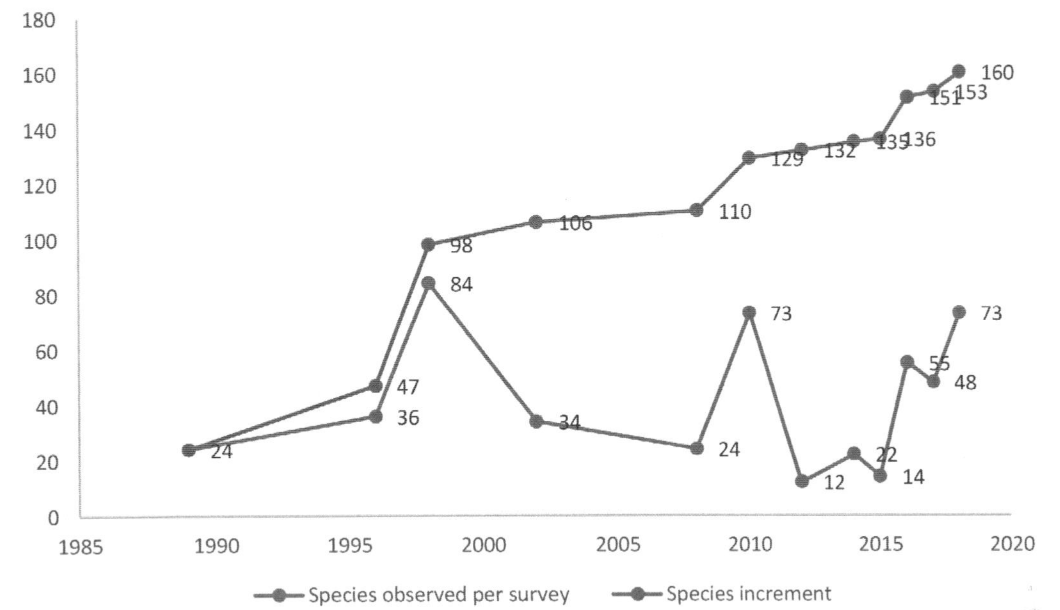

Figure 1. Total species in different ornithological studies in Limau Manis area and species increment contributed by each study

As shown by Figure 1, LMCC holds a considerable number of bird species, with certain potential to reach more species to be recorded, given the more study and observation conducted in this area in the future. Above figure also indicates that ornithological studies in human-interfered area, such as the studies on 2002, 2012 and 2014 [16, 17, 18] still yielded a handful of species which later pooling more new species recorded for LMCC. It is feasible to assume, from these observations, that most parts of the university establishment somehow synchronize with its surrounding nature, in which various bird species can live among these man-made structures. The record of new bird sighting, hence, can occur in every part of LMCC, while the forested area contributes more than human-interfered area [3]. This inventory list, furthermore, reasonably put LMCC as the most studied campus area for its avifauna diversity.

The LMCC area also harbors various birds with high conservation status. The inventory reveals that there are 31 birds protected by Indonesian regulation (Law no. 5 year 1990, Government Regulation no. 7 & 8 year 1999), 23 species listed as Critically Endangered (CR), Vulnerable (Vu) or Near Threatened (NT) species under IUCN Red List, and 16 species marked as Appendix I and II according to Convention on International Trade on Endangered Species or CITES (see Figure 2 below). Some species may have been assigned more than one conservation status, such as Helmeted Hornbill (*Rhinoplax vigil*) which has just raised its conservation status into CR level as well as included as Appendix I species. Furthermore, the LMCC's bird list also recorded 16 migrant species (all from northern hemisphere) and 6 endemics. In overall, the existence of high conservation status species has highlighted the importance of LMCC for avifauna conservation, recall that this area is more to accommodate the activity of scholars.

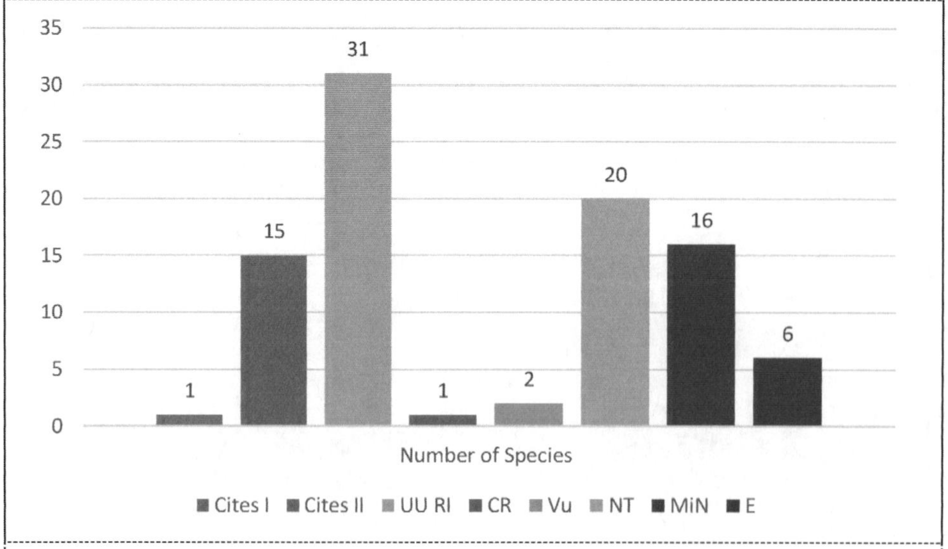

Figure 2. Number of birds with highlighted conservation and distributional status in LMCC

Birds within the proximity of LMCC may have been so far only regarded from its academic value, while its potential for avitourism yet far from being fully explored. The LMCC actually has what required for holding an avitourism activity, in exception of the management system for it. It has walking trails (semi- or non-permanent ones), as well as a progressively growing bird lists [19]. The endemic birds along with the migrant species are potential to provide attraction for specific birdwatchers for their exclusiveness on timing or geographical range. As most of the migrant species in LMCC area are from the northern hemisphere, they will be on their wintering ground around the period between as early as September up until late April on the subsequent year which give an amount of time for watching them. While the endemic species are more local and resident in this area. Regardless to its specificity, this type of avitourism promises potential significant revenue [20].

This study has been able to identify some potential sites for birdwatching in Limau Manis campus complex, either from ongoing bird survey or based on the literature studies. Green open spaces in between campus structures, the green shading along campus roads and its arteries provide habitat for high- to moderate-tolerance bird species, such as Yellow-vented Bulbul (*Pycnonotus goiavier*), Scarlet-backed Flowerpecker (*Dicaeum cruentatum*), Orange-bellied Flowerpecker (*D. trigonostigma*) and Green Iora (*Aeguthina viridissima*), although some uncommon species had been as well recorded from these types of habitat [16, 17, 18]. Lately, the campus authority has been actively conducting the landscape arrangement, adding some artificial ponds and dams into the complex, which later facilitate the kingfishers and other aquatic species to forage more into the campus complex. Meanwhile, a more challenging habitats for birdwatching are provided in the forested area in the proximity of campus complex. These habitats mainly consisted of secondary forest which characterized with dense ground cover mixed with leftover tree stands and a permanent concreted pathway access [14, 15] or primary forest that bordered with campus area with dense tree community and limited access [10-13]. Each of this forested area has rivers or tributary network within, as result from being the water catchment area with dense vegetation standing. In both habitat types, many forest interior bird species can be observed, such as hornbills (Bucerotidae), flycatchers (Muscicapidae) or babblers (Timaliidae). Birds identified as migrant and endemic species were mostly observed from these habitats.

Adversely to above facts, the LMCC area also faces some serious threats which impact not only birds, but also other biodiversity, environment and the people within. The threats are categorized into direct and indirect impacting bird community; the first comes from hunting and poaching many species for caged birds or as mere food item. The indirect threats may be in the form of illegal logging, poor campus waste management, and introduced feral predators that specifically prey upon birds that live near human settlement. Managing birdwatching activity in a human-impacted environment may come collided with classical aforementioned environmental issues, yet the trade-off from implementing it can present much larger benefit for conservation of birds and their habitat [21].

3.2. Suggestion for birdwatching activity in campus area

The LMCC area possesses most of the requirements needed in setting up a professional birdwatching activity inside the campus area, except for the lack of management system obliged in conducting the avitourism. Some favourable efforts, such as reforestation of barren land in campus, prohibition of bird hunting and bird poaching inside campus area will need to be continuously emphasized; while the disadvantageous activities toward bird communities should be minimized through systematic efforts. Campus authority has to start involving various stake holders in this regard, such as outsourcing manpower for managerial structure, preparing a continuous training system for students and other manpower who will potentially become the manpower for the avitourism. Meanwhile, the loggers, bird poachers or hunters should be controlled with and embraced through some sort of university's Corporate Social Responsibility program, where their experience with nature (including with birds) can be optimally in use for avitourism. The prospectus of professional birdwatching activity in LMCC is in line with the entrepreneurship spirit embedded in the vision and missions of Unand, hence it should be among the potential items of future campus development plan. Ecotourism through birdwatching activity in LMCC can be an alternative to reduce impact of human exploration (such from hunting, poaching, even from logging activities), as the doers of these adverse activities will be shifted into eco-friendlier activities through community-based conservation which also serve as income-generating for any involved party [22].

4. Conclusion

There have been recorded 160 bird species belong to 44 families and 14 orders, observed from the surrounding of Unand with possibility for finding more new species in more given time. The Limau Manis campus complex harbors birds with conservation status, as well as migrant and endemic species, which suggest the importance of this area for birds and their habitat conservation. The study identified major bird habitats in this area, namely ponds, rivers or tributaries, green open spaces, secondary and primary forest. Recent development activities in campus were in favor for bird community; some threats were, unfortunately, also recorded. Despite having both opportunity and threats over the possibility for avitourism activity, Unand needs to develop specific management system to run avitourism activity in LMCC, especially by empowering some stake holders, such as students and various parties in the surrounding community.

Acknowledgement

The author would like to thank the members of KCA-LH Rafflesia FMIPA Unand that provided assistance during the 2016-2018 survey. The gratitude also due to Welly Wendri, Aqil Fadhlullah who helped photographing some species.

5. Appendix

Total bird inventory in Limau Manis campus complex per July 2018.

No	Taxonomy (Order, Family, Species)	Year of Observation												Status
		89	96	98	02	08	10	12	14	15	16	17	18	
A	**Ciconiformes**													
i	Ardeidae													
1	*Gorsachius melanolophus*												x	M
2	*Ixobrychus cinnamomeus*			x										

IOP Conf. Series: Earth and Environmental Science **327** (2019) 012025 doi:10.1088/1755-1315/327/1/012025

No	Taxonomy (Order, Family, Species)	Year of Observation												Status
		89	96	98	02	08	10	12	14	15	16	17	18	
B	**Falconiformes**													
Ii	Pandionidae													
3	*Pandion haliaetus*		x											C2, AB, M
Iii	Accipitridae													
4	*Elanus caeruleus*			x										C2, AB
5	*Spilornis cheela*	X	x				x				x	x	x	C2, AB
6	*Ictinaetus malayensis*			x										C2, AB
Iv	Falconidae													
7	*Microhierax fringillarius*											x	x	C2, AB
C	**Galliformes**													
V	Phasianidae													
8	*Coturnix chinensis*						x							
9	*Gallus gallus*	x												
10	*Argusianus argus*	x												C2, AB, NT
D	**Gruiformes**													
Vi	Turnicidae													
11	*Turnix suscicator*	x	x	x							x	x	x	
Vii	Rallidae													
12	*Amarournis phoenicurus*	x		x		x						x	x	
E	**Charadriiformes**													
Viii	Rostratulidae													
	Rostratula bengalensis												x	
F	**Columbiformes**													
Ix	Columbidae													
13	*Treron oxyura*		x											NT, E
14	*Treron capellei*					x								Vu
15	*Treron vernans*		x	x		x	x			x	x	x	x	
16	*Ptilinopus jambu*			x								x		NT, E
17	*Columba livia domestica*										x			
18	*Ducula badia*						x							
19	*Macropygia unchall*	x		x			x						x	
20	*Macropygia ruficeps*			x	x									
21	*Streptopelia chinensis*	x		x		x	x				x	x	x	
22	*Geopelia striata*						x				x	x	x	
23	*Calcophaps indica*			x			x				x	x	x	
G	**Psittaciformes**													
X	Psittacidae													
24	*Loriculus galgulus*			x			x			x	x		x	C2
H	**Cuculiformes**													
Xi	Cuculidae													

No	Taxonomy (Order, Family, Species)	Year of Observation												Status
		89	96	98	02	08	10	12	14	15	16	17	18	
25	*Clamator coromandus*			x										M
26	*Cacomantis merulinus*		x	x		x	x				x	x	x	M
27	*Cacomantis sepulcralis*			x										
28	*Cacomantis sonneratii*						x							
29	*Chrysococcyx maculatus*										x			
30	*Chrysococcyx minutillus*										x			
31	*Surniculus lugubris*						x						x	
32	*Eudynamis scolopaceus*						x					x		
33	*Rhopodytes diardi*			x				x						NT
34	*Rhopodytes sumatranus*						x							NT
35	*Rhinortha chlorophaeus*	x		x			x						x	
36	*Zanclostomus javanicus*	x	x	x							x			
37	*Rhampococcyx curvirostris*			x	x			x					x	
38	*Centropus sinensis*	x	x	x		x	x						x	
39	*Centropus bengalensis*		x	x		x	x						x	
I	**Strigiformes**													
xii	Tytonidae													
40	*Tyto alba*										x		x	C2
xiii	Strigidae													
41	*Ketupa ketupu*										x			C2
42	*Otus lempiji*												x	C2
J	**Caprimulgiformes**													
xiv	Caprimulgidae													
43	*Eurostopodus temminckii*			x			x						x	
44	*Caprimulgus affinis*	x										x	x	
K	**Apodiformes**													
xv	Apodidae													
45	*Hydrochous gigas*						x				x			NT
46	*Collocalia fuciphagus*			x								x	x	
47	*Collocalia maximus*			x										
48	*Collocalia esculenta*	x	x	x		x	x				x	x	x	
49	*Apus nipalensis*	x	x	x		x								
xvi	Hemiprocnidae													
50	*Hemiprocne longipennis*		x				x				x	x	x	
51	*Hemiprocne comata*						x							
L	**Coraciiformes**													
xvii	Alcedinidae													
52	*Alcedo atthis*	x			x							x		AB
53	*Alcedo meninting*				x		x		x		X		x	AB
54	*Alcedo euryzona*			x										Vu, AB
55	*Ceyx erithaca*				X						x			AB, M

No	Taxonomy (Order, Family, Species)	Year of Observation												Status
		89	96	98	02	08	10	12	14	15	16	17	18	
56	*Ceyx rufidorsa*			X			X			X		X	x	AB
57	*Lacedo pulchella*		X	X										AB
58	*Halcyon coromanda*								x					AB
59	*Halcyon smyrnensis*		x	x			X				x	x	x	AB
60	*Todirhamphus chloris*										x	x	x	AB
xviii	Meropidae													
61	*Merops phillippinus*		x	x										M
62	*Merops viridis*			X			X				X		x	
63	*Nyctyornis amictus*						x							
xix	Coraciidae													
64	*Eurystomus orientalis*												x	
xx	Bucerotidae													
65	*Berenicornis comatus*										x			NT, C2, AB
66	*Anorrhinus galeritus*		x								x			C2, AB
67	*Rhyticeros undulatus*			x							x			C2, AB
68	*Anthracoceros albirostris*			x										C2, AB
69	*Buceros rhinoceros*			x		x	x							NT, C2, AB
70	*Rhinoplax vigil*				x									CR, C1, AB
M	**Piciformes**													
xxi	Capitonidae													
71	*Megalaima mystacophanos*			x			x					x	x	NT
72	*Megalaima australis*		x				x					x	x	
73	*Megalaima haemacephala*										x	x	x	
xxii	Picidae													
74	*Sasia abnormis*						x							
75	*Micropternus brachyurus*			x									x	
76	*Picus canus*					x								E
77	*Picus flavinucha*			x										E
78	*Dinopium javanense*										x			
79	*Meiglyptes tukki*								x				x	NT
80	*Dendrocopos canicapillus*						x					x	x	
81	*Hemicircus concretus*										x	x	x	
N	**Passeriformes**													
xxiii	Eurylaimidae													
82	*Cymbirhynchus macrorhynchos*			x			x					x	x	
83	*Eurylaimus javanicus*												x	
84	*Eurylaimus ochromalus*							x				x	x	NT
xxiv	Pittidae													
85	*Pitta moluccensis*										x			AB, M

No	Taxonomy (Order, Family, Species)	Year of Observation												Status
		89	96	98	02	08	10	12	14	15	16	17	18	
xxv	Hirundinidae													
86	*Hirundo rustica*										x	x	x	M
87	*Hirundo tahitica*	x		x		x	x				x	x	x	M
88	*Delichon dasypus*					x								
xxvi	Motacillidae													
89	*Motacilla flava*			x										
90	*Motacilla cinerea*			x			x		X		X		x	
91	*Anthus novaeseelandiae*		x	x									x	
xxvii	Campephagidae													
92	*Pericrocotus divaricatus*												x	
93	*Pericrocotus flammeus*	x												
94	*Hemipus hirundinaceus*			x	x		x				x		x	
xxviii	Aegithinidae													
95	*Aegithina tiphia*		x	x	x		x	x						
96	*Aegithina viridissima*			x	x		x	x	x		x	x	x	NT
xxix	Chloropsidae													
97	*Chloropsis cyanopogon*						x							NT
98	*Chloropsis cochinchinensis*		x											
xxx	Pycnonotidae													
99	*Pycnonotus melanoleucos*							x						NT
100	*Pycnonotus atriceps*	x	x	x	x		x						x	
101	*Pycnonotus melanicterus*		x	x										
102	*Pycnonotus aurigaster*			x		X					x	x	x	
103	*Pycnonotus eutilotus*			x										NT
104	*Pycnonotus goiavier*	x	x	x	x	x	x	x	x	x	x	x	x	
105	*Pycnonotus simplex*			x	x	x	x					x	x	
106	*Pycnonotus brunneus*		x	x	x		x	x		x			x	
107	*Pycnonotus erythropthalmos*			x			x							
108	*Tricholestes criniger*						x		x	x				
xxxi	Oriolidae													
109	*Irena puella*			x										
xxxii	Laniidae													
110	*Lanius tigrinus*			x	x		x	x	x		x	x	x	M
111	*Lanius cristatus*			x	x									M
xxxiii	Turdidae													
112	*Copsychus saularis*	x	x	x	x		x							
113	*Enicurus ruficapillus*			x			x							NT
xxxiv	Timaliidae													
114	*Pellorneum buettikoferi*				x									NT, E
115	*Trichastoma bicolor*								x					NT
116	*Malacocincla malaccense*			x					x					

No	Taxonomy (Order, Family, Species)	Year of Observation												Status
		89	96	98	02	08	10	12	14	15	16	17	18	
117	*Stachyris poliocephala*			x			x		x					
118	*Stachyris erythroptera*				x									
119	*Macronous gularis*			x	x		x			x				
120	*Macronous ptilosus*						x							NT
xxxv	Sylviidae													
121	*Cisticola juncidis*	x												
122	*Prinia familiaris*		x	x	x	x	x		x		x	x	x	E
123	*Prinia flaviventris*		x	x			x		x		x	x	x	
124	*Orthotomus atrogularis*			x										
125	*Orthotomus sericeus*			x			x				x			
126	*Orthotomus ruficeps*		X	x	x	x	x		x	x	x	x	x	
xxxvi	Muscicapidae													
127	*Rhinomyias olivacea*									x	x			M
128	*Muscicapa sibirica*			x					x					M
129	*Muscicapa dauurica*										x	x	x	M
130	*Muscicapa ferruginea*										x		x	M
131	*Eumyias thalassina*			x	x									
132	*Ficedula zanthopygia*												x	M
xxxvii	Monarchidae													
133	*Terpsiphone paradisi*			x										
xxxviii	Dicaeidae													
134	*Prionochilus maculatus*			x			x		x					
135	*Prionochilus percussus*			x	x		x							
136	*Dicaeum agile*											x		
137	*Dicaeum trigonostigma*		x	x	x		x	x			x	x	x	
138	*Dicaeum concolor*				x		x			x				
139	*Dicaeum ignipectus*				x									
140	*Dicaeum cruentatum*		x	x	x		x	x	x	x	x	x	x	
141	*Dicaeum trochileum*		x											E
xxxix	Nectarinidae													
142	*Anthreptes simplex*				x		x						x	AB
143	*Anthreptes malacensis*			x	x		x				x		x	AB
144	*Anthreptes rhodolaema*						x							NT, AB
145	*Anthreptes singalensis*			x	x		x					x	x	AB
146	*Hypogramma hypogrammicum*						x		x					AB
147	*Cinnyris jugularis*						x			x	x	x	x	AB
148	*Aetophyga siparaja*	x	x	x	x		x							AB
149	*Arachnothera longirostra*		x	x		x	x		x					AB
150	*Arachnothera chrysogenys*												x	AB
xl	Estrildidae													

No	Taxonomy (Order, Family, Species)	Year of Observation												Status
		89	96	98	02	08	10	12	14	15	16	17	18	
151	*Lonchura striata*			x	x	x	x		x	x	x	x	x	
152	*Lonchura malacca*					x								
153	*Lonchura punctulata*	x	x	x	x	x	x		x		x	x	x	
154	*Lonchura maja*	x	x	x		x	x		x	x	x	x	x	
xli	Ploceidae													
155	*Passer montanus*	x	x	x	x	x	x				x	x	x	
156	*Ploceus philippinus*			x									x	
xlii	Sturnidae													
157	*Aplonis panayensis*							x			x	x	x	
158	*Sturnus sturninus*						x				x	x	x	
xliii	Dicruridae													
159	*Dicrurus paradiseus*			x										
xliv	Artamidae													
160	*Artamus leucorhynchus*										x	x	x	

Remarks of the table: capital Roman numeral signifies bird orders, lower case Roman numeral points out families and Arabic numbers are for the species. Year of observation: 89, 96 and 98 from before 2000, the rest are from afterward. The conservation status based on IUCN Red List, where CR = Critically Endangered, Vu = Vulnerable, NT = Near Threatened; CITES (Convention on International Trades on Endangered Species, where C1 = Appendix 1 organism, C2 = Appendix 2 organism; Indonesian Regulation, where A = Indonesian Law No. 5 Year 1990, B = Government Decree No. 7&8 Year 1999. The letter M = Migrant species, E = Endemic species.

References

[1] Anonim. (2010). *Salingka Unand.* Unand Publisher. Padang.

[2] Anonim. (2017). Universitas Andalas – Sejarah Ringkas. Diakses pada http://www.unand.ac.id/id/tentang-unand/selayang-pandang/sejarah pada 30 Agustus 2017.

[3] Janra, M.N., SME. Ananta, D.A. Saragih, AAA. Faturrahman, R. Septiavi & G. Komonichi. (2017). Ekologi burung di kawasan Kampus Limau Manis Padang, dengan catatan beberapa jenis baru untuk Kawasan Kampus Unand. *Proceeding Konferensi Peneliti dan Pemerhati Burung Indonesia 3, 2-4 Februari 2017.* Universitas Udayana, Bali.

[4] Wood, M.E. (2002). Ecotourism: Principles, practices & policies for sustainability. The International Ecotourism Society. United Nation Environment Program. Burlington, USA.

[5] Bibby, C., M. Jones & S. Marden. (2000). *Teknik-Teknik Ekspedisi Lapangan Survey Burung.* BirdLife. Bogor.

[6] MacKinnon, J., Phillips, K., 1993. A Field Guide to the Birds of Borneo, Sumatra, Java and Bali, the Greater Sunda Islands. Oxford University Press, Oxford.

[7] MacKinnon J, K. Phillips & S. van Balen. (2010). Burung-Burung di Sumatera, Jawa, bali dan Kalimantan. Puslitbang Biologi-LIPI, Bogor.

[8] Eaton, J.A., S. van Balen, N.W. Brickle & F.E. Rheindt. (2016) *Birds of the Indonesian Archipelago. Greater Sundas and Wallacea.* Lynx Edicions, Barcelona.

[9] Sukmantoro, W., M. Irham, W. Novarino, F. Hasudungan, N. Kemp & M. Muchtar. (2007). *Daftar Burung Indonesia No. 2.* Indonesian Ornithologists' Union. Bogor.

[10] Rahman, M., R. Tamin & A. Salsabila. (1989). Komposisi flora dan fauna di hutan hujan tropika Bukit Limau Manis, Padang.

[11] Aswad, D. (1996). Aspek avifauna di HPPB Universitas Andalas. Laporan Kuliah Lapangan

Singkat Taksonomi Hewan Vertebrata. [Unpublished]
[12] Azmardi. (1998). *Jenis-jenis Burung di Hutan Pendidikan dan Penelitian Biologi.* Skripsi Sarjana Biologi FMIPA Universitas Andalas, Padang.
[13] Sari, N.F. (2008). Jenis-jenis burung yang ditemukan di sekitar sarang buatan dan yang memanfaatkan sarang buatan di hutan pendidikan dan penelitian Biologi (HPPB) Universitas Andalas. Skripsi Sarjana Biologi. Universitas Andalas. Padang. Unpublished.
[14] Sukmawati, S. (2010). Jenis-jenis burung di kawasan kebun tanaman obat farmasi (KTOF) dan arboretum Kebun Raya Universitas Andalas. Skripsi Sarjana Biologi. Universitas Andalas. Padang. Unpublished.
[15] Mardianti, F.S. (2015). Interaksi burung dengan tumbuhan benalu di Kebun Raya Andalas. Skripsi Sarjana Biologi. Universitas Andalas, Padang. [Unpublished]
[16] Afriyeni, V. (2002). Jenis-jenis burung yang memanfaatkan Macaranga javanica (Bi.) M.A. yang sedang berbuah di Kampus Universitas Andalas Limau Manis Padang. Skripsi Sarjana Biologi. Universitas Andalas. Padang. Unpublished.
[17] Surya, D.C., W. Novarino & A. Arbain. (2013). Jenis-jenis burung yang memanfaatkan *Euryia acuminata* DC di kampus Universitas Andalas Limau Manis. *Jurnal Biologi Universitas Andalas* 2(2): 90-95.
[18] Andira, A, J. Nurdin & W. Novarino. (2014). Struktur komunitas burung pada tiga tipe habitat di kampus Universitas Andalas, *Jurnal Biologi Universitas Andalas* 3(3): 227-230.
[19] Conradie, N. (2015). Profiling the international avitourist: preferences of avitourists at the British and Dutch birdwatching fairs. *African Journal of Hospitality, Tourism and Leisure* 4(1): 1-26.
[20] Callaghan, C.T., M. Slater, R.E. Major, M. Morrison. J.M. Martin & R.T. Kingston. (2017). Travelling birds generate eco-travellers: The economic potential of vagrant birdwatching. *Human Dimensions of Wildlife*: 1-12.
[21] Steven, R. 2015. The relationships between birders, avitourism and avian conservation. Doctoral Thesis. Griffith University, Australia.
[22] Sekercioglu, C.H. (2002). Impacts of birdwatching on human and avian communities. Environmental Conservation 29(3): 282-289.

Analysis of Relationship Between Production Factors of Citra Water Apple Business in Hamlet II Paya Salit, Langkat District

J Marisa[1] and S A Sitepu[2]

[1] Department of Agroecotechnology, Faculty of Sains and Technology, Pembangunan Panca Budi University, Medan , Indonesia

[2] Department of Animal Husbandry, Faculty of Sains and Technology, Pembangunan Panca Budi University, Medan, Indonesia

Corresponding author's e-mail: juliamarisa@pancabudi.ac.id

Abstract. The purpose of this study is to analyze the relationship between land area, fertilizer, insecticide, and labor to the production of citra water apple business in Hamlet II Paya Salit Langkat. This research type is descriptive research that is researcher describe condition of variable of cultivation business of citra water apple. Population in this research is all farmer of citra water apple that exist in Hamlet II Paya Salit. The analytical method used is Chi-Square analysis using SPSS 18.0 software. The results showed that there was a significant relationship between land area, fertilizer, insecticide and labor on the production of citra water apple business in Hamlet II Paya Salit Langkat District with the correlation of the relationship between the business area, fertilizer to the production of 91.4%, and the correlation of labor relations with production of 87.4% and the correlation of the insecticide relationship with the production of 52.9%.

Keywords: Citra water apple, Production Factor, Production.

1. Introduction

Farming is a process of combining production factors such as area, labor, fertilizer and insecticide to produce agricultural products. The success of farming depends on three elements such as seeds, fertilizer, and management. Management includes the management of seeding, fertilizing, pruning, and control of plant diseases. In addition, management in farming is inseparable from the socio-economic characteristics of farmers so that it will affect the results obtained later.

Hamlet II Paya Salit, Langkat District is an area where most of the people work as farmers. One of the agricultural commodities that is widely cultivated by the community is citra water apple. Citra water apple is one of the leading commodities in Langkat Regency. Citra water apple farming is much in demand by farmers as a livelihood which is a source of farmers' income. The prospect of the farmer working on citra water apple farming can increase farmers' income. The advantage of the citra water apple is that it has a large size weighing around 200 grams to 350 grams, having a sweet and fresh taste. In addition, the content of vitamin C is also very high which is good for health. This results in a very high demand for citra water apple and has an expensive price compared to other types of water guava.

The total production of water apple in Langkat Regency in 2015 was 2,456 tons and increased in 2016 to 3,845 tons. Production of water apple which is quite high in Langkat Regency is not a guarantee that productivity is also high. Decrease in productivity is thought to result from the use of production factors that are not efficient at the level of cultivators of the citra water apple. In the people's farming business, farmers usually function as decision makers who try to make effective and efficient decisions in running and managing their farming. The socio-economic characteristics of farmers can influence farmers in making decisions that can benefit their farming. So that the socio-economic characteristics will affect the income obtained by farmers so that the relationship between the factors of production and the production of citra water apple should be identified.

The use of appropriate and efficient production factors influences the income of farmers in the citra water apple. If production increases, it can eventually increase farmers' income in the citra water apple. The relationship between land area, fertilizer, insecticide, and labor to the production of citra water apple business in Hamlet II Paya Salit Langkat District.

2. Research Methods

2.1. Location Selection

The study was conducted in Pulau Sembilan Hamlet II Paya Salit langkat District. The location selection in this study was conducted purposively because Hamlet II Paya Salit is a place for cultivating citra water apple which has the largest harvest and production area in Langkat District.

2.2. Data Collection

The method of data collection was conducted in a survey of 45 citra water apple farmers. The data used in this study include primary data and secondary data. Primary data was obtained through interviews and filling out questionnaires by respondents, which aimed to obtain information and input on constraints and efforts that must be made in developing the cultivation of citra water apple on the Hamlet II Paya Salit Langkat District. Secondary data was obtained from the Langkat District Agricultural Service and related institutions as well as the literature supporting the research.

2.3. Data Analysis Method

2.3.1. Production Function. To see the relationship between input and output production, it would require a form of production function. According to the production function is a mathematical relationship between input and output [1]. Whereas according to the production function is a physical relationship between input and production [2].

Presentation of production functions can be done in various ways, among others in the form of graphs, tables or in systematic equations. Systematically, the production function can be indicated by the equation:

$$\hat{Y} = f(X_1, X_2, X_3, X_4, X_5) \quad (1)$$

Description:

$\hat{Y}$: Production (*output*)
X_1 : Area
X_2 : Labor
X_3 : Fertilizer
X_4 : Insecticide

The production function shows that the amount of production depends on the factors of production which are a function of the factors of production used in the production process. In conducting an agricultural business, a farmer will always try to allocate the inputs he has as efficiently as possible to be able to produce maximum output.

2.3.2. Chi-Square Analysis. Data were analyzed quantitatively using the SPSS 18.0 for Windows (Spread Sheet For Statistic) computer program with Chi-square analysis test model. Determining the significance of a value equal to the calculated value is estimated using Table C in the statistical table with db = k-1 and set = 0.05. If the probability associated with the calculated value is greater than the Table value, then reject H0 and accept H1.

Chi Square analysis is a non parametic statistical analysis, used to test whether the frequency of observed data from a categorical variable corresponds to the frequency of expectations. Chi Square Test Formula:

$$X^2 = \sum\left[\frac{(f0 - ft_{ax})^2}{ft_{ax}}\right] \qquad (2)$$

Description:
X^2 = Chi Squared
f_0 = Frequency of Observation Results
ft_{ax} = Expected frequency in the study population, by sharing the number of subjects in the sample and subject categories.

The Chi Square test results in the value of Asympyotic Significance (Asymp. Sig.) Which shows whether there is a relationship between the two factors studied, and then compared with the value (0.05). Benchmark decision making Based on the value of Asymp. Sig. is Asymp. Sig. smaller than value (0.05), then Ho is rejected.
The decision making criteria are:
H_0 accepted if $X^2 \leq X^2$
H_1 accepted if $X^2 \geq X^2$

3. Results And Discussion

3.1. Analysis of the Relationship between Production Factors and Production.
Production of citra water apple has a relationship with several factors such as the use of production facilities consisting of area, labor, fertilizer and insecticide. The citra water apple area can be planted with 200 trees per hectare. Fertilizers used by farmers are compost, urea and NPK. Using one compost, the use of urea and NPK is 2 grams. Insecticide is used to prevent loss of citra water apple flowers and fruit. In addition, it is also used to prevent white lice that cause stems to dwarf due to the presence of parasites that feed on plant nutrients. Labor used in the business of citra water apple are workers and family members. The results of the research processed using Chi-Squre analysis using SPSS version 18.0 soft ware can be seen in table 1.

Table 1. Results of Chi-Square Analysis of The Relationship Between The Factors of Image Production of Apple Water

Variable	Chi-Squre Value	Significant Value	*Contingency Coefficient*
Area	267	0,000	0,914
Labor	162	0,000	0,874
Fertilizer	267	0,000	0,914
Insecticide	43	0,000	0,529

Source: Primary (processed) data, 2018.

3.1.1. Relationship between Area and Production. Chi-Square test analysis showed an area of 267 by the significance value of 0.000. These results indicate that H0 is rejected and H1 is accepted because the significance value is smaller than α0.05. It can be concluded that there is a real relationship between the area with the production of a citra water apple on the Hamlet II Paya Salit Langkat District. The strength of the relationship between the two variables can be done by looking at the value of the contingency coefficient [3]. The value of the contingency coefficient obtained in the study is 0.914. This shows that there is a closeness in the relationship between the area and the production of a citra water apple of 91.4%.

The results of the study are the same as the research of cashew nut [4] which states that cashew farming area has a positive effect on $\alpha = 5\%$, meaning that farmers with large land will have the opportunity to adopt high technology. With increasing land area, the level of technology adoption by farmers will be higher. This is due to the wider farming land that will increase the number of plants that farmers can cultivate with the possibility of production which also increases. Expectations of benefits and revenues from an innovation are one of the determinants of technology adoption rates [5].

3.1.2. Relationship between Labor and Production. The Chi-Square labor analysis test showed a value of 162 with a significance of 0,000. These results indicate that H0 is rejected and H1 is accepted because the significance value is smaller than α0.05. It can be concluded that there is a real relationship between labor and the production of a citra water apple on the Hamlet II Paya Salit Langkat District. The strength of the relationship between the two variables can be done by looking at the value of the contingency coefficient [3]. The value of the contingency coefficient obtained in the study was 0.874. This shows that there is a closeness in the relationship between labor and the production of a citra water apple of 87.4%.

Labor is one of the factors that influence the quantity of products produced by a company. Product quality is influenced by labor because the higher the skills and expertise of labor, will produce better products [6]. The income obtained by a company is determined by the efficiency of labor use, because the more labor is used, the more expenditure will be spent [7]. Many citra water apple farmers use labor from family members. Because of using family members, farmers do not count labor as expenditure because they do not pay wages.

3.1.3. Relationship between Fertilizer and Production. The Chi-Square fertilizer analysis showed a value of 267 with a significance of 0,000. These results indicate that H0 is rejected and H1 is accepted because the significance value is smaller than α0.05. It can be concluded that there is a real relationship between fertilizer and the production of a citra water apple on the Hamlet II Paya Salit Langkat District. The strength of the relationship between two variables can be done by looking at the value of the contingency coefficient [3]. The value of the contingency coefficient obtained in the study is 0.914. This shows that there is a closeness between the relationship between fertilizer and citra water apple production of 91.4%.

Plant needs for nutrients greatly affect crop production. Nutrients absorbed by plants will be used in the process of growth and development as well as the process of forming food reserves. Nutrient intake in plants can be given through fertilization [8]. Fertilizers are one of the determining factors for increasing the production of citra water apple. There are several things that must be considered to increase production through fertilization, namely the type of fertilizer, fertilizer dosage, fertilization method, fertilization time, and fertilization intensity. The development and progress of fertilization technology and the occurrence of changes in nutrient status in the soil cause the fertilization process to be considered and refined [9]. Continuous fertilizer treatment includes the type of fertilizer used, the amount of fertilizer, the time of fertilizer application, and the method of fertilizer application which will result in optimal quality growth of the citra water apple. Provision of 200 gram NPK fertilizer (15:15:15) every 2 months for plants less than 2 years old will produce maximum production.

3.1.4. Relationship between Insecticide and Production. The insecticide Chi-Square analysis showed a value of 43 with a significance of 0,000. These results indicate that H0 is rejected and H1 is accepted because the significance value is smaller than α0.05. It can be concluded that there is a real relationship between insecticide and the production of a citra water apple on the Hamlet II Paya Salit Langkat District. The strength of the relationship between the two variables can be done by looking at the value of the contingency coefficient [3]. The value of contingency coefficient obtained in the study was 0.529. This shows that there is a closeness between the relationship between insecticide and the production of a citra water apple of 52.9%.

Pest and disease control that attacks citra water apple plants is only done when it has caused losses. The use of insecticide production factors is the most widely used method in controlling pests and diseases in plants [10]. The use of insecticides is the easiest and most effective way, with the use of effective insecticides that will produce satisfactory results. The main pest water apple is young leaf rollers, white fleas, stem borer and the most common are fruit fly pests which can cause fruit damage. Fruit wrapping can be done to prevent pests other than using methyl eugenol and trapping fly trapping glue [11].

4. Conclusion

The area, fertilizer, insecticide, and labor used are significantly related to the production of citra water apple cultivation in Langkat Regency. The closeness of the relationship between area, fertilizer with production of 91.4%, and the closeness of the relationship between insecticides and production of 52.9% and the closeness of labor relations with production of 87.4%.

Reference

[1] Nicholson, W. 2004. *Intermediate Microeconomics And Its Application Eigh*t. Edition. Harcourt Inc.

[2] Soekartawi. 2003. *Agribisnis Teori Dan Aplikasinya.* Penerbit Pt. Raja Grafindo Persada , Jakarta.

[3] Supriana, T Dan Barus, R. 2010. *Statistik Nonparametrik: Aplikasi Dalam Bidang Sosial Ekonomi Pertanian*. Universitas Sumatera Utara Press, Medan.

[4] Sudjarmoko, B. (2016). Analisis Adopsi Teknologi Jambu Mete Di Nusa Tenggara Timur. *Buletin Penelitian Tanaman Rempah Dan Obat*, *21*(1), 69-79.

[5] Feder, R.G., R.E. Just, And D. Zilberman. 1985. Adoption Of Agricultural Innovation In Developing Countries. A Survey. American Journal Of Agricultural Economics 33 : 255 – 298.

[6] Tajerin Dan M. Noor. 2005. *Analisis Efisiensi Teknis Usaha Budifaya Pembesaran Ikan Kurapu Dalam Keramba Jaring Apung Diperairan Teluk Lampung*. Jurnal Ekonomi Pembangunan.

[7] Fathoni, A. H, 2004. Manajemen Sumber Daya Manusia. Penerbit Pt. Rineka Cipta, Jakarta.

[8] Hernawan, F., & Santosa, P. B. (2013). *Analisis Faktor-Faktor Yang Mempengaruhi Usaha Jambu Air Merah Delima (Kasus Desa Betokan, Kabupaten Demak)* (Doctoral Dissertation, Fakultas Ekonomika Dan Bisnis).

[9] Suheli, M., Hastuti, D., & Nurjayanti, E. D. (2013). Analisis Kelayakan Usahatani Jambu Air Merah Delima (Syzygium Samarangense (Blume) Merr. & Perry.) Di Kabupaten Demak (Studi Kasus Di Kelurahan Betokan Kecamatan Demak Kabupaten Demak). *Mediagro*, *9*(2).

[10] Fauzia, L., Dan H. Tampubolon., 1991. Pengaruh Keadaan Sosial Ekonomi Petani Terhadap Keputusan Petani Dalam Penggunaan Sarana Produksi.

[11] Candra, D., Sutikno, A., & Salbiah, D. (2014). Uji Daya Tahan Beberapa Bahan Pembungkus Dalam Mengendalikan Lalat Buah (Bactrocera Spp.) Pada Tanaman Jambu Biji (Psidium Guajava L.) Di Sentra Pengembangan Pertanian Universitas Riau. *Pest Tropical Journal*, *1*(2), 1-12.

Percentage Value of Membrane Integrity and Acrosome Integrity Spermatozoa in Simmental Liquid Semen with Addition Penicillin and Sweet Orange Essential Oil

S A Sitepu[1] and J Marisa[2]

[1] Department of Animal Husbandry, Faculty of Sains and Technology, Pembangunan Panca Budi University, Medan

[2] Department of Agroecotechnology, Faculty of Sains and Technology, Pembangunan Panca Budi University, Medan

Corresponding author's e-mail: sukmaaditya@dosen.pancabudi.ac.id

Abstract. Growth of bacteria in the Simmental liquid semen is is the cause of the quality of liquid cement to be bad. The addition of antibiotics has been done to inhibit the growth of bacteria in the liquid semen. One of antibiotic that often used is penicillin. The addition of sweet orange essential oil is expected to further suppress the growth of bacteria because it contains antibacterial. The research objective was to determine the value of the percentage Membrane Integrity and Acrosome Integrity on Simmental liquid semen with the addition penicillin and sweet orange essential oil. The method used was experimental using Completely Randomized Design with 5 treatments and 5 replications. The treatment was addition 0%, 0,25%; 0.5%; 0.75% and 1% sweet orange essential oil on tris yolk and penicillin extender. The results showed that addition sweet orange essential oil and tris yolk and penicillin extender resulted the best percentage value of the Membrane Integrity and Acrosome Integrity in Simmental liquid semen was the addition of sweet orange essential oil 1%.
Keywords: Essential Oil, Liquid Cement, Simmental, Sweet Orange, Penicillin..

1. Introduction

The low population and genetic quality of local cattle is one of the causes of the demand for beef in Indonesia is still not sufficient.The characteristics of local cows in Indonesia such as low weight growth and low body weight result in not much meat produced. Demand for meat continues to increase along with population growth, public awareness of the importance of nutritious food for the body and increasing community income. Therefore it is necessary to think about how to improve genetic quality and cattle population.

To improve the genetic quality of local beef cattle can be done by mating with the kind of superior cattle. One of the superior cows is Simmental Cow because it has high body weight, high growth and high carcas. However Simmental Cows are hard to find in Indonesia. Besides that the price of good quality Simmental Cows is very expensive and to maintain it also requires a lot of costs. To optimize the use of Simmental cattle for mating with limited quantities, can be done with Artificial Insemination.

Increased population and productivity of Simmental Cows can be done using Artificial Insemination. Mating with Artificial Insemination can optimize the use of a superior bull to mating many cows. Artificial Insemination has several advantages such as preventing the occurrence of diseases caused by natural mating, optimizing the use of Simmental bull studs for mating programs because in one semen shelter can be used for many cows as well as improving the genetic quality of livestock because it inherits good properties from Simmental Cows.

There are several factors that influence the success percentage of artificial insemination program, including inseminator skills, the timing of Artificial Insemination, and the factors of cattle farmers. In addition, the quality of liquid semen Simmental Cattle used is also very influential to determine the success of Artificial Insemination. The quality of liquid cement is determined by the raw materials used such as fresh cement and the material of semen extender. Liquid cement thinner material must be able to provide protection, especially at the time of cold shock [1]. Liquid cement must be used immediately for Artificial Insemination because the longer the storage period the quality will continue to decline if it is not immediately used [2].

Simmental Cattle liquid semen quality may be lower if contaminated with bacteria in large quantities. Bacterial contamination can occur during the process of fresh semen storage, preparation of liquid semen extender, equipment and the environment. What can be done to inhibit bacterial growth is the addition of antibiotics in liquid semen extender of Simmental Cattle. The antibiotics commonly used in liquid semen extender are penicillin, streptomycin and gentamicin. Penicillin is an antibiotic that can be used as an antibacterial. The addition of the antibiotics less effective because only active against certain types of bacteria. Efforts have been made to combine the use of antibiotics such as GTLS (Gentamisin, Tylosin and Linco-Spectin) [3], penicillin and streptomycin [4] and Andromed® (gentamicin sulfate, spectinomycin, and linkomycin) [5].

The addition of other materials that contain antibacterial can be used to minimize the growth of bacteria in the liquid semen Simmental Cattle. Besides giving antibiotics such as penicillin, can also add sweet orange essential oil. Addition of sweet orange essential oil has been carried out by [6] on Boer Goat semen, which shows increased quality with additions of up to 1%. Additionally sweet orange essential oil can also be added to the Simmental Cattle semen extender to improve the quality [7].

The content of linalool and limonane is very high in sweet orange essential oil so that it can be used as an antibacterial [8]. The results of the study by [7] show that the addition of sweet orange essential oil to the tris yolk extender can improve the quality of Simmental Cows semen. The addition of sweet orange essential oil and penicillin is expected to further improve the quality of Simmental Cows liquid semen. To see the quality of liquid semen was carried out by microscopic analysis, namely observations made by observing spermatozoa using a microscope. The aim of the study was to determine the effect of addition of sweet orange peel essential oil on tris yolk and penicillin extender to the percentage value Membrane Integrity and Acrosome Integrity in Simmental liquid semen.

2. Methods

2.1 Research Method

This study used a completely randomized design with 5 treatments and 5 replications. The treatments are:

P_0 = Penicillin + Sweet Orange Essential Oil 0%
P_1 = Penicillin + Sweet Orange Essential Oil 0.25%
P_2 = Penicillin + Sweet Orange Essential Oil 0.5%
P_3 = Penicillin + Sweet Orange Essential Oil 0.75%
P_4 = Penicillin + Sweet Orange Essential Oil 1%

2.2. Evaluation of microscopic sperm parameters

Membrane integrity: Evaluation of the integrity of the plasma membrane of the spermatozoa. The evaluation was done using methods hypoosmotic swelling test (HOST). Tests done by mixing 0.1 ml to 9.9 ml cement hypoosmotic medium. Once mixed, the preparations are incubated in a water bath temperature of 37 ° C for 30 minutes.
Acrosome integrity: Evaluation of the integrity of the sperm acrosome hood characterized by head spermatozoa black thick cement when exposed in physiological saline solution containing 1% formalin [9]. The evaluation was done at a minimum of 200 spermatozoa by using a light microscope magnification of 400 times.

3. Results and Discussion

The results of membrane integrity and acrosome integrity of spermatozoa in liquid semen Simmental cattle before equlibration and after equlibration can be seen in Table 1.

Table 1. Recapitulation of Membrane Integrity and Acrosome Integrity of Spermatozoa Simmental Cattle Before Equlibration and After Equlibration.

Parameters	Treatment	Observation	
		Before equlibration	After equlibration
Membrane Integrity	P_0	70	68
	P_1	75	72
	P_2	77	73
	P_3	78	75
	P_4	81	78
Acrosome Integrity	P_0	66	61
	P_1	70	65
	P_2	72	67
	P_3	73	70
	P_4	75	72

Note: Different superscripts on the column show very significant differences (P <0.01)

3.1. Membrane Integrity

Observation of membrane integrity spermatoza is very important to test the quality of liquid cement in Simmental Cattle. The lowest percentage value without the addition of sweet orange essential oil and its value increases with increasing sweet orange essential oil given. The results of various membrane integrity analysis showed the effect of sweet orange essential oil on the extender material had a very significant effect (P <0.01) both on observation before and after equality. The highest average value of membrane integrity of Simmental cattle before equality is shown in treatment P4, which is 81% and after equality of 78%. The presence of toxic substances originating from dead spermatozoa or from substances contained in the extender that have undergone oxidation due to storage causes high levels of free radicals which can damage the integrity of the plasma membrane of the spermatozoa. If spematozoa membrane integrity has been damaged, then the sperm metabolism will be disturbed and begin to lose motility, resulting in the death of spermatozoa [10].

The results of the study showed a decrease in the quality of liquid semen before the equlibration until after equlibration. The decrease in the value of membrane integrity percentage is related to the motility of spermatozoa. The motility of spermatozoa after equilibration has decreased due to the egg-yolk coagulating enzyme factor in toxin plasma semen and cold shock. The decrease in motility is caused by the reduced energy supply of spermatozoa needed to maintain the life and movement of the spermatozoa. Decreased spermatozoa motility Simmental cattle are also caused by treatments that cause damage and death of spermatozoa. During the spermatozoa equilibration process Simmental cattle are very susceptible to cell damage due to changes in osmotic pressure. Only spermatozoa have a strong plasma membrane that can survive to fertilize the egg cell [11].

The difference in percentage of membrane integrity in each treatment was caused by differences in the concentration of sweet orange essential oil in each treatment. The membrane integrity is very necessary for spermatozoa because damage to the plasma membrane will affect the metabolic process and is related to the viability of spermatozoa. If the membrane integrity is damaged then the cell metabolism process will be disrupted and result in the death of spoermatozoa. Acrosomes contain many enzymes that are needed during fertilization, therefore the success of acrosome fertilization process must be intact, because the enzymes in it function to lyse the oocyte protective layer [12].

Cell metabolism will take place well if the plasma membrane of the cell is in an intact state so that it is able to properly regulate traffic in and out of the cells of all the substrates and electrolytes needed in the metabolic process [13]. The metabolic process will be disrupted if the plasma membrane is damaged. ATP synthesis does not run normally so that the sperm will become damaged so that the value of the percentage of viability and motility of the sperm in the liquid cement of the Simmental Cow decreases.

Spermatozoa with a high percentage value of viability indicate that the plasma membrane is still intact, so that spermatozoa cell organelles will be protected and ions for metabolic processes and food requirements are available [14]. The high lactose concentration causes changes in osmotic pressure on the extender in the hypertonic direction which indicates more particles outside the cell than in the cell so that water is released from the cell so that the cell will shrink [15].

During the maturation process of spermatozoa in the epididymis, there is also a change in the composition of the components of the constituent compounds of spermatozoa plasma membranes. The plasma membrane of spermatozoa cells will lose some cholesterol, so there is an increase in the ratio between unsaturated fatty acids and cholesterol. This causes the plasma membrane to become more fragile. Such a condition of plasma membrane is physiologically needed to facilitate spermatozoa undergoing the capacitation process in the uterus and at the time of fusion with the oocyte plasma membrane during fertilization [16].

3.2. Acrosome Integrity

Acrosome Integrity percentage value is very important because it plays a role in the process fertility. The results of the variance analysis showed that the addition of sweet orange essential oil to extender had a very significant effect ($P < 0.01$) on the percentage value of Acrosome Integrity before equlibration and after equlibration. The highest percentage value of Acrosome Integrity is in treatment P4 before equlibration which is 75% and 72% after equlibration.

The results showed that the percentage of the Acrosome Integrity would be higher if there was an increase in the level of sweet orange essential oil in the Simmental semen extender. Damage to the Acrosome Integrity is usually accompanied by damage to the acrosome hood so that the observation of the Whole Acrosome Hood is important because damage to the acrosome can result in the loss of proteolytic enzyme enzymes and most occur during dilution of cement which causes failure of Artificial Insemination [17].

The Acrosome Integrity is the layer that covers the nucleus, inside it is a collection of enzymes that function to help the nucleus enter the cytoplasm of the egg at fertilization, by damaging the layer of wrapping the egg through an acrosome reaction. The percentage of Acrosome Integrity shows a significant correlation with fertility. The most important aspect in evaluating semen quality is the investigation of the ability to experience an acrosome reaction. This process is related to activation and release of proteolytic enzymes from acrosomes which are very crucial during the penetration of eggs [12]. In the Acrosome Integrity there are hyaluronidase enzymes, acrosine and Corona Penetrating Enzyme which are needed to penetrate the culumus oophorus and pellucida zone. Therefore this part of the Acrosome Integrity is an important part to consider in evaluating liquid semen [14].

Viability of spermatozoa is also influenced by the pH of semen. Changes in pH to a more acidic direction occur because of the accumulation of lactic acid which is the result of cell metabolism, namely the breakdown of fructose. Lactic acid is toxic to spermatozoa. With the percentage of Acrosome Integrity decreasing, this results in decreased spermatozoa fertilization [12]. Physiologically there is a relationship between the motility and integrity of the plasma membrane and the survival of spermatozoa. Damage to the plasma membrane will cause the loss of necessary enzymes [18].

4. Conclusion

Addition of essential oils of sweet orange and penicillin in Simmental liquid semen diluent can increase the percentage value of membrane integrity and acrosome integrity. The best results obtained is the addition of essential oils as much as 1% and the lowest is without the addition of essential oils.

References

[1] Mumu, M.I. 2009. Viabilitas Semen Sapi Simental Yang Dibekukan Menggunakan Kriopektan Gliserol. Journal Agroland 16 (2) : 172-179.

[2] Sitepu, S. A., Udin, Z., Jaswandi, J., & Hendri, H. (2018). Quality Differences of Boer Liquid Semen During Storage with Addition Sweet Orange Essential Oil In Tris Yolk And Gentamicin Extender. *Journal of Community Research and Service*, *1*(2), 78-82.

[3] Hasan, S., S.M.H. Andrabi, R. Munir, M. Jehangir, P. Shafique, M. Anzar and N. Ahmad. 2000. Effect Of New Antibiotic Combination On Post-Thaw Semen Quality Of Buffalo And Sahiwal Bulls. 33rd Annual Meet. Soc. Study Reprod., 62: 157. Abstract.

[4] Paulenz, H., Söderquist, L., Ådnøy, T., Fossen, O. H., & Berg, K. A. 2003. Effect of milk-and TRIS-based extenders on the fertility of sheep inseminated vaginally once or twice with liquid semen. *Theriogenology*, *60*(4), 759-766.

[5] Minitub. 2001. Certificate Andromed. Minitub Abfullund Labortechnik GmbH andCo KG. Germany.

[6] Sitepu, S. A., & Zaituni, U. (2018, February). Improved quality of frozen boer goat semen with the addition of sweet orange essential oil on tris yolk and gentamicin extender. In *IOP Conference Series: Earth and Environmental Science* (Vol. 122, No. 1, p. 012125). IOP Publishing.

[7] Sitepu, S. A., & Putra, A. (2017). Influence of Addition Sweet Orange Essential Oil on Tris Yolk Extender to Quality of Post-Thawing Semen Simmental Cattle. *Jurnal Peternakan Indonesia*, *19*(3), 145-151.

[8] Fisher, K., and C. A. Phillips. 2006. The effect of lemon, orange and bergamot essential oils and their components on the survival of Campylobacter jejuni, Escherichia coli O157, Listeria monocytogenes, Bacillus cereus and Staphylococcus aureus in vitro and in food systems. J Appl Microbiol. 2006 Dec;101(6):1232-40.

[9] Saacke, R. G., and J. M. White. 1972. Semen Quality Tests and Their Relationship to Fertility. Proceeding 4th Tech Conf on AI and Reprod NAAB.

[10] Yulnawati and Setiadi MA. (2005). Motilitas dan keutuhan membran plasma spermatozoa epididimis kucing selama penyimpanan pada suhu 4 C. *Media Kedokteran Hewan*, *21*(3).

[11] Maxwell W. M. C. and P. F. Watson. 1996. Recent Progress in the Preservasion of Ram Semen. Animal reproduction.

[12] Sugiarti, T., Triwulanningsih, E., Situmorang, P., Sianturi, R. G., & Kusumaningrum, D. A. (2004). Penggunaan katalase dalam produksi semen dingin sapi. In *Pros. Seminar Nasional Teknologi Peternakan dan Veteriner. Bogor* (pp. 4-5).

[13] Surachman, M., Herdis, Yulnawati, M. Rizal, dan H. Maheshwari. 2009. Kualitas Semen Cair Asal Epididimis Kerbau Belang dalam Bahan Pengencer Andromed yang Mendapat Penambahan Sukrosa. Media Peternakan: Vol. 32 No. 2.

[14] Tambing, S.N. 1999. Efektivitas berbagai dosis gliserol dan waktu ekuilibrasi terhadap kualitas semen beku kambing Peranakan Etawah. Thesis Pascasarjana IPB-Bogor.

[15] Supriatna, I. dan F, H. Pasaribu. 1992. In Vitro Fertilisasi, Transfer Embrio dan Pembekuan Embrio. Pusat Antar Universitas, Institut Pertanian Bogor, Bogor.

[16] Rizal, M. dan Nasrullah. 2004. Pemanfaatan Spermatozoa Epididimis Dalam Teknologi Reproduksi. Wartazoa: Vol. 14 No. 1.

[17] Valcarcel, A., M.A. De Las Heras, L. Perez, D.F. Moses and H. Baldassarre. 1997. Assessment ofthe acrosomal status of membrane-intact ram spermatozoa after freezing and thawing by simultaneous lectin/Hoechst 33258 staining. Anim. Reprod. Sci. 45: 299-309.

[18] Salisbury, G. W. and N. L. Van DeMark. 1985. Fisiologi Reproduksi dan Inseminasi Buatan pada Sapi. Terjemahan R. Djanuar. Gadjah Mada University Press, Yogyakarta.

IOP Conf. Series: Earth and Environmental Science **327** (2019) 012028 doi:10.1088/1755-1315/327/1/012028

The Influence of Preliminary Treatment on Cassava (*Manihot utilissima*) on the Quality of Analog Rice Produced

M Harni[1], S K Putri[2], F Violalita[2], Evawati[2], and N Tazar[2]

[1, 2] Food Technology, Agricultural Polytechnic of Payakumbuh, Jl Raya Negara Km 7 Tanjung Pati, Payakumbuh, Indonesia

Corresponding author's e-mail address: mimiharni2009@gmail.com

Abstract. The research of analog rice cassava has been conducted in the Laboratory of Food Processing Agricultural Polytechnic State of Payakumbuh, and the Laboratory of Agricultural Technology Andalas University, Padang. Research conducted with the aim to search for materials that can shorten the fermentation time of cassava which is usually by natural fermentation, so as to produce analog rice with the best quality. The design used in this research is Complete Random Design (*CRD*) with 5 treatments and 3 replications. One treatment as a control is cassava that is soaked naturally and 4 treatments with the addition of material in the form of curd, yeast of bread, yeast of *tempe* and yeast of *tape*. Advanced test carried out by means Duncan's New Multiple Range Test (DNMRT) on the real level of 5%. Based on the research that has been done the treatment chosen is the treatment with the addition of bread yeast in fermentation (C). The results of the chemical test of analog rice with the treatment of the addition of bread yeast (C) are: water content 13,42%, ash content of 0,75%, a protein content of 1,30%, a fat content of 4,53% and starch content 70,25%.

1. Introduction

Analog rice is one of the efforts to reduce dependence on one staple food and to vary food consumption. Analog rice using raw material source other than rice, but is derived from raw materials which contain other carbohydrates such as cassava, corn, taro, sago and others. Usually analog rice has a value of carbohydrates is approaching even exceeds that of rice.

In the process of analog rice manufacture with granulation derived from cassava through a process of immersion for a few days viewing of the age of the cassava used. Immersion makes the process long and followed with the replacement of the water every day as well as cause a bad odor.

Efforts to speed up the immersion process is performed by means of adding microbes such as curd, bread yeast, yeast of tempe and tape yeast. A variety of pretreatment in the research is expected to shorten the fermentation time and produces analog rice with good quality. The purpose of the research that has been done is to determine the effect of some pretreatment on the quality of analog rice that is produced.

2. Material and Methods

2.1. Time and place

This research has been conducted on the Labor of Food Processing Agricultural Polytechnic State of Payakumbuh, and Labor of Agricultural Technology Andalas University, Padang.

2.2. Materials and tools

The materials used in this research are cassava, filter cloth, cooking oil, curds, bread yeast, yeast of tempe and tape yeast. The tools used in this research are pan, stove, mortar, scales, a plastic container, cheese grater, tray, a cauldron and a spoon of fried.

2.3. Implementation of the research

1. Fresh cassava peeled, cleaned and inserted into a plastic container (basin)
2. The cassava is then soaked in clean water according to the treatment until the texture of the cassava is soft and dumped the water until dry.
3. The process continued with cassava grating that are not destroyed in the process of soaking while destroyed cassava in the process of soaking directly done washing and the disposal of the pith of the cassava as well as rough fiber.
4. Crushed cassava is next pulverized with the use of wood to further smooth the texture of cassava. The next printed above the filter cloth or paper in bamboo tray. then analog rice that has been printed dried by the wind.
5. Further roasting by using as much oil as 10 ml/200 grams for 8 minutes. The rice is then dried under sunlight until dry.

2.4. Research observation

The observations made on the analog rice of cassava are: water content test, ash content test, protein content test, fat levels test and levels of starch test. The method of analysis as follows:

1. Water content [1]
   Determination of water content was performed by the method of termogravimetri by using the oven
2. Ash content [1]
   Determination of ash content was done by weighing the material remaining from the combustion of organic material
3. Protein content [1]
   Determination of protein content determined by the method of Kjeldahl
4. Fat content [1]
   The fat content is determined by extraction method using Soxhlet
5. Starch content [1]
   Starch content determined by the method of Luff Schrool

3. Results and discussion

3.1. Water content

The results of the variety investigation showed that the treatment of addition of material in the form of curd, yeast of bread, yeast of tempe and yeast of tape in pretreatment in the manufacture of analog rice significantly affect to the water content (real level of 5%). This can be seen in Table 1 below:

Table 1. The average of water content

Treatment	The average of water content
B (adding of curd)	14,35 a
D (adding of tempe)	13,56 b
C (adding of yeast of bread)	13,42 b
E (adding of tape)	13,04 c
A (Natural fermentation)	12,43 d
KK = 1,96%	

Note: The numbers followed by the same small letters on the same row and the same uppercase letters on the same line is not real different according to DNMRT advanced test on real level of 5%.

From the table above it is seen that the value of the highest water content value derived from treatment of the addition of curd (B) and the lowest from natural fermentation (A). Based on [2] the water content of the rice is 14-15%, thus all treatments meet the quality standards of rice.

The water content of analog rice depends on the process of reduction of the water after fermentation. The process of water reduction is performed manually by pressing and the end limits to be printed. This value turned out to affect the water content of analog rice although it roasted to reduce the water content in the material. According to [4] that the longer of the cooking time, the water content of the material will decrease thus causing the evaporation of more water. This will cause the water content in the materials will be smaller. Evaporation is also caused due to the occurrence of the difference in vapor pressure between the water on the material with water vapor in the air. The water vapor pressure on the material in general is greater than the vapor pressure of water in the air so that the occurring mass transfer of water from the material into the air.

3.2. Ash content

On the variety investigation showed that the treatment of the addition of curd, bread yeast, yeast of tempe, yeast tape on soaking cassava for the manufacturing of analog rice was real effect on ash content (the real level of 5%). This can be seen on Table 2 below:

Table 2. The average of ash content

Treatments	The average of ash content
A (Natural Fermentation)	1,40 a
C (Adding of bread yeast)	1,01 ab
D (Adding of tempe yeast)	0,75 bc
B (adding of dadih)	0,72 bc
E (adding of tape yeast)	0,57 c
KK = 26,1%	

Note: The numbers followed by the same small letters on the same row and the same uppercase letters on the same line is not real different according to DNMRT advanced test on real level of 5%.

Based on [2] the value of the ash content of analog rice is higher than rice of 0,19. This is due to the influence of different raw materials and the process of fermentation. The ash content of treatments are real different, value of ash content decreased with the addition of material in the form of tape yeast, curd, bread yeast and yeast of tempe when compared with the natural fermentation. The results of the fermentation produced organic acids in the form of lactic acid. The amount of lactic acid produced depends on the material which is added that of tape yeast, curd, bread yeast and yeast of tempe. The amount of organic acid is inversely proportional to the ash content because the ash is inorganic material.

According to [9] ash is the inorganic substances remaining results of combustion of organic material. Ash content and its composition in food material depending on the type of material and how its incirenation. The ash content of a material is closely related to the mineral content of the material. Minerals contained in a material can be a two kinds of salt, that organic salt and inorganic salt. Furthermore, according to the [8] on the fermentation time 96 hours, *Lactobacillus plantarum* is able to produce lactic acid by (0,895 % (wt/vol)) followed by tempe yeast (0,878 %(wt/vol)) and bread yeast (0,552% (wt/vol)).

3.3. Fat content

On the variety investigation showed that the treatment of the addition of curd, bread yeast, yeast of tempe, yeast tape on cassava fermentation for the manufacturing of analog rice was real effect on fat content (the real level of 5%). This can be seen on Table 3 below:

Table 3. The average content of fat

Treatments	The average of fat content
D (adding tempe)	5,53 [a]
E (adding of tape yeast)	5,11 [b]
C (adding of bread yeast)	4,53 [c]
B (adding of curd)	4,07 [d]
A (natural fermentation)	3,38 [e]
KK = 5,18%	

Note: The numbers followed by the same small letters on the same row and the same uppercase letters on the same line is not real different according to DNMRT advanced test on real level of 5%.

On the Table 3 above can be seen that the fat content between treatments are mutually real different. Based on [2] the fat content of analog rice is higher than rice rice of 0,19.

This is caused by the addition of oil in the process of pre-gelatinization. The oil used as much as 10 ml per 200 grams of the ingredients in the roasting in this research. High fat in the process of roasting will speed up the process of rancidity in storage so it is not worth consumed. This should be handled in the process of roasting lest too much. In addition, much oil not only speed up the damage but will also affect the visibility of the analog rice so that oily in the dry state. The difference of each treatment is caused by the absorption of the oil in the process of roasting.

Oil absorption is influenced by the water content of the material. On roasting that uses heat to cause the water will evaporate and water position will be replaced by oil. According to [6] on the process of ripening takes place by the penetration of heat from the oil that goes into food. In this section the empty space that was initially filled by water will be filled by oil.

3.4. Protein content

On the variety investigation showed that the treatment of the addition of curd, bread yeast, yeast of tempe, yeast tape on cassava fermentation for the manufacturing of analog rice was real effect on protein content (the real level of 5%). This can be seen on Table 4 below:

Table 4. The average of protein content

Treatments	The average of protein
E (adding of tape yeast)	1,30 [a]
C (adding of bread yeast)	1,29 [a]
D (adding of tempe yeast)	1,22 [b]
B (adding of curd)	1,21 [bc]
A (Natural Fermentation)	1,18 [c]
KK = 2,5%	

Note: The numbers followed by the same small letters on the same row and the same uppercase letters on the same line is not real different according to DNMRT advanced test on real level of 5%.

On the Table 4 above can be seen that the protein content of analog rice is lower than the standard composition of sosoh rice namely 7,39 [3]. The low value is because the raw material used is a low protein content of cassava. The value of the protein increased with the onset of the fermentation process a result of enzymes produced by microorganisms.

According to [7] the protein content is high due to the ability of *Sacharomyces cerevisae* and *Rhizopus oryzae* to secrete several enzymes of extracellular (protein) in cassava during the process of fermentation, or the development of *Sacharomyces cerevisae and Rhizopus oryzae* in cassava in the form of single cell proteins during the fermentation process.The low protein on the analog rice can be coupled with other ingredients that contain carbohydrates such as beans. According to [3] analog rice is made from rice non rice plant with the content approach or exceed the rice and originated from a combination of food materials.

3.5. Starch content

On the variety investigation showed that the treatment of the addition of curd, bread yeast, yeast of tempe, yeast tape on cassava fermentation for the manufacturing of analog rice was real effect on starch content (the real level of 5%). This can be seen on Table 5 below:

Table 5. The average of starch

Treatments	The Average of starch
E (adding of tape yeast)	71,19 [a]
D (adding tempe yeast)	70,25 [ab]
C (penambahan bread yeast)	69,49 [b]
B (adding of curd)	68,15 [c]
A (natural fermentation)	67,40 [d]
KK = 0,82%	

Note: The numbers followed by the same small letters on the same row and the same uppercase letters on the same line is not real different according to DNMRT advanced test on real level of 5%.

From the table above it is seen that the highest starch content is in the treatment of addition tape yeast (E), this happens because in the process of fermentation of cassava for 5 days the texture of cassava do not change that is still hard.

Hard texture occured because of the absence of erosion of the walls of cassava by water which causes the number of starch dissolved in water immersion so that the starch is wasted. According to [9] starch is the carbohydrate which is a polymer of glucose, consists of amylose and amylopectin. The magnitude of the ratio of amylose and amylopectin varies depending on the type of starch. Amylose has the structure of straight bonds of (1,4) D-glycoside, is easily soluble in water because it contains a lot of hydroxyl groups. Based on the standard sosoh rice the amount of carbohydrates in rice is 79,64 [3]

The amount of starch in the treatment is lower than the standard because on observations is the amount of starch which is the part of carbohydrates. But with this fermentation, the starch obtained already modified so it is more easily digested.

According to [5] starch modification is through a process of fermentation naturally or with the addition of microbes that help the fermentation process. In the fermentation process, the microbes will also produce enzymes that will hydrolyze the starch so that produced modified starch with similar characteristic with starch that modified by enzymes.

References

[1] Apriyantono A, Fardiaz D , Puspitasari N L, Yasni S, Budijanto S 1989 Petunjuk laboratorium analisis pangan Departemen Pendidikan dan Kebudayaan Pusat Antar Universitas Insititut Pertanian Bogor

[2] BSN 2017 Mutu beras Menurut SNI 6128-2015 dan SNI 6128-2008. https://www.penggilinganpadi.com/2017/11/14/mutu-beras-menurut-sni-6128-2015-dan-sni-6128-2008/ (Didownload tanggal 20 Oktober 2018)

[3] Budijanto S, Muaris H J 2013 Beras analog pangan alternatif mirip beras dari non padi. Penerbit PT Gramedia Pustaka Utama Jakarta

[4] Fitriani S 2008 Pengaruh suhu dan lama pengeringan terhadap beberapa mutu manisan belimbing wuluh (*Averrhoa bilimbi* L) kering Jurnal Sagu Volume 7 No. 1

[5] Julianti E 2013 Modifikasi pati ubi kayu melalui proses fermentasi dan pemanfaatannya sebagai alternatif terigu. Laporan Penelitian. Perpustakaan Universitas Indonesia (Didownload tanggal 20 Oktober 2018)

[6] Ketaren 2005 Minyak dan lemak pangan Cetakan 2 Penerbit Universitas Indonesia. Jakarta.

[7] Okafor 1998 *An integrated bio-systemfor the disposal of assava wastes, integrated bio-systems in zero emissions applications*. In Proceeding of the Internet Conference on Integrated Bio-Systems Available: http://www.ias.unu.edu/proceedings/icibs (Didownload tanggal 24

Oktober 2018)
[8] Pratama A Y, Febriani R N, Gunawan S 2013 Pengaruh ragi roti, ragi tempe dan *Lactobacillus plantarum* terhadap total asam laktat dan pH pada fermentasi singkong. Jurnal Teknik POMITS Vol 2 No.1
[9] Winarno F G 2004 Kimia pangan dan gizi PT Gramedia Pustaka Utama Jakarta.

IOP Conf. Series: Earth and Environmental Science **327** (2019) 012029 doi:10.1088/1755-1315/327/1/012029

POME: An Alternative Nutrient Source for Bio-Organic Plant Tissue Culture Media

M N A Nadirah[1], A Awal[1,2], Y N Azma[1,2] and A Sulaiman[1]

[1] Faculty of Plantation and Agrotechnology, Universiti Teknologi MARA (UiTM), 40450 Shah Alam, Selangor, Malaysia

[2] Agricultural Biotechnology Research Group, Faculty of Plantation and Agrotechnology, Universiti Teknologi MARA (UiTM), 40450 Shah Alam, Selangor, Malaysia.

Corresponding author's e-mail address: asmah138@uitm.edu.my

Abstract. One of the methods in sustainable agriculture is transforming crop wastes into beneficial products such as bio-fertilizer, which will improve the soil fertility, preserving the natural sources, and cost-effective practices. Currently, these practices also need to be applied in plant tissue culture technology to enhance an eco-friendly and sustainable resources which are abundant and locally available. To date, the new bio-organic media derived from different types of agricultural wastes, especially from fruit wastes were successfully established. The objective of this study was to determine the efficiency of agricultural waste, i.e. Palm Oil Mill Effluent (POME) through a natural fermentation process as a potential nutrient source to substitute the existing standard and synthetic plant tissue culture media (PTC) such as Murashige and Skoog's (1962) (MS) Medium, Woody Plant Medium (WPM), N6 (Chu) Medium and Gamborg's (B-5) Medium, available in the market. 100 mL (v/v) POME solution was added with 100 mL (v/v) molasses and then were fermented naturally for 4 weeks in dark condition at room temperature. The results showed that the nutritional value of the fermented POME solution was higher than the synthetic MS basal salt medium. The new POME PTC medium was successfully formulated as a new bio-plant tissue culture media for *in vitro* regeneration of *Musa acuminata*.

1. Introduction

Plant tissue culture industry benefits many agricultural and plantation sectors as the technology able to produce a large amount of a high quality with disease free and true-to-type offspring derived from a single explant. Plant tissue culture or micropropagation is a method used to grow the entire plants from one cell (i.e. totipotency cell), which is genetically identical to a parent plant, on a formulated nutrient medium, under *in vitro* conditions. The formulated plant tissue culture media such as Murashige and Skoog's (MS) (1962) Medium, Woody Plant Medium (WPM) (1981), Gamborg's B-5 Medium (1981), and Chu (N6) (1975) medium are generally made up of macronutrients, micronutrients, vitamins, amino acids, sugar, other undefined organic supplements, solidifying agents and plant growth regulators as stated in Table 1. In plant tissue culture technology and micropropagation process, the macronutrients and micronutrients are compulsory for plant development and the plants require an accurate combination of macronutrients and micronutrients in order to live, grow and reproduce. The dynamic factor that contributed to the effectiveness of plant tissue culture medium is the presence of various types of ion rather than the compound itself [1] because, too few nutrients will cause the deficiency of plant uptake and too many nutrients will cause toxicity to the plant [2].

The ultimate aim in plant tissue culture technology is for the mass propagation of elite and uniform tissue cultured plant within scarce land available. The plant tissue culture is defined as an *in vitro* culture method for mass production of selected plant at a competitive price [3]. Even though, the benefit of plant tissue culture technology is widely applied over the countries, the cost of commercialising still incurred high operational cost. The chemicals requirement in plant tissue culture practices was identified as the main causal of high production cost [4]. This was supported by [5] where the expensive cost in the plant tissue culture method restrict its technology as a good advantage to a few institutions. Therefore, the cost effective approach in plant micropropagation techniques is required to produce an elite clonal seedling.

Table 1. The Standard Plant Tissue Culture Medium Components.

Medium Components (mg/L)	MS [6]	G_5 [7]	WPM [8 & 9]	N6 [10 & 11]
Macronutrients				
$Ca_3(PO4)_2$				
NH_4NO_3	1650.0		400.0	
KNO_3	1900.0	2500.0		2830.0
$CaCl_2 \bullet 2H_2O$	440.0	150.0	96.0	166.0
$MgSO_4 \bullet 7H_2O$	370.0	250.0	370.0	185.0
KH_2PO_4	170.0		170.0	400.0
$(NH_4)_2SO_4$		134.0		463
$NaH_2PO_4 \bullet H_2O$		150.0		
$CaNO_3 \bullet 4H_2O$			556.0	
Na_2SO_4				
KCI				
K_2SO_4			990.0	
Micronutrients				
KI	0.83	0.75	-	0.8
H_3BO_3	6.20	3.0	6.2	1.6
$MnSO_4 \bullet 4H_2O$	22.30		-	4.4
$MnSO_4 \bullet H_2O$		10.0	29.43	3.3
$ZnSO_4 \bullet 7H_2O$	8.6	2.0	8.6	1.5
$Na_2MoO_4 \bullet 2H_2O$	0.25	0.25	0.25	
$CuSO_4 \bullet 5H_2O$	0.025	0.025		
$CoCl_2 \bullet 6H_2O$	0.025	0.025		
$CO(NO_3)_2 \bullet 6H_2O$				
Na_2 EDTA	37.3	37.3	37.3	37.3
$FeSO_4 \bullet 7H_2O$	27.8	27.8	27.8	27.8
$MnCl_2$				
$Fe(C_4H_4O_6)_3 \bullet 2H_2O$				
Vitamins and other supplements				
Inositol	100.0	100.0	100.0	0.0
Glycine	2.0	2.0	2.0	2.0
Thiamine HCI	0.1	10.0	1.0	1.0
Pyridoxine HCI	0.5		0.5	0.5
Nicotinic acid	0.5		0.5	0.5
Ca-panthothenate				
Cysteine HCI				

Riboflavin
Biotin
Folic acid

Prakash, Hoque and Brink [12] highlighted there are several options to minimize the production cost either by turning into cheaper source or improving the efficiency in method and materials without affecting the plant quality. Alternate for inexpensive resources which performed a similar function like laboratory sucrose in substitute with local commercialize sugar for *in vitro* micropropagation of potato (*Solanum tuberosum* L.) were reported [13], or the previous study reported the potential of Isabgol-husk to substitute laboratory solidifying agent for *in vitro* rooting of *Balanites aegyptica* where the Isabgol-husk decreased the production cost by ~44% and proposed that table sugar can be used in commercial scale as both of the plantlet; *Balanites aegyptica* and *Phyllanthus emblica* responded well for *in vitro* rooting [14]. [15] also identified the effectiveness of coconut water, papaya, tomato and banana juices in regenerating shoots of *Celosia* sp. Whereas, [16] found that the addition of coconut water and banana homogenate produced high number of shoot and early plantlet development of *Cymbidium pendulum*.

Ogero *et al.* [17] used vegetative fertilizer containing both macro and micronutrients were successful on *in vitro* regeneration of *Ipomea batatas* (L.) Lam to substitute MS media. In addition, [18] found that *in vitro* growth of *Mentha* sps. on low cost plant tissue culture media KFA and KFA-plus derived from fly ash was almost similar to the artificial MS media as the clonal plants successfully produced in large numbers and had high rate of plant survival during acclimatization. Furthermore, in a recent study by [19], it was found that replacing the whole standard MS media preparation where MS salt, sucrose, and gelling agent changed to much cheaper sources resolved the cost ineffective ingredients in producing *in vitro* banana plantlets and the waste disposal issues.

Therefore, the present study was done to determine the effectiveness of fermented palm oil mill effluent solutions to substitute the MS medium for *in vitro* regeneration as the fertilizing properties in palm oil mill effluent are desirable for plant growth. According to previous studies conducted by [20], where the treated POME culture medium potentially promoted the growth of *H. callitrichoides* compared the full strength MS medium due to its ability in subdue the plant morphological changes. Since 2002-2011, these effluent wastes increased 1.6 times and 60 million tons of POME were generated in 2012, later, the waste will be overflowing to approximately 70-110 million tons in 2020 if improperly managed [21]. Embrandiri *et al.* [22] pointed out POME is a free and non-toxic material which contains valuable macronutrients required for plant growth. POME also consists of high total nitrogen sources which increase the composting period efficiency by degrading the cellulose and hemicellulose components of empty fruit bunch (EFB). In addition, the organic nutrient constitution in POME and EFB are desirable by plant which it is able to reduce the usage of inorganic fertilizer as part of environment protection [23]. However, there are few reports on POME utilization as a nutrient substitution for plant micropropagation techniques. Thus, this study was conducted to substitute the synthetic plant tissue culture media, with agricultural wastes such as POME, which is locally abundant in the palm oil mill and this will also minimizing the production cost of clonal plant without compromising the quality of regenerant produced.

2. Materials and Method

2.1. Establishment of POME Solution for Fermentation Process

bnnv POME sample was obtained from the FELDA palm oil mills in Pahang, Malaysia. POME sample was collected from the main receiving tank as a main nutrient source. Molasses was purchased from small entrepreneur in Raub, Pahang, Malaysia. The POME sample and molasses were measured approximately 100 (v/v) mL and 100 (v/v) g, respectively at a ratio of 1:1. Then, the molasses were added into the POME solution. The container (24cm x 13cm x11cm) was sealed with a mesh cloth and wrapped with an aluminum foil to begin the fermentation process. The fermentation process was

IOP Conf. Series: Earth and Environmental Science **327** (2019) 012029 doi:10.1088/1755-1315/327/1/012029

carried out for 4 weeks in dark condition (24 hours darkness) at room temperature. The initial pH value and final pH values were weekly recorded.

2.2. Preparation of Fermented Solution for Nutrient Analysis
Some text. The fermented POME solution was extracted manually by squeezing using filter to remove soils and debris from the crude solution. Then, the fermented crude solution was filtered using filter paper to remove any fine particles. Subsequently, the filtered fermented crude nutrient and mineral content was analyzed using Inductively Coupled Plasma Optical Emission Spectrometer (ICP-OES) instrument at the Soil Laboratory, Faculty of Plantation and Agrotechnology, Universiti Teknologi MARA, Malacca, Malaysia.

2.3. Preparation of Bio-Plant Tissue Culture Medium
Some text. The fermented POME solution was taken accordingly to the composition for the establishment culture media as mention in Table 2. Each formulation was supplemented with 30.00 g/L laboratory sucrose, 2.00 g/L solidifying agent, 1.00 g/L activated charcoal, and 2.00 mg/L Benzylamonipurine (BAP). The MS basal salt medium was served as standard control medium. All culture media were adjusted to pH 6.4 prior to autoclaving at temperature of 121˚C for 20 minutes. Totally, there are 6 treatments including control.

Table 2. Treatments for bio-organic plant tissue culture media formulation derived from fermented POME solution.

Treatment	Composition
MS_0	MSBS + Sucrose + Agar + Plant Hormone
PM_1	1% concentration + Sucrose + Agar + Plant Hormone
PM_2	2% concentration + Sucrose + Agar + Plant Hormone
PM_3	3% concentration + Sucrose + Agar + Plant Hormone
PM_4	4% concentration + Sucrose + Agar + Plant Hormone
PM_5	5% concentration + Sucrose + Agar + Plant Hormone

2.4. In Vitro Regeneration of Musa acuminata
Some text. Healthy suckers of *Musa acuminata* were obtained from the banana farm at Tanjung Karang, Selangor, Malaysia. The leaves and roots were removed and washed with tap water. Then, the suckers were trimmed to the sizes of 10.0 cm length and 5.0 cm diameter before washed with liquid detergent and later, rinsed off the liquid detergent with tap water. The clean suckers were placed under running tap water for 30 minutes before being transferred to the laminar air flow chamber. The explants were surface sterilized by treated the explants with 50% (v/v) chlorox (a.i. 5.25% sodium hypochlorite) solutions added with a few drops of Tween20 for 30 minutes and then rinsed thoroughly with sterile distilled water once. The exposed explants area with chlorox were excised and dipped into 95% (v/v) alcohol for 3 minutes, followed by the solutions of 100% chlorox (v/v) (a.i. 5.25% sodium hypochlorite) with a few drops of Tween20 for 5 minutes. Lastly the explants were rinsed three times with sterile distilled water and dried on autoclaved tissue paper for 15 minutes to remove the excess moisture before the explants were trimmed to the final size of 2.0 cm length and 2.0 cm diameter. The half cut explants were cultured on the MS basal salt media and the formulated POME PTC media. All the explants were incubated for 3 months under the temperature at 25±2˚C, photoperiod condition of 16 hours with light intensity of 2000 lux per day for shoot initiation. The day for explants changed into green and explant responses on different concentrations of formulated POME PTC media were observed and recorded.

3. Results and Discussion

3.1. Fermentation of POME

ccvv The fermentation process was successfully completed after 4 weeks in the dark condition at room temperature. The weekly pH values were measured as provided in Figure 1. Visual and physical changes in the fermented solution for 4 weeks were observed. The first week fermentation, the solution released a sweet and sour odor almost like a rotten fruit. Subsequently, the solution exhibited bubble/foam on top of the liquid and disappeared at the end of the fermentation period. This finding was also supported by the conducted experiment where the fermented solution of the banana peel form bubbles at the second days, the banana peel floated, liquid settle at the bottom, and lastly the sweet and sour smell produced [24]. This indicated that the microorganism actively decomposed and transformed all the sugars, nutrient and oxygen efficiently into ionic macro and micro nutrients. There was no change in term of physical colour whereby the fermented solution is still maintained black in colour as similar with the original POME.

Figure 1 shows that the fermented solution recorded an alkaline pH value during the early phase of fermentation which is pH 7. The pH value of the fermented solution started to decrease rapidly to pH 5.61 after being fermented for one week. It is indicated that during the first week of fermentation, the microorganism using the food source provided. Subsequently, the pH value decreased in week 2 and week 3 to pH 4.23 and pH 4.19 respectively. Then, the final (week 4) pH slightly increased to pH 4.45. The fermentation process was considered completed as the pH begins to increase due to the microorganism had achieved stationary and death phase. Hence, the fermented POME solution was an acidic solution with pH 4.45 at the end of fermentation.

The reduction in pH proved that the POME solution was succeeded in the fermentation process. Nanson and Fields [25] reported that the fermentation caused a decreasing in pH value. Cokgor *et al.* [26] highlighted that the significant factors affecting the efficiency of sludge fermentation are pH and temperature. According to the Sundberg, Smårs, and Jöhnson [27], the final stage of composting causes the pH to decline to the below neutral value. This is because the formation of organic acid discharges a large amount of carbon dioxide (CO_2). In addition, the pH reduction is caused by the acidification process where the carbohydrates are being converted into alcohols or acids. This trend is also supported by [28] whom reported that during the glucose fermentation, the pH values decreased due to the higher production of acetic acid or butyric acid as the pH values remains constant at pH 4.68-4.85 over the time of fermentation period. Therefore, the fermented solutions of the POME may offer high amounts of nutrient content similar as the decomposition of organic materials.

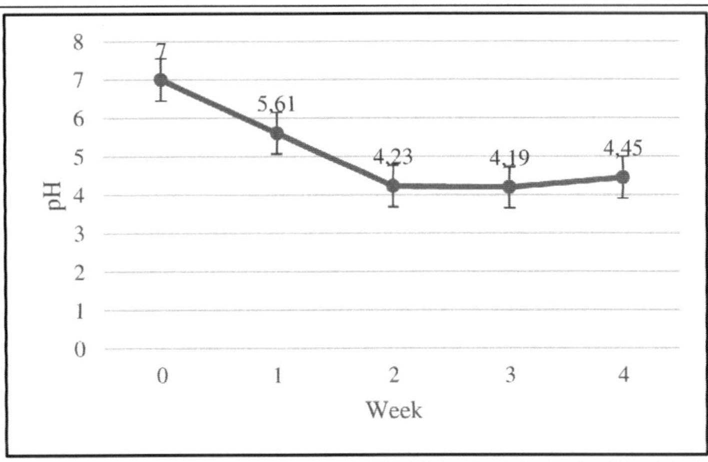

Figure 1. The pH profile of the POME solution after 4 weeks of fermentation process.

3.2. Macro and Micro Nutrient Analysis

The fermented POME solution was analysed to determine the presence of macronutrient by using Inductively Coupled Plasma Optical Emission Spectrometer (ICP-OES) instrument. The macronutrients are potassium (P), phosphorus (K), magnesium (Mg), calcium (Ca), and sulphur (S). Meanwhile, the determination of selected micronutrients is iron (Fe), copper (Cu), manganese (Mn), and sodium (Na).

Table 3 and Figure 2(a) showed that the concentration of macronutrients in fermented POME solution was relatively higher than the MS basal salt media. The total nutrient content of P and K in the fermented POME solution were found to be at 84.16 % and 55.82% respectively higher than in MS basal salt concentration. While, the total nutrient content of Mg, and Ca in fermented POME solution was also found to be at 87.97 %, and 74.99 % relatively higher than in MS basal salt concentration. All these macronutrients content were increased by an average of four times during the fermentation process.

Table 4 and Figure 2(b) showed that the concentration of micronutrient in fermented POME solution was relatively higher than the MS basal salt media. The total nutrient content of Fe, Cu, and Na were found to be at 95.23 %, 97.62 % and 89.27 %. However, the total Mn content is lower than the required concentration in MS basal salt. The Mn concentration was 21.68 % lower than the MS basal salt concentration. The lower concentrations of Mn in the fermented POME solution are probably caused by specific nutrients suppressed the Mn enhancement during the fermentation process. According to the Alam, Kamei, and Kawai [29] the significant decreased of Mn toxicity level in shoots and roots of barley (*Hordeum vulgare* L. cv. Minorimugi) by elevating the Fe concentration where the plant regained its constituents. Alam *et al.* [30] also conducted the relationship the effect of potassium (K) on absorption and translocation in Mn uptake where high K assist in preventing the plant to intake excessive Mn amount. These findings showed that iron (Fe) and potassium (K) attributed in repressing the plant toxicity especially for Mn level. Therefore, the Mn concentration below the level apparently caused by the presence of elevated K and Fe in fermented POME solution. All these micronutrients content were increased by an average of sixteen times during the fermentation process.

The conversion process occurs during fermentation synthesizes vitamins, minerals, nutrient, and other bioactive compounds better compared than the raw materials. The successful fermentation of agricultural waste, i.e. POME solution was proven by the declining of pH value and other characters present. Addition of molasses as substrate acts as a source of energy for the microorganism growth and survival during the fermentation process [31]. Kawaguchi and Minamisawa [32] highlighted the symbiotic interaction between the natural occurring microbes during the fermentation process improved some mineral intake by plants. These nutrients act as synthetic elements in cultures and it may differ for each plant species and genotype [33]. This is consistent with previous studies done by [34] reported the correlation between the fermentation process of agricultural waste and pH reduction directly enhanced the quality of fermented waste. The nutrient enhancement in fermented POME solution is vital in promoting the plant growth especially in sterile environment because the fresh POME lacking in primary macronutrients [35] nutrient composition in the fermented POME was found relatively higher compared to the MS medium. In addition, the stationary phase of fermentation process offers various metabolites for biotechnological interest [31]. The action of microorganisms during the fermentation process changed the structure, texture and added flavour and aroma of the product. In some cases, fermentation enhanced the contents of mineral, vitamins and better digestibility than the raw materials [36].

Table 3. Macro nutrient contents (mg/L) of the MS basal salt and the fermented POME solution after 4 weeks of fermentation.

Elements	MS basal salt	POME
Potassium (K)	695.88	1575.20
Phosphorus (P)	21.88	138.16
Calcium (Ca)	107.78	431.04
Magnesium (Mg)	19.38	161.20

Note: Slight instability of mineral content reading may happen due to the manual extraction method which may cause instability pressure during extraction. However, the subject used the pressure is the same to decrease potential error.

Table 4. Micro nutrient contents (mg/L) of the MS basal salt and the fermented POME solution after 4 weeks of fermentation.

Elements	MS basal salt	POME
Iron (Fe)	4.34	90.96
Copper (Cu)	0.05	1.68
Manganese (Mn)	4.47	3.92
Sodium (Na)	9.99	93.12

Note: Slight instability of mineral content reading may happen due to the manual extraction method which may cause instability pressure during extraction. However, the subject used the pressure is the same to decrease potential error.

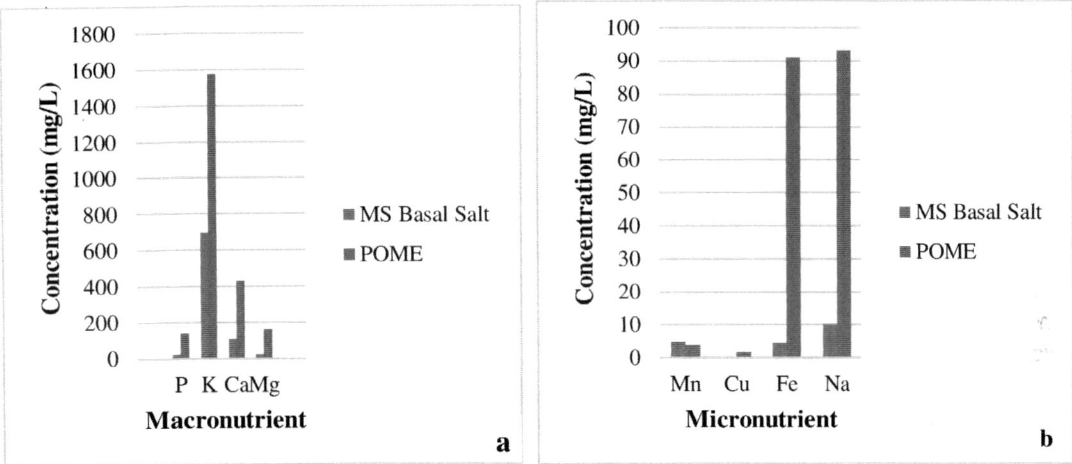

Figure 2. The Comparison on nutrient content between the fermented OME solution and MS basal salt for (a) and (b)

3.3. The Effects of Fermented POME Solution on In Vitro Regeneration of Musa acuminate

Some text. Bio-organic plant tissue culture medium was formulated according to the different concentration requirements of the fermented POME solutions. The concentration levels started with 1%, 2%, 3%, 4%, to 5% respectively. The formulated bio-organic media displayed a light yellowish medium to a thick yellowish medium as the concentration level increased compared to the translucent and clear MS basal salt medium. Even though the formulated bio-organic media were yellow in color, the media still translucent which was similar to MS basal salt medium.

In addition, the pH of formulated bio-organic media was set to pH 6.40 prior to the autoclaving instead pH 5.80 as optimum pH in most standard plant tissue culture media. This is because, the pH of formulated bio-organic media tends to drop in average of 0.9±1 after autoclaving as the pH condition of formulated media became an acidic again (pH 4.30±0.5). An acidic plant tissue culture media is not preferable in plant microporpagation techniques to promote plant growth and development where optimum pH for plants to grow is pH 5.5-6.5.

After 3 months of culture initiation, the aseptic plantlets in the formulated bio-organic media ranges from 1%, 2%, 3%, 4%, and 5% showed a positive response. The days of plantlet to green was observed in different concentration levels and standard MS basal salt media as mention in Table 4. The result showed that the explants in the MS_0 took 20-25 days to transform into green colour, while the explants in PM_3, PM_4, and PM_5 took 18-25 days compared to PM_1 and PM_2. Hence, this indicated that the high concentration level of formulated POME media had similar days to MS basal salt for the explants to transform into green colour than the lower concentration. In addition, each explant cultured into the formulation POME PTC media and MS media were replicated into 10 and the total of responsive explants were recorded and it is shown that more than 50% of the culture positively responded on the formulated POME PTC media. Further statistical analysis on the mean number of plantlet height and number of shoot produced will be analysed and recorded.

Table 5. Days for explants of Musa *acuminata* responded to morphogenesis process and exhibited green colour after being cultured into MS basal salt media and formulated POME PTC media within 3 months of culture initiation.

Treatment	Concentration	Explant response (day)
MS_0	-	25
PM_1	1%	45-60
PM_2	2%	30-36
PM_3	3%	18- 25
PM_4	4%	18-25
PM_5	5%	18-25

Figure 3. Establishment of bio-organic POME PTC media, according to its composition and supplemented with 30.0 g/L sucrose, 2.00 g/L gelling agent, and 1.0 g/L active charcoal.

Several researchers had identified the usage of agricultural wastes such as bio-organic plant tissue culture media to promote plant growth. A study conducted by [37] whom using two types of fruit waste which were banana and papaya through a fermentation process had resulted a better *in vitro* seed germination of pitaya than the synthetic media. In addition, the outcome of research conducted by [38] also proved that the commercial cooking agar and cocopeat has successfully reduced solidifying agent cost by 97.289% and 99.937% respectively to promote pitaya growth. The potential use of banana and papaya as nutrient replacement was continued by [34] by adding pineapple, calamansi lime, and key lime peels as new fruits waste combination to promote growth of different plant species. The formulated fruits waste solutions media showed the significant interaction of *Lycium barbarum* (Goji) *and Aquilaria malaccensis* Lamk. (Agarwood) in promoting the plant growth. Nevertheless, the application of formulating fruits waste solutions was proven to promote plant growth as both of the plant positively responsive to the formulated fruits waste media. Subsequently, [39] improved the usage of fruit waste by choosing the fruit peel waste in developing bio-organic plant tissue culture media. The fruit peel waste was selected as a main nutrient source because the fruit peel wastes have a high content of bioactive compound which can be transformed into potentially valuable products [40]. The low concentration of the formulated fruits peel media showed a greater performance in plant height than the higher concentration. Hence, *Musa acuminata* cv. Berangan (AAA) were successfully regenerated on the developed media of fruit peel waste. The constrained in former researches were the availability of fruit wastes. Even though it is easy to find the fruit waste due to the fruit is a daily consumption routine, but they are not available in abundant amount. Hence, the agricultural waste that produces in mass volume and easy accessible are required and POME meets all the requirements indicated. Biotransformation of POME into valuable products especially in plant tissue culture techniques was a profitable industry and more approachable sources which in agreement with [40] point out that three key elements to achieve good development in sustainable agriculture which were economic thriving, eco-friendly, and favorable to human.

Figure 4. Green explants of *Musa acuminata* after 3 months cultured into different media formulation (a) MS_0 (b) PM_1 (c) PM_2 (d) PM_3 (e) PM_4 (f) PM_5

4. Conclusion

The fermented POME solution was successful in enhancing the nutrient and mineral content and supported the theory of increasing nutrient quality through fermentation process. Hence, the fermented POME solution fulfills in providing important macronutrient, micronutrient and essential vitamins for the satisfactory plant growth in the aseptic condition. Furthermore, the preliminary study showed that POME PTC media were succeeding in regeneration of *Musa acuminta* in *in vitro* condition The fermented POME solution contained nutrient needed by the tissue cultured plant as the nutrient uptake by the tissue cultured plant physiologically similar to the optimum nutrient of the standard plant tissue culture media, depending on the different needs of plant species. The collected data on plantlet height and number of shoot will be further analysed using one-way analysis of variance (ANOVA) by SPSS programme to determine its significant difference for *in vitro* regeneration of *Musa acuminata*.

Acknowledgement

This study was financially supported by the 600-IRMI/DanaKCM 5/3/GIP (089/2017) and 600-IRMI/DANA 5/3/BESTARI (0009/2016) facilitated by Research Management Institute Universiti Teknologi MARA (UiTM) Shah Alam, Selangor, Malaysia.

References

[1] Saad AI and Eishahed AM. Plant Tissue Culture Media. In *Recent Advances in Plant In Vitro Culture*. 2012 Oct 17. Intech Open. doi:10.5772/50569

[2] Ann M, Clain J and Jeff J. *Plant Nutrient Functions and Deficiency and Toxicity Symptoms.* In Nutrient Management (Self-study course). 2009. MSU. Extension Service Continuing Education Series. Montana State University. 4449.9.

[3] Debergh PC and Read PE. Micropropagation. In *Micropropagation.* 1991 (pp.1-13). Springer, Dordrecht.

[4] Savangikar VA. Role of Low Cost in Tissue Culture. In *Low Cost Options for Tissue Culture*

Technology in Developing Countries. Proceedings of a Technical Meeting Organized by the Joint FAO/IAEA Division of Nuclear Techniques in Food and Agriculture. 2004 Feb (pp. 11-15).

[5] Kodym A and Zapata-Arias FJ. Low Cost Alternatives for the Micropropagation of Banana. In *Plant Cell, Tissue and Organ Culture*. 2001 Jul 1:66(1) (pp. 67-71)

[6] Murashige T and Skoog F. A Revised Medium for Rapid Growth and Bio Assays with Tobacco Tissue Culture. In *Physiologia Plantarum*. 1962 Jul 1:15(3) (pp. 473-97).

[7] Gamborg OL. Nutrition Media and Characteristics of Plant Cell and Tissue Cultures (Thorpe TA Ed.). In *Plant Tissue Culture Methods and Application in Agriculture*. 1981. Academic Press, New York.

[8] Lloyd G and McCown B. Commercially-feasible Micropropagation of Mountain Laurel, *Kalmia latifolia*, by Use of Shoot-tip Culture. *Commercially-feasible Micropropagation of Mountain Laurel, Kalmia latifolia, by Use of Shoot-tip Culture*. 1980:30 (pp. 421-7).

[9] Nas MN and Read PE. A Hypothesis for the Development of a Defined Tissue Culture Medium of Higher Plant and Micropropagation of Hazelnuts. *Scientica Horticulture*. 2004 May 3;101(1-2) (pp. 189-200).

[10] Chih-ching C, Ching-chu W, Ching-san H, Chen H, Kwang-chu Y, Chih C and Feng-yun B. Establishment of an Effiecient Medium for Anther Culture of Rice Through Comparative Experiments on the Nitrogen Sources. Scientica Sinica. 1975 May 20;18(5) (pp. 659-68).

[11] Talebi R, Rahemi MR, Arefi H, Nourozi M and Bagheri N. In Vitro Plant Regeneration Through Anther Culture of Some Iranian Local Rice (*Oryza sativa* L.) Cultivars. *Pak J Biol Sci*. 2007 Jun 15;10(12) (pp. 2056-60). doi: 10.3923/pjbs.2007.2056.2060

[12] Prakash S, Hoque MI and Brinks T. Culture Media and Containers. Low Cost Options for Tissue Culture Technology in Developing Countries, FAO/IAEA Division of Nuclear Techniques in Food and Agriculture, Vienna. 200 Feb (pp. 29-40).

[13] Demo P, Kuria P, Nyende AB and Kahangi EM. Table Sugar as an Alternative Low Cost Medium Component for In Vitro Micro-propagation of Potato (*Solanum tuberosum* L.) *African Journal of Biotechnology*. 2008;7(15).

[14] Gour VS and Kant T. Efficacy of Low Cost Gelling Agents an Carbon Source Alternatives During In Vitro Rooting of *Balanites aegyptiaca* and *Pyllanthus emblica* Microshoot. *Tree For Sci Biotechnol*. 2011;5 (pp.58-60).

[15] Daud N, Taha RM, Noor NN and Alimon H. Effects of Different Organic Additives on In Vitro Shoot Regeneration of *Celosia sp. Pak J Bio Sci.* 2011 May1;14(9) (pp. 546-51).

[16] Kaur S and Bhutani KK. Organic Growth Supplement Stimulants for In Vitro Multiplication of *Cymbidium pendulum* (Roxb.) Sw. *Horticultural Science*. 2012 Feb 16;39(1) (pp.47-52).

[17] Ogero KO, Mburugu GN, Mwangi M, Mgugi MM and Ombori O. Low Cost Tissue Culture Technology in the Regeneration of Sweet Potato (*Ipomea batatas* (L) Ram). *Research Journal of Biology,* 2012;2(2) (pp. 51-8).

[18] Biswas K, Biswas R and Negi P. Novel Low Cost Culture Media KFA and KFA plus for Micropropagation of Mentha sps. *Int. J. Curr. Microbiol Ap. Sci.* 2014;3(4) (pp. 172-82).

[19] Kadam DD, Chhatre AA, Lavale SA and Shinde NA. Low-cost Alternatives for Conventional Tissue Culture Media. *Int. J. Curr. Microbiol. App. Sci.* 2018;7(4) (pp.2523-29).

[20] Ng YS, Lim CR and Chan DJ. Development of treated Plam Oil Mill Effluent (POME) Culture Medium for Plant Tissue Culture of *Hemianthus callitrichoides*. *Journal of Environmental Chemical Engineering*. 2016;4(4) (pp. 4890-6).

[21] Melssen B. National Biomass Strategy 2020: New Wealth Creation for Malaysia's Biomass Industry. 2013. *Agensi Inovasi Malaysia, Malaysia.*

[22] Embrandiri A, Ibrahim MH and Singh RP. Palm Oil Mill Wastes Utilization, Sustainability in the Malaysian Context. *International Journal of Scientific and Research Publication*. 2013;24(3) (pp. 63-8).
[23] Nordiana I, Wan Noranida WMN, Asmah A, Muhammad Zahid M, Nur Nabila Huda A and Husien AG. Comparativeness Cost of Production for Sugar's Substitute Through Different Amended Planting Media. *International Journal of Engineering Science and Management Research.* 2016;3(4) (pp. 1-4).
[24] Kadir AA, Rahman NA and Azhari NW. *The Utilization of Banana Peel in the Fermentation Liquid in Food Waste Composting*. In IOP Conference Series: Materials Science and Engineering. 2016 Jul (Vol. 136, No. 1, p. 012055). OI Publishing.
[25] Nanson NJ and Fields ML. Influence of Temperature of Fermentation on the Nutritive Value of Lactic Acid Fermented Cornmeal. *Journal of Food Science*. 1984;49(3) (pp. 958-9).
[26] Cokgor EU, Oktay S, Tas DO, Zengin GE and Orhon D. Influence of pH and Temperature on Soluble Substrate Generation with Primary Sludge Fermentation. *Bioresource Technology*. 2009;100(1) (pp. 380-6).
[27] Sundberg C, Smårs S and Jönsson H. Low pH as an Inhibiting Factor in the Transition from Mesophilic to Thermophilic Phase in Composting. *Bioresource Technology.* 2004; 95(2) *(pp.* 145-150.)
[28] Li Z, Wang H, Tang Z, Wang X and Bai J. Effects of pH Value and Substrate Concentration on Hydrogen Production from the Anaerobic Fermentation of Glucose. *International Journal of Hydrogen Energy. 2008;33(24)* (pp. 7413-18).
[29] Alam S, Kamei S and Kawai S. Amelioration of Manganese Toxicity in Barley with Iron. *Journal of Plant Nutrition.* 2001;29(4) (pp. 1421-33).
[30] Alam S, Akiha F, Kamei S, Imamul Huq S, and Kawai S. Mechanism of Potassium Alleviation of Manganese Phytotoxicity in Barley. *Journal of Plant Nutrition.* 2005;28(5) (pp. 889-901).
[31] Pumphrey B, Julien C, and BV NBSB. (1996). An introduction to fermentation. In *Fermentation Basics,* New Brunswick Scientific (UK and The Netherlands) May.1996 (pp.1-25)
[32] Kawaguchi M and Minamisawa K. Plant-microbe Communication for Symbiosis. *Plant and Physiology*. 2010; 51(9) (pp 1377-80).
[33] George EF, Hall MA, and De Klerk GJ. The Components of Plant Tissue Culture Media I: Macro- and Micro-nutrients. In *Plant Propagation by Tissue Culture.* 2008 (pp. 65-113). Springer, Dordrecht.
[34] Jaafar F, Awal A, Abdullah S, Sidik NK and Yusuf NA. Bio-potential of Fermented Fruit Waste Solutions on In Vitro Seed Germination and Regeneration of *Lycium barbarum* and *Aquilaria malaccensis* Lamk. Malaysian Applied Biology. 2015;44(3) (pp.127-34).
[35] Mohd Nasir AR, Morad N, Ismail N. Omar WM and Razali RM. Utilization of *Chlorella vulgaris* in the Removal of Ammonical Nitrogen and Phosphorus from Palm Oil Mill Effluent. *Journal of Industrial & Research Technology.*2013;3(1). (pp.6-11).
[36] Chojnacka K. *Fermentation products*. K. Chojnacka, Chemical Engineering and Chemical Process Technology. 2010;12. Wroclaw, Poland: Wroclaw University of Technology.
[37] Yusuf NA. *Bio-organic Tissue Culture Media for Pitaya's (Hylocereus polyrhizus)*

Shoot Growth (Master-degree thesis). 2012. Universiti Teknologi MARA (UiTM), Shah Alam, Selangor, Malaysia.

[38] Rasuddin R. *Potential of cooking agar and cortpith as an alternative agar for tissue culture media.* (Master degree thesis).2012. Universiti Teknologi MARA (UiTM), Shah Alam, Selangor, Malaysia.

[39] Khirul Anuar NA, Mohd Shaain UH, Awal A, and Yusuf NA. *Selected Fruits Peel as an Alternative Low Cost Medium Component for In Vitro Micropropagation of Musa acuminata cv. Berangan (AAA).* Paper presented at regional Conference on Science, Technology and Social Sciences. Copthorne Hotel, Cameron Highlands, Brinchang, Malaysia. 2016 Dec 4-5.

[40] Duda-Chodak A, and Tarko T. Antioxidant Properties of Different Fruit Seeds and Peels. *Acta Scientiarum Polonorum Technologia Alimentaria.* 2007;6(3) (pp. 29-36).

[41] Hamid MN, and Abdullah MY. Contribution of Mathematical Model for the Development of Sustainable Agriculture. *Malaysian Journal of Mathematical Sciences*. 2008; 2(2) (pp. 83-91).

AUTHOR INDEX